Optical Path Theory

Fundamentals to freeform adaptive optics

IOP Series in Emerging Technologies in Optics and Photonics

Series Editor

R Barry Johnson, a Senior Research Professor at Alabama A&M University, has been involved for over 50 years in lens design, optical systems design, electro-optical systems engineering, and photonics. He has been a faculty member at three academic institutions engaged in optics education and research, has been employed by a number of companies, and has provided consulting services.

Dr Johnson is an IOP Fellow, an SPIE Fellow and Life Member, an OSA Fellow, and was the 1987 President of SPIE. He serves on the editorial board of *Infrared Physics & Technology* and *Advances in Optical Technologies*. Dr Johnson has been awarded many patents, has published numerous papers and several books and book chapters, and was awarded the 2012 OSA/SPIE Joseph W Goodman Book Writing Award for Lens Design Fundamentals (second edition). He is a perennial co-chair of the annual SPIE Current Developments in Lens Design and Optical Engineering Conference.

Foreword

Until the 1960s the field of optics was primarily concentrated in the classical areas of photography, cameras, binoculars, telescopes, spectrometers, colorimeters, radiometers, etc. In the late 1960s optics began to blossom with the advent of new types of infrared detector, liquid crystal display (LCDs), light emitting diode (LEDs), charge coupled device (CCDs), laser, holography, and fiber optics along with new optical materials, advances in optical and mechanical fabrication, new optical design programs, and many more technologies. With the development of the LED, LCD, CCD, and other electro-optical devices, the term 'photonics' came into vogue in the 1980s to describe the science of using light in the development of new technologies and the operation of a myriad of applications. Today optics and photonics are truly pervasive throughout society and new technologies are continuing to emerge. The objective of this series is to provide students, researchers, and those who enjoy self-education with a wide-ranging collection of books, each of which focuses on a topic relevant to the technologies and applications of optics and photonics. These books will provide knowledge to prepare the reader to be better able to participate in these exciting areas now and in the future. The title of this series is *Emerging Technologies in Optics and Photonics*, in which 'emerging' is taken to mean 'coming into existence', 'coming into maturity', and 'coming into prominence'. IOP Publishing and I hope that you will find this series of significant value to you and your career.

Optical Path Theory

Fundamentals to freeform adaptive optics

Rafael G González-Acuña
Huawei Technologies, Tampere Camera Lab, Finland

Héctor A Chaparro-Romo
Independent researcher

IOP Publishing, Bristol, UK

ISBN 978-0-7503-4705-1 (ebook)
ISBN 978-0-7503-4703-7 (print)
ISBN 978-0-7503-4706-8 (myPrint)
ISBN 978-0-7503-4704-4 (mobi)

DOI 10.1088/978-0-7503-4705-1

Version: 20220601

IOP ebooks

British Library Cataloguing-in-Publication Data: A catalogue record for this book is available from the British Library.

Published by IOP Publishing, wholly owned by The Institute of Physics, London

IOP Publishing, Temple Circus, Temple Way, Bristol, BS1 6HG, UK

US Office: IOP Publishing, Inc., 190 North Independence Mall West, Suite 601, Philadelphia, PA 19106, USA

In memory of Jose Ángel Romo Diaz

To all those who have dedicated part of their lives to optical design.

To all those who deeply love optical design.

Contents

Preface

This treatise is mostly based on an equation that we recently published. This equation is the general formula for adaptive optics mirrors, which was published in January 2021 in 'General mirror formula for adaptive optics' *Applied Optics* **60(2)**. We were so happy with this equation that we decide to expound its theoretical background; we have therefore written a book about it and its potential applications.

This treatise addresses the introductory topics, the solution (a general formula for adaptive optics mirrors), the implications, and the potential applications of this formula. We decided to do all this because we agree that the abovementioned equation is the analytical closed-form solution to the fundamental problem in adaptive optics.

We named this project 'optical path theory' because the derivation of this enigmatic equation was only necessary to study the path of light. So in the treatise in your hand, we are going to make in-depth use of Fermat's principle and its implications for the path of light. This treatise has three parts: the first is the introduction; the second part deals with the abovementioned equation in two dimensions and its direct implications for aspherical design; the third part consists of the generalization of the chapters of the second part to three dimensions, in which we move from aspherical designs to freeform designs.

With the exception of chapter 1, all the chapters have code for the respective examples and theory. The code is written for `Mathematica`. `Mathematica` is a package containing several built-in libraries that have applications in symbolic computing. This book does not have a `Mathematica` tutorial.

This treatise is specially designed to provide solid training in rigorous optical design. It is recommended to read this treatise in order.

The Authors.

Acknowledgements

Acknowledgements of Rafael G González-Acuña

I want to thank God, this work is for your glory.

I want to thank my mother Carmen Leticia Acuña Medellín and my father Rogelio González Cantú for their unconditional love and support.

Héctor Alejandro Chaparro-Romo: comrade, once again! we did it, comrade! your patience has been rewarded! Thanks for all the time you have dedicated to optical design and our endless discussions about optics.

To Professor Barry Johnson for your support. I also want to thank the IOP family, especially Ashley Gasque and Robert Trevelyan, for all the support and help.

I would also like to thank Adriana Mabel Serrano, Yoshio Catillejos, Israel Meléndez, Gustavo Medina, Daniel Lomas, Roberto Martinez, Andres Roaro, Hernan Izguerra, Adrian Lozano, Eliel Guadarrama, and Benjamin de Jesus Perez Garcia.

Acknowledgements of Héctor A Chaparro-Romo

I wish to deeply thank the Great Spirit, his love towards us allowed us to accept the humility of our hearts and therefore showed us the way to perfection; this work was done to remind us of his omnipotent and omnipresent presence. This book is an offering to your infinite goodness.

I want to thank all those optical designers who, through their contributions, have allowed us to understand the nature of optical design and its use for humanity.

To Dr Barry Johnson, and the IOP production team, in particular, Ashley Gasque and Robert Trevelyan, for their unconditional support; they made it possible for this work to be completed.

I wish to express my love to my parents, José Eleazar Chaparro-Villegas and Ma. del Refugio Romo-González, for their infinite aid and unconditional support, this book is also the result of their life's efforts, thank you, thank you, thank you very much.

I wish to express my gratitude to my brother Eleazar Enrique Chaparro-Romo, who has shared great things with me: his advice, example, and support have formed the character that defines us today. Thanks to my grandparents, my aunts and uncles, my cousins, my nephews and nieces, thanks to all of you for being a part of me.

To my eternal teachers Descartes, Maxwell, Marx, Fermi, Huygens, Leibniz, Euclides, Newton, Keynes, Friedman, Euler, Einstein, Boltzmann, Cantor, Perelman, Plato, Aristotle, Pythagoras, Gauss, Socrates, and all those who have contributed to what I am today: thank you very much, sincerely.

To Dr José Raúl Miranda-Tello, for your unconditional support and your great human quality; your teachings and advice are invaluable. This book is a sample of the professional commitment that you instilled in me with your example. Thank you very much, my friend.

I want to thank Dr Nicolas Dominguez-Vergara, his example and professionalism are an example to follow, his insightful observations and recommendations were helpful in not giving up. Many thanks.

To the professor of the Department of Electronics at the Autonomous Metropolitan University, Vicente Yuri Plata-Gárnica, who trained me as an electronics engineer and gave me the foundations to start my professional life. Thanks a lot.

To the professor of the Faculty of Economics of the National Autonomous University of Mexico, Miriam Berenice Lavarios-Arzate, for having given me the reason to build a new path in the economic sciences. Thank you very much for your commitment to Mexican education.

To the Marakame Alberto Bautista Lopéz and his family, Estela, Joél, Braulia, Luís, Melisa, and Jose Ignacio, thank you for showing me the spiritual path and love of life through our Father-Fire.

To Taita Pina Varinawa and his wife Memi Kamanawa from the Amazonian tribe of Brazil, Noke Koi, for their great love for Mother Earth and our brothers on the way, thank you for your light and your guide.

To my friends Rodolfo Antulio Luna Rodriguez, Martín Hernández, Piero Rangel Fonseca, Antonio and Magdalena Moreno Hernández, Pablo and Gabriela Sanchez de la Rosa, Martín and Pedro Ramírez, Héctor Lugo, Ismael Bernardino, Alfonso Alcantara, Diana Goméz, David Hernandez Piedra, Indira Lara, Fabiola Segobia, Marco Vinicio Beltrán, Ingird and Maru Arroyo, Moises Jardínes, Linda Aceves, Marevi Santos, Gabriel Villegas, Eduardo Ordoñez, Juan Arevalo, Edgar Fragoso, Francisco Delgado, Marcelo Vaca, Alejandro Arredondo -Billy- and to all those that I do not mention but with whom I have shared and enjoyed many things that have left a mark on my life, thank you very much friends for all those beautiful and unforgettable moments

To the invaluable women who have marked their history and left their mark on my life, Romina Julieta Arevalo Mendez, Karina Gabriela Flores Goméz, Lidia Marquez Osco, Dafne Elisabeth Reyes Ramos, Elizabert Martinez, Diana Jardines, Norma Barrios, Laura Casillas, Blanca Mendez, Cony Peréz, Olivia Reyes, Luz Arreola Medrano, and Monica Paez, they are and will be my guides on the way to coexistence with the feminine spirit!

I wish to deeply thank my colleague and friend Rafael Guillermo González-Acuña; having shared part of this adventure with him is an invaluable gift, the hard work and the long hours of solitude allowed our ideas to be written in the history of science, thank you very much again, comrade.

Finally, I want to thank all those who dedicated time from their lives to understand our theory. This work is my last scientific contribution in this direction, my path ends here and I close my cycle as an optical designer.

I wish you the best for your future challenges in the field of optical design!

Author biographies

Rafael G González-Acuña

Rafael G González-Acuña studied industrial physics engineering at the Tecnológico de Monterrey and studied for the masters degree in optomechatronics at the Centro de investigaciones en Óptica, A.C. He is currently studying for his PhD at the Tecnológico de Monterrey. His doctoral thesis focuses on the design of spherical-aberration-free lenses. He is co-author of the solution to the problem of designing bi-aspheric singlet lenses free from spherical aberration. He is co-author of the book *Analytical Lens Design.*

Héctor A Chaparro-Romo

Héctor A Chaparro-Romo is a Mexican scientist born in 1985 in Mexico City. He obtained his degree in electronic engineering at the Universidad Autónoma Metropolitana and is currently studying for a degree in Economics at the Universidad Nacional Autónoma de México. He is co-author of the solution to the problems of spherical aberration and deformable mirrors for adaptative optics in the lens design field. He is also co-author of several peer-reviewed scientific articles and books on the analytical design of optical systems. Héctor is an independent and self-employed researcher in his home office, where he fully focuses his capabilities in the complex field of computer networks and the Internet. As a pioneer, his main goal is to develop http://www.biaspheric.com as a reference portal for all those who want to achieve a deep understanding of the theory of optical design that has been developed from a rigorous analytical paradigm. He is co-author of the books *Analytical Lens Design* (2019), *Stigmatic Optics* (2020), and *Optics and Artificial Vision* (2021).

Part I

Introduction to optical path theory

Chapter 1

The path of light

In this chapter, we cover the preliminary concepts necessary to understand the general formula for adaptive optics mirrors. These concepts are Fermat's principle, aberration theory, stigmatism, adaptive optics, and null lenses.

1.1 Purpose and introduction to this treatise

This treatise was written in order to explain a recently-published equation in greater detail; this equation is the general formula for adaptive optics mirrors, which was published in January 2021 and highlighted with the 'editor's pick' distinction. As a result of this publication, the authors decided to write a treatise in which an explanation of the abovementioned formula could be set out in more detail. Thus, a book on the topic of the published article would be quite useful. The book would have to address the introductory topics, the solution, the implications, and the potential applications. We decided to do all this because we agreed that the abovementioned equation is the analytical closed-form solution to the fundamental problem in adaptive optics.

We have called this project 'optical path theory' because the derivation of this enigmatic equation was only necessary to study the path of light. In other words, we only use Fermat's principle and its implications. For us, Fermat's principle is the only axiom of optical path theory and all its implications are the theorems. Thus, the aforementioned equation is just one more theorem derived from Fermat's principle. This is something that we explore in the following chapters.

This treatise is written in three parts: the first is the introduction, consisting of this single chapter. In this chapter, we will discuss the issues necessary to understand the equation this book is about. In addition, this chapter will study other concepts related to possible applications, such as adaptive optics and null lenses. These concepts will only be covered in a superficial way and the reader is recommended to check the bibliography of this chapter on these topics for more in-depth and detailed reading.

doi:10.1088/978-0-7503-4705-1ch1

The second part deals with the abovementioned equation in two dimensions and its direct implications for aspherical design. The third part consists of the generalization of the chapters of the second part to three dimensions, i.e. we move from aspherical designs to freeform designs.

A characteristic of this treatise is that from chapter 2, each equation is directly explained step by step and the code is later presented in the `Mathematica` coding language, ready to be executed, so that the reader can easily reproduce the content of this book. `Mathematica`, officially called Wolfram Mathematica, is a package containing several built-in libraries that have applications in symbolic computing, matrices, plotting, implementing algorithms, creating user interfaces, and interacting with programs written in other programming languages.

1.2 The optical path and Fermat's principle

This treatise studies optics using the paradigm of geometric optics. Geometric optics is the study of the behaviour of light according to geometry. How would light relate to geometry? Well, everything makes sense if and only if you have a postulate that justifies it. That postulate is Fermat's principle. Fermat's principle is the only axiom of the whole theory that we are going to develop.

An axiom is a truth in itself that does not need to be proved. What we ask ourselves is: if this is true, what else is true? What are the implications if Fermat's principle is true? The most important implication is that under this paradigm, the behaviour of light tends to be geometric in nature. We mean that the propagation of light tends to be geometric in nature.

Rays are typically used to describe the propagation of light in geometric optics. Rays are useful abstractions that approximate the paths along which light propagates under certain conditions. The first premise of the propagation of light in straight lines is only valid when light travels within a homogeneous medium. A homogeneous medium has a constant refractive index, which it is a dimensionless number that describes how fast light travels through it. The refractive index is defined by

$$n = \frac{c}{v} \tag{1.1}$$

where c is the speed of light in a vacuum and v is the phase velocity of light in the medium. In this book, we only work with homogeneous mediums. The premise of 'in straight lines' comes, in fact, from Fermat's principle.

According to Fermat's principle, a ray of light can pass between two points in the shortest possible time when travelling between them.

Fermat's principle can be expressed as follows:

> The optical length of the path followed by light between two fixed different points is the global minimum. The optical length is the physical length multiplied by the refractive index of the medium.

It is important to remark on the word *global minimum*. Generally, the global minimum refers to the smallest value of a set. Consider for a moment that we have a set whose elements are all those possible optical paths between two points. There is an optical length for each of these paths. According to Fermat's principle, the only path in our set that is physically valid is the one that has the shortest optical path length (OPL). The OPL is the physical path multiplied by the refractive index of the medium.

Fermat's principle can be mathematically described as the amount of time it takes for a ray point to travel between two points, T, as follows:

$$T = \int_{t_0}^{t_1} dt = \frac{1}{c}\int_{t_0}^{t_1} \frac{c}{v}\frac{ds}{dt}dt = \frac{1}{c}\int_{A}^{B} nds, \tag{1.2}$$

where the points are A and B, ds is an infinitesimal displacement along the ray, $v = ds/dt$ is the speed of light in the medium, $n = c/v$ is the refractive index of that medium, t_0 is the starting time (the ray is in A), and t_1 is the arrival time at point B. The optical path length of a ray from A to B is given by the following integral:

$$S = \int_{A}^{B} nds, \tag{1.3}$$

which is related to the travel time by $S = cT$. The optical path length is a purely geometrical quantity, since time is not considered in its calculation. The global minimum of the light's travel time between two points A and B is equivalent to the global minimum of the optical path length between A and B.

Fermat's principle not only predicts the straight path of the light across a homogeneous medium but also describes the change of the light's direction when it touches a mirror or passes through another medium. The first phenomenon is called reflection and the second is called refraction. We will explore them in the following subsections.

1.3 The law of reflection

Reflection is the change in the direction of a ray at an interface between two different media, so that the ray returns into the medium from which it originated. This media is usually a polished surface or a mirror.

When light is reflected, it behaves according to Fermat's principle. Please see figure 1.1; a light beam is shown that is incident to a surface. Reflection occurs at the surface. We can see that the ray, if it originates at x_1, y_1, travels in a straight line to the point where it is reflected in the surface and then reaches point x_2, y_2. The time it takes for light to travel from point x_1, y_1 to point x_2, y_2 can be written as follows:

$$T(x) = \frac{1}{c}\left[\sqrt{(x - x_1)^2 + y_1^2} + \sqrt{(x_2 - x)^2 + y_2^2}\right], \tag{1.4}$$

where x is the point on the plane in the horizontal direction where the light strikes. By computing $T(x)$, we find that light moves in straight lines in places with constant refractive indices, just as Fermat's principle predicts.

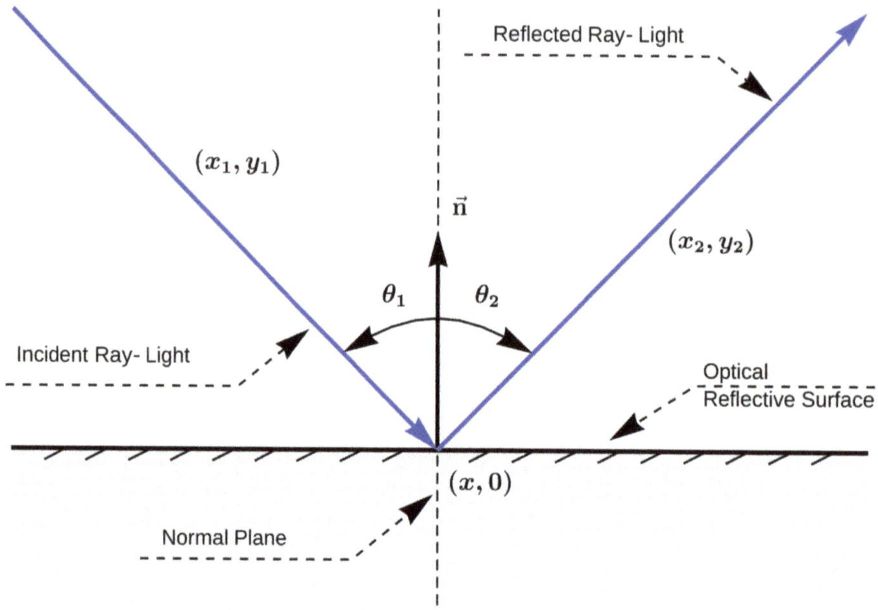

Figure 1.1. Schematic diagram of ray tracing for the phenomenon of light ray reflection. A ray of light falls on the reflective surface at the point $(x, 0)$ and is deflected at the same angle as the one at which it arrived; both rays coexist within the same optical medium.

However, we are unaware of how the light reflects. There is a critical point at $(x, 0)$. The slopes of the straight lines are determined by the value of x. In accordance with Fermat's principle, light moves within the shortest amount of time. Therefore, we can derive the following with respect to x and equal to zero:

$$\frac{\partial T(x)}{\partial x} = 0, \tag{1.5}$$

and by computing the derivative

$$\frac{x - x_1}{\sqrt{(x - x_1)^2 + y_1^2}} = \frac{x_2 - x}{\sqrt{(x_2 - x)^2 + y_2^2}}, \tag{1.6}$$

from figure 1.1, we can relate the angles θ_1 and θ_2 using the last expression:

$$\sin\theta_1 = \sin\theta_2, \tag{1.7}$$

to finally obtain

$$\boxed{\theta_1 = \theta_2}. \tag{1.8}$$

A simple representation of reflection in a flat mirror can be seen in figure 1.1, where the incident angle θ_1 is equal to the reflected angle θ_2, which leads to $\theta_1 = \theta_2$. In angular notation, this is Snell's law for reflection.

By computing the second derivative, we can show that this is a real minimum:

$$c\frac{\partial^2 T}{\partial x^2} = \frac{1 - \sin^2\theta_1}{\sqrt{(x - x_1)^2 + y_1^2}} + \frac{1 - \sin^2\theta_2}{\sqrt{(x_2 - x)^2 + y_2^2}}$$
$$= (1 - \sin^2\theta_2)\left[\frac{1}{\sqrt{(x - x_1)^2 + y_1^2}} + \frac{1}{\sqrt{(x_2 - x)^2 + y_2^2}}\right]. \tag{1.9}$$

Since $\sin\theta_2^2 \leqslant 1$ is positive for all values of x, then equation (1.9) is an absolute minimum.

1.4 The law of refraction

Light is refracted when passing through an area with a changing refractive index. A simple case of refraction occurs at a point where there is an interface between a uniform medium with an index of refraction of n_1 and another medium with an index of refraction of n_2. Snell's Law, also known as the law of refraction, describes how light is deflected in such a situation. This is a result of Fermat's principle.

Refraction is illustrated in figure 1.2, in which two transparent media with equal refractive indexes interact at a flat surface. Based on the figure, we can deduce that light travels as follows:

$$R(x) = n_1\sqrt{d_1^2 + x^2} + n_2\sqrt{d_2^2 + (L - x)^2}\,. \tag{1.10}$$

The height of the light ray at its initial position is defined as d_1. In a coordinate system, x is the horizontal distance between the light ray's origin and its initial position. In the diagram below, L represents the horizontal distance between the starting and finishing positions. Finally, in the diagram above, d_2 is the final height of the light ray.

By differentiating equation (1.10) with respect to x, we can find the path that the light covers in the shortest amount of time.

$$\frac{dR(x)}{dx} = n_1\frac{x}{\sqrt{d_1^2 + x^2}} - n_2\frac{(L - x)}{\sqrt{d_2^2 + (L - x)^2}}. \tag{1.11}$$

As we can see in figure 1.2, $\sin\theta_1$ is equal to

$$\sin\theta_1 = \frac{x}{\sqrt{d_1^2 + x^2}}. \tag{1.12}$$

As we can also see from the same figure, $\sin\theta_2$ is

$$\sin\theta_2 = \frac{L - x}{\sqrt{d_2^2 + (L - x)^2}}. \tag{1.13}$$

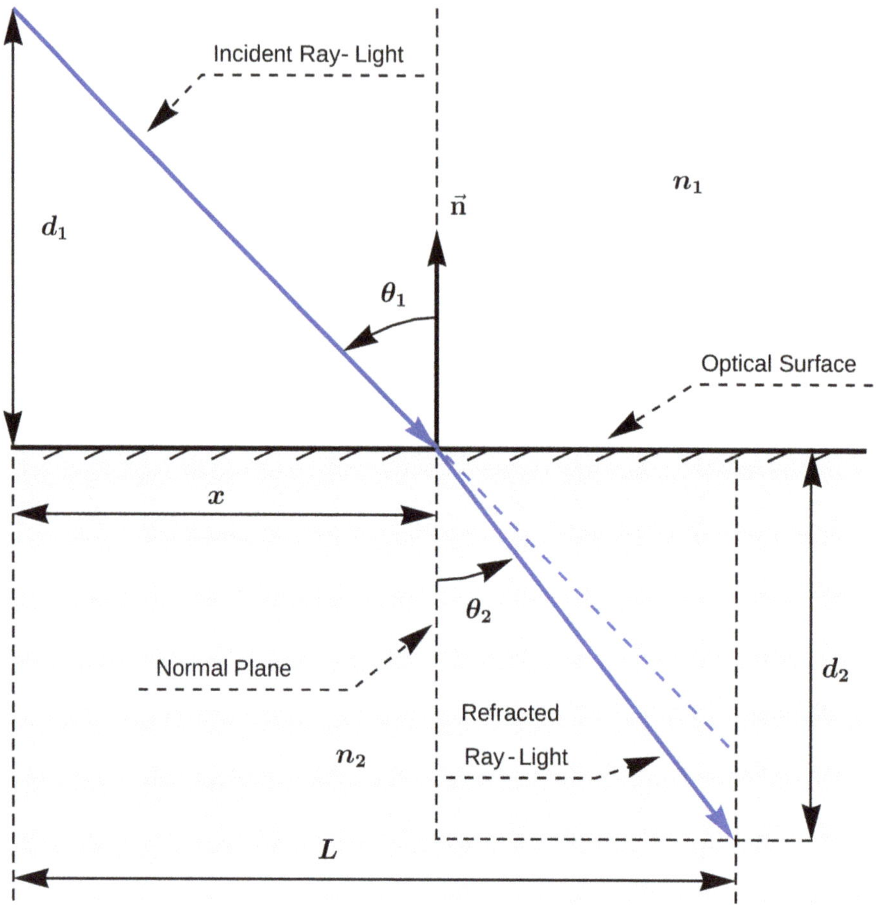

Figure 1.2. Schematic diagram of ray tracing for the phenomenon of light ray refraction. A ray of light falls on the refractive surface and is deflected at a different angle than the one it exhibited on entry; the participating rays are in different optical media.

By replacing equation (1.11) and (1.12) in equation, we obtain

$$\boxed{n_1 \sin \theta_1 = n_2 \sin \theta_2.} \tag{1.14}$$

If the light rays are moving through a homogeneous medium, their paths are straight. A change of homogeneous medium will result in refraction (deviation) of the light. This is expressed by the equation above. In angular notation, this is Snell's law for refraction.

1.5 The vector form of Snell's law

In vector form, Snell's law can be expressed in a more manageable manner. To obtain Snell's law in vector form, let $\vec{a}_1$ be the incident unit vector, let $\vec{a}_2$ be the refracted unit vector, and let $\vec{n}$ be the normal unit vector of the surface; θ_1 is the incident angle with respect to $\vec{n}$ and θ_2 is the refracted angle with respect to $-\vec{n}$.

From figure 1.2 we can now express the refracted vector, which is given by

$$\vec{\mathbf{a}}_2 = \frac{n_1}{n_2}\vec{\mathbf{a}}_1 + \left(\vec{\mathbf{a}}_2 \cdot \vec{\mathbf{n}} - \frac{n_1}{n_2}\vec{\mathbf{a}}_1 \cdot \vec{\mathbf{n}}\right)\vec{\mathbf{n}}, \tag{1.15}$$

and using the definition of the dot product between vectors,

$$\vec{\mathbf{a}}_2 = \frac{n_1}{n_2}\vec{\mathbf{a}}_1 + \left(\cos\theta_2 - \frac{n_1}{n_2}\cos\theta_1\right)\vec{\mathbf{n}}, \tag{1.16}$$

equation (1.16) can also expressed as

$$\vec{a}_2 = \frac{n_1}{n_2}\vec{a}_1 + \left(\frac{n_1}{n_2}|\vec{n}||\vec{a}_1|\cos\theta_1 - |\vec{n}||\vec{a}_2|\cos\theta_2\right)\vec{n}. \tag{1.17}$$

From figure 1.2 we deduce that the vertical component of the refracted blue ray $\vec{\mathbf{a}}_2$ is,

$$\vec{\mathbf{n}} \cdot \vec{\mathbf{a}}_2 = |\vec{\mathbf{n}}||\vec{\mathbf{a}}_2|\cos\theta_2. \tag{1.18}$$

Remember that we are working with unit vectors, so $|\vec{\mathbf{n}}| = |\vec{\mathbf{a}}_2| \equiv 1$,

$$\vec{\mathbf{n}} \cdot \vec{\mathbf{a}}_2 = \cos\theta_2, \tag{1.19}$$

and squaring both sides, we obtain

$$(\vec{\mathbf{n}} \cdot \vec{\mathbf{a}}_2)^2 = \cos\theta_2^2 = 1 - \sin\theta_2^2 = 1 - \frac{n_1^2}{n_2^2}\sin\theta_1^2. \tag{1.20}$$

From the definition of the cross-product, we know that

$$\vec{\mathbf{n}} \times \vec{\mathbf{a}}_1 = |\vec{\mathbf{n}}||\vec{\mathbf{a}}_1|\sin\theta_1(\vec{\mathbf{e}}_1 \times \vec{\mathbf{e}}_2), \tag{1.21}$$

and also that $|\vec{\mathbf{n}}| = |\vec{\mathbf{a}}_1| = 1$.

$$\vec{\mathbf{n}} \times \vec{\mathbf{a}}_1 = \sin\theta_1; \tag{1.22}$$

squaring both sides in the last expression, we obtain

$$(\vec{\mathbf{n}} \times \vec{\mathbf{a}}_1)^2 = \sin\theta_1^2. \tag{1.23}$$

By replacing the above expression in $(-\vec{\mathbf{n}} \cdot \vec{\mathbf{a}}_2)^2 = 1 - (n_1^2/n_2^2)\sin\theta_1^2$, we get

$$\vec{\mathbf{n}} \cdot \vec{\mathbf{a}}_2 = \sqrt{1 - \frac{n_1^2}{n_2^2}(\vec{\mathbf{n}} \times \vec{\mathbf{a}}_1)^2}. \tag{1.24}$$

On the other hand, the other terms of equation (1.17) are

$$\frac{n_1}{n_2}\left[\vec{\mathbf{a}}_1 + |\vec{\mathbf{n}}||\vec{\mathbf{a}}_1|\cos\theta_1\vec{\mathbf{n}}\right] = \frac{n_1}{n_2}\left[\vec{\mathbf{a}}_1 - (\vec{\mathbf{n}} \cdot \vec{\mathbf{a}}_1)\vec{\mathbf{n}}\right]. \tag{1.25}$$

As a consequence, Snell's law in vector form is

$$\boxed{\vec{\mathbf{a}}_2 = \frac{n_1}{n_2}[\vec{\mathbf{a}}_1 - (\vec{\mathbf{n}} \cdot \vec{\mathbf{a}}_1)\vec{\mathbf{n}}] - \vec{\mathbf{n}}\sqrt{1 - \frac{n_1^2}{n_2^2}(\vec{\mathbf{n}} \times \vec{\mathbf{a}}_1)^2}}. \tag{1.26}$$

Equation (1.26) will be very useful and will be applied in the following chapters.

The vector notation form of Snell's law for reflection is equation (1.27); we set $\frac{n_1}{n_2} = -1$, thus

$$\boxed{\vec{\mathbf{a}}_2 = -[\vec{\mathbf{a}}_1 - (\vec{\mathbf{n}} \cdot \vec{\mathbf{a}}_1)\vec{\mathbf{n}}] - \vec{\mathbf{n}}\sqrt{1 - (\vec{\mathbf{n}} \times \vec{\mathbf{a}}_1)^2}.} \tag{1.27}$$

1.6 The wavefront and the Malus–Dupin theorem

Light follows the shortest path, according to Fermat's principle. One can think of the refractive index as a measurement of a material's speed of light. Therefore, the incident and refracted rays are related equally in angular terms, according to Fermat's law. Although Snell's law and Fermat's principle share some information in common, they are not the same. Snell's law only relates to the angles at which light passes from one medium to another and is an implication of Fermat's principle.

An inherent property of geometric optics that connects it to wave optics is the Malus–Dupin theorem. According to the Malus–Dupin theorem, if one follows the same optical paths along all the rays emitted from a source, then there is a surface normal to all those rays. 'Wavefront' is the name given to the surface perpendicular to the rays. Fermat's principle provides proof of the Malus–Dupin theorem. The Malus–Dupin theorem holds, no matter how many reflections or refractions a ray may experience en route to its destination.

Two infinitesimally separate paths are taken, $[AB]$ and $[AB']$, where A is the focus and B and B' are the arrival points separated by equal optical paths. As a result, we designate the following optical paths:

$$L_B = \int_A^B n(\vec{\mathbf{r}})ds \tag{1.28}$$

and

$$L_{B'} = \int_A^{B'} n(\vec{\mathbf{r}}')ds', \tag{1.29}$$

where $\vec{\mathbf{r}}$ and $\vec{\mathbf{r}}'$ are the respective positional vectors, ds and ds' are the respective differentials of space, and $n(\vec{\mathbf{r}})$ is the refractive index. In this case, we admit that the refractive index can be differentiable. We also consider the following relationship: $\vec{\mathbf{r}'} = \vec{\mathbf{r}} + \varepsilon\vec{\mathbf{b}}$, in which $\vec{\mathbf{dr}}' = \vec{\mathbf{dr}} + \varepsilon\vec{\mathbf{db}}$ is true if and only if $\varepsilon \approx 0$. See figure 1.3.

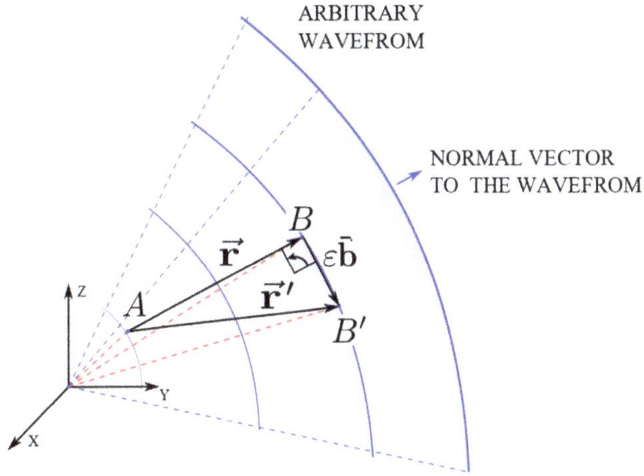

Figure 1.3. Schematic diagram of the Malus–Dupin theorem, an abstraction of the relationship between the vectors **r** and **r**′ that describes the relationship between the propagation of the wave and the light rays.

Our derivation is based on the first-order Taylor series,

$$f(\vec{\mathbf{r}}) = f(\vec{\mathbf{r}}_0) + \nabla f(\vec{\mathbf{r}}_0) \cdot (\vec{\mathbf{r}} - \vec{\mathbf{r}}_0). \tag{1.30}$$

We can obtain the refractive index in terms of a first-order Taylor series development,

$$n(\vec{\mathbf{r}}') = n(\vec{\mathbf{r}}) + \nabla n(\vec{\mathbf{r}}) \cdot (\vec{\mathbf{r}}' - \vec{\mathbf{r}}) = n(\vec{\mathbf{r}}) + \nabla n(\vec{\mathbf{r}}) \cdot \varepsilon\vec{\mathbf{b}}. \tag{1.31}$$

We do the same for the positional vector:

$$|\vec{\mathbf{r}}'| = |\vec{\mathbf{r}}| + \nabla|\vec{\mathbf{r}}| \cdot (\vec{\mathbf{r}}' - \vec{\mathbf{r}}) \tag{1.32}$$

$$= |\vec{\mathbf{r}}| + \frac{\partial}{\partial r} r\vec{\mathbf{u}} \cdot (\vec{\mathbf{r}}' - \vec{\mathbf{r}}) \tag{1.33}$$

$$= |\vec{\mathbf{r}}| + \vec{\mathbf{u}} \cdot (\vec{\mathbf{r}}' - \vec{\mathbf{r}}) \tag{1.34}$$

$$= |\vec{\mathbf{r}}| + \vec{\mathbf{u}} \cdot \vec{\mathbf{r}}' - \vec{\mathbf{u}} \cdot \vec{\mathbf{r}} \tag{1.35}$$

$$= |\vec{\mathbf{r}}| + \vec{\mathbf{u}} \cdot \vec{\mathbf{r}}' - |\vec{\mathbf{r}}| \tag{1.36}$$

$$= \vec{\mathbf{u}} \cdot \vec{\mathbf{r}}', \tag{1.37}$$

where $\vec{\mathbf{r}} = \vec{\mathbf{u}}$, so $\vec{\mathbf{u}}$ is the vector in the direction of the ray. Then, for the differential ds',

$$ds' = |\vec{\mathbf{dr}}'| = \vec{\mathbf{u}} \cdot \vec{\mathbf{dr}}' = \vec{\mathbf{u}} \cdot \vec{\mathbf{dr}} + \vec{\mathbf{u}} \cdot \varepsilon\vec{\mathbf{db}} = ds + \varepsilon\vec{\mathbf{u}} \cdot \vec{\mathbf{db}}. \tag{1.38}$$

The last expression is in equation (1.29),

$$L_{B'} = \int_A^{B'} n(\vec{\mathbf{r}'})ds' = \int_A^{B'} \left[n(\vec{\mathbf{r}}) + \nabla n \cdot \varepsilon\vec{\mathbf{b}}\right](ds + \varepsilon\vec{\mathbf{u}} \cdot \vec{\mathbf{db}}); \tag{1.39}$$

expanding the above equation, we obtain

$$\begin{aligned} L_{B'} = & \int_A^{B'} n(\vec{\mathbf{r}})ds + \int_A^{B'} \nabla n \cdot \varepsilon\vec{\mathbf{db}} + \int_A^{B'} n(\vec{\mathbf{r}})\varepsilon\vec{\mathbf{u}} \cdot \vec{\mathbf{db}} + \\ & \int_A^{B'} \nabla n \cdot \varepsilon\vec{\mathbf{b}}\varepsilon\vec{\mathbf{u}} \cdot \vec{\mathbf{db}}. \end{aligned} \tag{1.40}$$

In the last expression we set $\varepsilon^2 \to 0$, because we considered that $\varepsilon \approx 0$, so,

$$L_{B'} = \int_A^{B'} n(\vec{\mathbf{r}})ds + \int_A^{B'} \nabla n \cdot \varepsilon\vec{\mathbf{db}} + \int_A^{B'} n(\vec{\mathbf{r}})\varepsilon\vec{\mathbf{u}} \cdot \vec{\mathbf{db}}. \tag{1.41}$$

We calculate the optical path difference ΔL using equation (1.28) minus equation (1.41):

$$\Delta L = \int_A^{B'} n(\vec{\mathbf{r}})ds + \int_A^{B'} \nabla n \cdot \varepsilon\vec{\mathbf{b}}ds + \int_A^{B'} n(\vec{\mathbf{r}})\varepsilon\vec{\mathbf{u}} \cdot \vec{\mathbf{db}} - \int_A^{B} n(\vec{\mathbf{r}})ds. \tag{1.42}$$

In the last equation, we now collect the terms multiplied by ε:

$$\Delta L = \int_A^{B'} n(\vec{\mathbf{r}})ds - \int_A^{B} n(\vec{\mathbf{r}})ds + \varepsilon\left[\int_A^{B'} \nabla n \cdot \vec{\mathbf{b}}ds + \int_A^{B'} n(\vec{\mathbf{r}})\vec{\mathbf{u}} \cdot \vec{\mathbf{db}}.\right] \tag{1.43}$$

By differentiating the trajectories, we find that $d(n\vec{\mathbf{u}} \cdot \vec{\mathbf{b}}) = d(n\vec{\mathbf{u}}) \cdot \vec{\mathbf{b}} + n\vec{\mathbf{u}} \cdot \vec{\mathbf{db}} = \nabla n \cdot \vec{\mathbf{b}}ds + n\vec{\mathbf{u}} \cdot \vec{\mathbf{db}}$, so equation (1.43) becomes

$$\Delta L = \int_A^{B'} n(\vec{\mathbf{r}})ds - \int_A^{B} n(\vec{\mathbf{r}})ds + \varepsilon\left[\int_A^{B'} d(n\vec{\mathbf{b}} \cdot \vec{\mathbf{u}})\right]. \tag{1.44}$$

If we assume that we have chosen $B \equiv B'$ then

$$\Delta L = \int_A^{B'} n(\vec{\mathbf{r}})ds - \int_A^{B'} n(\vec{\mathbf{r}})ds + \varepsilon\left[\int_A^{B'} d(n\vec{\mathbf{b}} \cdot \vec{\mathbf{u}})\right] = 0 \tag{1.45}$$

turns into

$$\Delta L = +\varepsilon\left[\int_A^{B'} d(n\vec{\mathbf{b}} \cdot \vec{\mathbf{u}})\right] = 0. \tag{1.46}$$

Thus,

$$\int_A^{B'} d(n\vec{\mathbf{b}} \cdot \vec{\mathbf{u}}) = 0, \tag{1.47}$$

which implies

$$n\vec{\mathbf{b}} \cdot \vec{\mathbf{u}} = 0. \tag{1.48}$$

Dividing by n in both sides, we obtain

$$\vec{\mathbf{b}} \cdot \vec{\mathbf{u}} = 0, \tag{1.49}$$

which also implies

$$\vec{\mathbf{b}} \perp \vec{\mathbf{u}} \tag{1.50}$$

and since $\vec{\mathbf{u}}$ is the vector in the direction of the ray and $\vec{\mathbf{b}}$ is the vector tangential to the wavefront, we can conclude that rays are perpendicular to waves.

As a final remark, in this book we also call the wavefront predicted by the Malus–Dupin theorem 'the eikonal.'

1.7 Optical path difference and phase difference

At this point we have shown that the trajectory of light is described by rays and that there is a surface perpendicular to the rays. In reality, that surface, the wavefront, is a surface with a constant phase. To consider what we mean by 'phase,' we first consider a point source of light emitting rays in all possible directions around it. The OPL of a ray is given by the length travelled multiplied by the refractive index where the ray travels. For our example, the OPL is a radius multiplied by the refractive index. These rays form a surface perpendicular to themselves that expands as the light travels further from the source; this surface is the wavefront. The wavefront is a surface with equal phase, so it is also called a phase front. The concept of phase comes from the electromagnetic nature of the wave phenomenon. An electric field can be represented as

$$u = u_o \sin \Phi = u_o \sin\left(\frac{2\pi(x - ct)}{\lambda}\right), \tag{1.51}$$

where u_o is the amplitude of the field, x is a variable representing special propagation, t is the time, c is the speed of light, and λ is the wavelength. The term Φ is the phase, which is a point in the cycle of the sine function. So if $\Phi = 0$ then $u = 0$ or if $\Phi = 90°$ then $u = u_o$. Going back to our point source, the light emitted from it has the same phase Φ_o, no matter the direction of the ray. So along the ray path, the cyclic nature of light is presented in the number of wavelengths necessary to fit the OPL. See figure 1.4. Thus, at the surface of the wavefront, the rays have the same phase Φ_o. For this example, the wavefront is stigmatic, since it comes from a single point source and in fact, it is a spherical wavefront. If this point source is far away, at infinity, then we can see that the spherical wavefront turns into into a flat wavefront.

1.8 Stigmatism and aberrated wavefronts

Once we know how light propagates, it is time to manipulate it to get artifacts of interest.

Let us suppose that one has a solid body and that the refractive index of the environment is different from that of the body. When the physical body focuses or

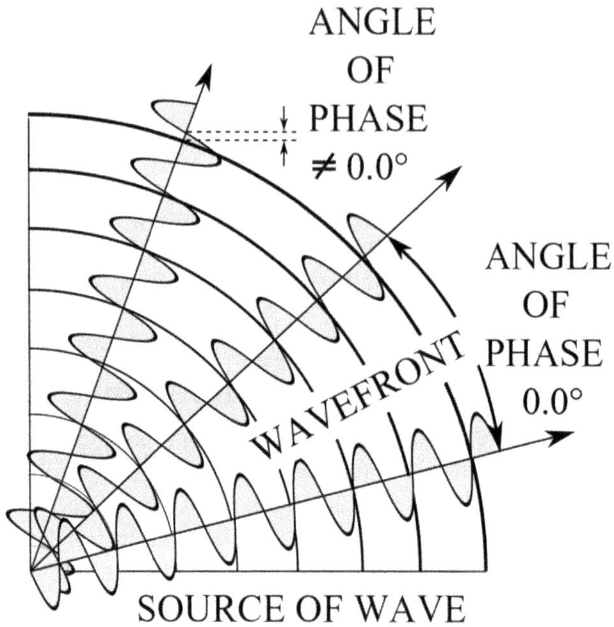

Figure 1.4. Schematic diagram showing the phase between different waves that propagate from the same source. Two possible cases are shown: (1) when the waves travel in phase, (2) when there is a phase shift between them.

disperses rays due to the difference of refraction indices between it and the medium, it is a lens. The study of lenses is central to geometric optics and in this treatise we are going to study them in an interesting way.

Light is focused and scattered by lenses. Mirrors also possess such features. All this happens due to the laws of reflection and refraction. Surface normals determine both the law of reflection and the law of refraction. Consequently, the shape of the lens/mirror plays a crucial role in determining its function.

A basic principle of the focusing of light is that after rays are emitted by a point object, they converge into an image. Consequently, stigmatic lenses and mirrors are required. The stigmatism of an optical system describes the process of focusing a point object into a point image. Such points are called a stigmatic pair of the optical system. Mirrors and stigmatic lenses need to have very specific shapes.

Stigmatic mirrors are formed from conic sections of reflective surfaces.

Rays reflected by mirrors with parabolic surfaces converge into a common focus when parallel rays hit them. When a mirror has a spherical surface, all rays that emerge from a point object are reflected back to the same location if and only if the object is situated at the centre of the circumference. Rays reflected from a focus of an elliptical mirror appear as a point image at the other focus. Last but not least, rays that originate from a point object are reflected in a single virtual point image when they are reflected by mirrors with hyperbolic surfaces. Similarly, other curved surfaces may also focus light, but not on a single point. See figure 1.5.

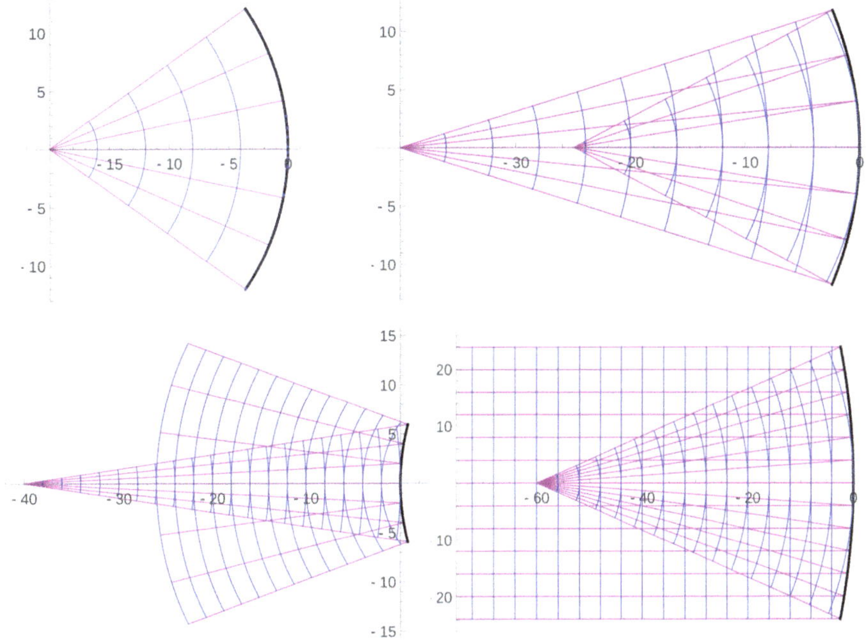

Figure 1.5. Schematic diagrams of the reflective behaviours of conical mirrors. Top right: a spherical mirror; top left: an elliptical mirror; bottom right: a hyperbolic mirror; bottom right: a parabolic mirror.

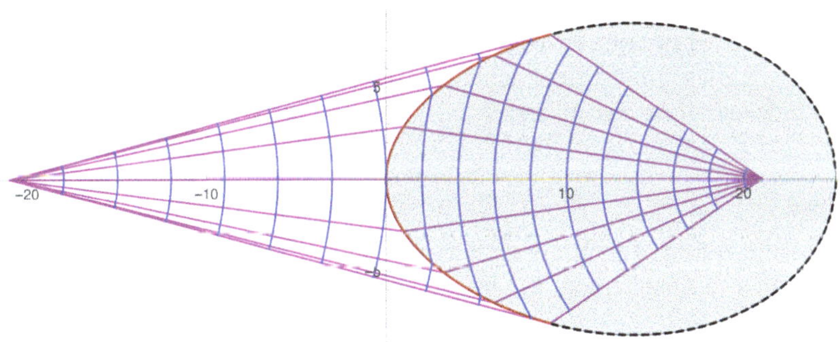

Figure 1.6. Schematic diagram of the refractive behaviour of a Cartesian oval surface. By definition, the Cartesian oval is the stigmatic surface.

In terms of refractive surfaces, the stigmatic surface is the Cartesian oval, which is a fourth-order function. Another way of explaining this is that when the rays emitted from a point object are refracted, they focus on a single image point. See figure 1.6.

So, recalling the Malus–Dupin theorem that rays are perpendicular to the wavefront, observe that the spherical wavefronts are generated by point objects, and flat wavefronts are also generated by point objects but this time, the objects are far away. Stigmatic optics deals with the optical artifacts that convert spherical wavefronts to other spherical wavefronts or plane wavefronts.

Optical path theory is more general than stigmatic optics; it contemplates the path of light described by its pure nature, the Fermat principle. It also may study the path of light in optical systems with aberrations. A system has aberrations if it is not stigmatic for all points of an object. It is not always possible to resolve the point object at a single point on the image space, in which case, the optical system has aberrations. Depending on the type of aberration, the region of space where the image is formed differs.

The two main types of errors are chromatic and monochromatic. Chromatic aberrations are caused by variations in the refractive index of a lens with regard to the wavelength. Monochromatic aberrations occur because of optical system geometry. There are three laws associated with it: the reflection law, Snell's law, and Fermat's principle (note that the first two laws are implications of Fermat's principle). Information about the imaging system geometry is available to Fermat's principle. It is crucial to remove monochromatic aberrations by designing the optical system to have a specific shape. The most common monochromatic aberrations are spherical aberration, along with coma, astigmatism, field curvature, and image distortion.

1.8.1 Spherical aberration

The term spherical aberration refers to a situation in which a point object located on an optical axis lacks a stigmatic correlation with a point image. Rays leaving a point object on an optical axis do not converge to a point image. Because spherical lenses exhibit this phenomenon, it is called spherical aberration. Aspherical lenses are lenses that are non-spherical in shape.

In most cases, they aim to reduce spherical aberrations. Figure 1.7 shows an example of spherical aberration generated by a spherical surface with a constant refractive index n along the material.

1.8.2 Coma

Coma is the aberration of an image caused by an object that is outside the optical axis of the system. As a result of coma, the image appears distorted, with a tail like a comet.

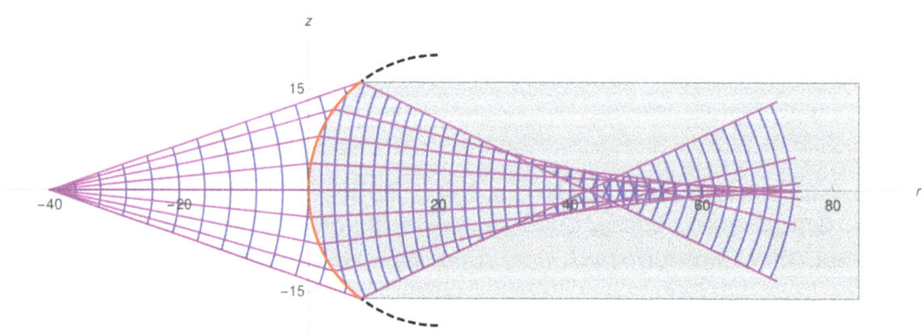

Figure 1.7. Schematic diagram of the behaviour of spherical aberration at a refractive interface. The light source is in a different medium from the one in which the phenomenon of spherical aberration (typical of refractive surfaces) is generated.

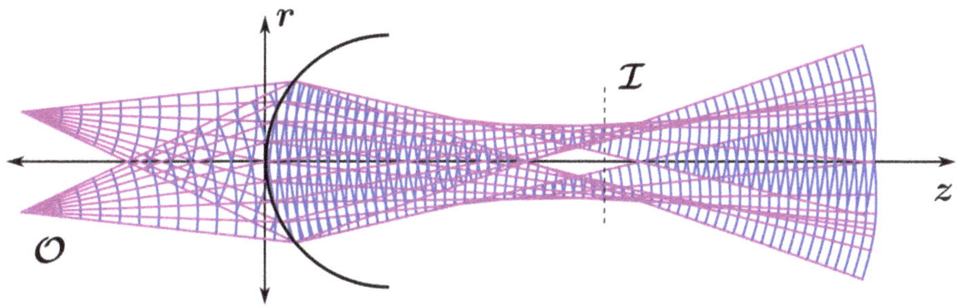

Figure 1.8. Schematic diagram of the behaviour of coma aberration at a refractive interface. The light source is in a different medium from the one in which the coma aberration phenomenon is generated; this phenomenon is typical of the relationship between a refractive surface and positions outside the optical axis where the light sources are located.

Thus, when an optical system has coma, a point outside the optical axis does not have a stigmatic relationship with a point image, since the image is not a point, but a region.

The coma is shown in figure 1.8, in which a ray crosses the surfaces of index refraction n and is refracted; the ray then crosses the interface of interposition. Some of the rays are inverted due to the refraction, as seen in the above figure. These photos depict an inversion that appears to be that of a coma or a comet.

1.8.3 Astigmatism

When two perpendicular planes have different image points, an image becomes astigmatic. These planes are usually called the meridional and sagittal planes. In a three-dimensional model using the x, y, and z coordinate system, the meridional plane is the $y - z$ plane and the sagittal plane is the $x - z$ plane. The direction z is called the optical axis; this is the axis where the light travels, usually from $-z$ to z. An illustration of astigmatism is presented in figure 1.9.

1.8.4 Distortion

A rectilinear projection is used to measure distortion in an image-forming system. When the system is functioning as it should, a rectilinear projection (a projection that renders straight lines in a scene as straight lines in the image) passes through. Figure 1.10 shows the most common distortions.

1.8.5 Field curvature

The Petzval field curvature, named for Joseph Petzval, is an optical aberration in which an object is focused as a curved image. Figure 1.11 shows the field curvature due to a Cartesian oval.

1.9 Adaptivc optics

Adaptive optics is a technique that makes it possible to correct the evolving and non-predictive deformations of a wavefront in real time using a deformable mirror.

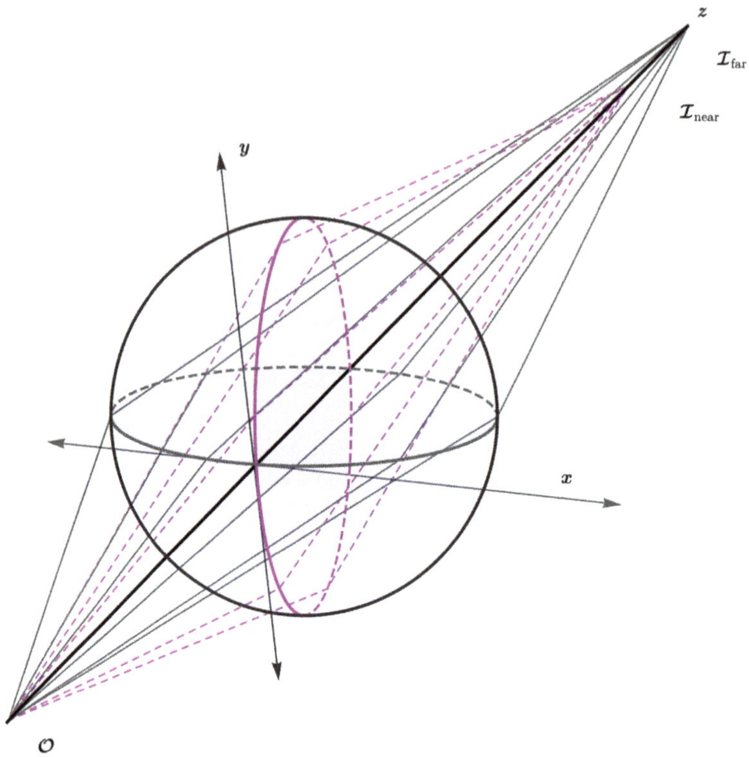

Figure 1.9. Schematic diagram of the aberration of astigmatism at a refractive surface. Note that two image points are formed, one corresponding to the vertical axis and the other to the horizontal axis.

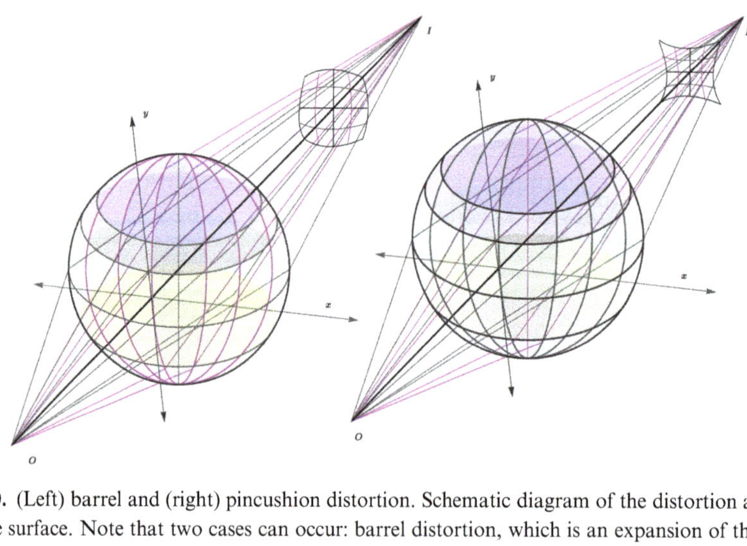

Figure 1.10. (Left) barrel and (right) pincushion distortion. Schematic diagram of the distortion aberration at a refractive surface. Note that two cases can occur: barrel distortion, which is an expansion of the image and pincushion distortion, which is image compression.

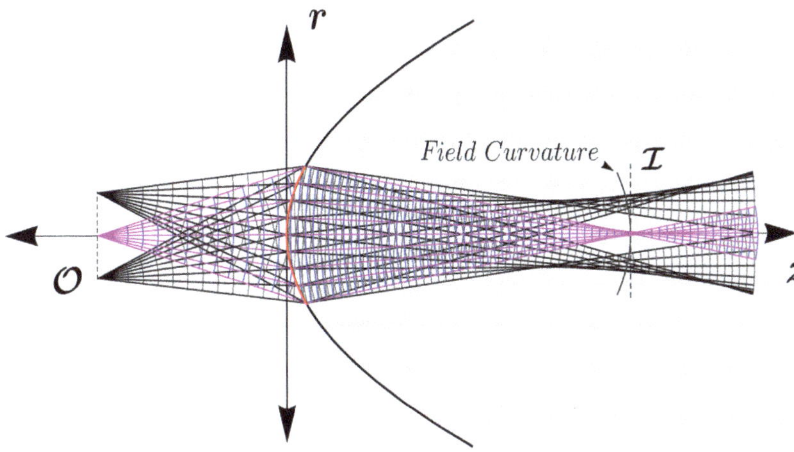

Figure 1.11. Schematic diagram of field curvature aberration or Petzval curvature in a refractive surface. Note that the field in the image plane curves as the object moves away from the optical axis.

Deformable mirrors compensate for non-predictive deformations in real time. In this section, we introduce the general concept of adaptive optics, and in the following subsection, we explore their main components.

Adaptive optics was first developed in the 1950s; its main field of use is astronomy, but it has started to extend to many other fields such as ophthalmologic medicine and telecommunications. Here, we are only going to focus on its astronomical applications.

Astronomers have found that the atmosphere introduces aberrations into the input beams of light that enter telescopes. These aberrations affect image quality and they are very non-predictive. Thus, the main limitation of the quality of astronomical observations is no longer the physical dimensions of the mirrors but atmospheric disturbances.

This observation prompted the creation of high-altitude observatories or even the sending of telescopes into space, since space telescopes are free from atmospheric problems. The problem of atmospheric turbulence can also be solved by adaptive optics, which use computer-controlled rapidly deformable mirrors to compensate for wavefront distortions.

Adaptive optics is an optical technique that allows us to counteract, in real time, the effects of the Earth's atmosphere on the formation of astronomical images. To achieve this, a deformable mirror supported by a set of computer-controlled actuators is inserted into the optical path of the telescope. In astronomy, this technique is used in particular by terrestrial telescopes to correct star observations, among other things.

In the case of stars, if we have the impression that a star is twinkling, it is not because it emits light in a non-constant way, but because of the atmospheric turbulence which distorts the image we have of it; such turbulence distorts the wavefront and thus the phase. Indeed, a star, assumed to be a point in the visible sky and located at a very great distance compared to the scale of the Earth, emits light with a spherical wavefront which, on our scale, can be considered a plane. However, if we consider the case of a telescope that has a primary mirror with a diameter of several tens of meters, the wavefront incident upon the surface of the primary mirror undergoes random deformations when it passes through the atmosphere because of variations in refractive index caused by atmospheric turbulence.

This is explained by the dependence of the refractive index on the temperature and the local pressure of the atmosphere encountered. The optical path traversed by a light ray is defined as the integral of nds, where n is the refractive index and ds is the elementary displacement along the path. The light rays do not travel the same optical path: the wavefront that is observed is then no longer plane, and the image is distorted. In adaptive optics, a wavefront analyser is used to estimate the disturbance due to the atmosphere, and a mirror is deformed so as to exactly compensate for this disturbance. Thus the image after reflection by the mirror is almost as if there had been no degradation by atmospheric turbulence.

In order to use this technique, a reference star is needed in the stellar field: the analysis of its aspect makes it possible to evaluate the disturbances to which the image is subjected in real time. The computer reacts by sending commands to the deformable mirror actuators many times every second; the mirror then takes a shape that compensates for image defects.

This system can also use a reference to an artificial star produced by a laser beam that passes through the layers of air causing poor image quality. Any point or large object in the field of view, such as a galaxy, thus improves its sharpness. It should be noted that adaptive optics not only compensates for the variable disturbances induced by the atmosphere, but is also capable of correcting a good part of the intrinsic aberrations due to telescope optics.

Although it is a technique that finds its natural field of application in astronomy, it is currently also being investigated in adaptive optics applied to human vision. To do this, we proceed to analyse the images of point objects (artificial 'stars') projected onto the retina through the front lenses of the eye (cornea and crystalline). The study of these images makes it possible to evaluate the aberrations due to the organ of vision, and to act on external adaptive systems that introduce the necessary corrections into the incident light beam so that the images projected on the retina are as sharp as possible.

In practice, the implementation of an adaptive optics system begins with the construction of a control matrix. This matrix represents the actuators to be modified to reproduce each of the optical aberrations of the base of the Zernike polynomials.

Zernike polynomials are a series of polynomials that are orthogonal to the unit disk. From the analysis of the disturbance of the wavefront by the atmosphere via a wavefront analyser, we can decompose the wavefront fault on the basis of Zernike polynomials to compensate for the delays using the deformable mirror. In practice, only a limited number of Zernike orders need to be corrected to obtain a sufficiently small residual defect.

If we consider a light wave that has passed through an imperfect system, the wavefront at the output of the system is not completely flat: we define the phase shift function Φ, which at any point of a plane front denotes the phase shift between the theoretical light wave in the geometrical optics model and the real light wave, while taking into account the defects; it would be equal to the null function if the system were perfect. It is then possible to approximate this so-called aberrant phase using a linear combination of Zernike polynomials; each of the polynomials of the base is considered to correspond to a different category of aberration.

Thus, in adaptive optics, it is possible to use a wavefront analyser coupled to a computer system capable of calculating Φ and its decomposition into Zernike polynomials in real time. This allows us to know the nature of the aberrations of the system under study at any time and possibly correct them using a deformable mirror. It is important to remark that in this treatise we are not going to directly use the Zernike polynomials, since the equations that describe the adaptive optical system that we are going to study are simpler if we study them in Cartesian coordinates instead of working in a unitary disk such as that of the Zernike polynomials. However, Zernike polynomials have a direct transformation or translation to Cartesian coordinates. So all the mathematical models that we will present in the following chapter can ultimately be expressed in terms of Zernike polynomials.

1.9.1 The main configuration in adaptive optics

The general configuration of adaptive optics is presented in figure 1.12.

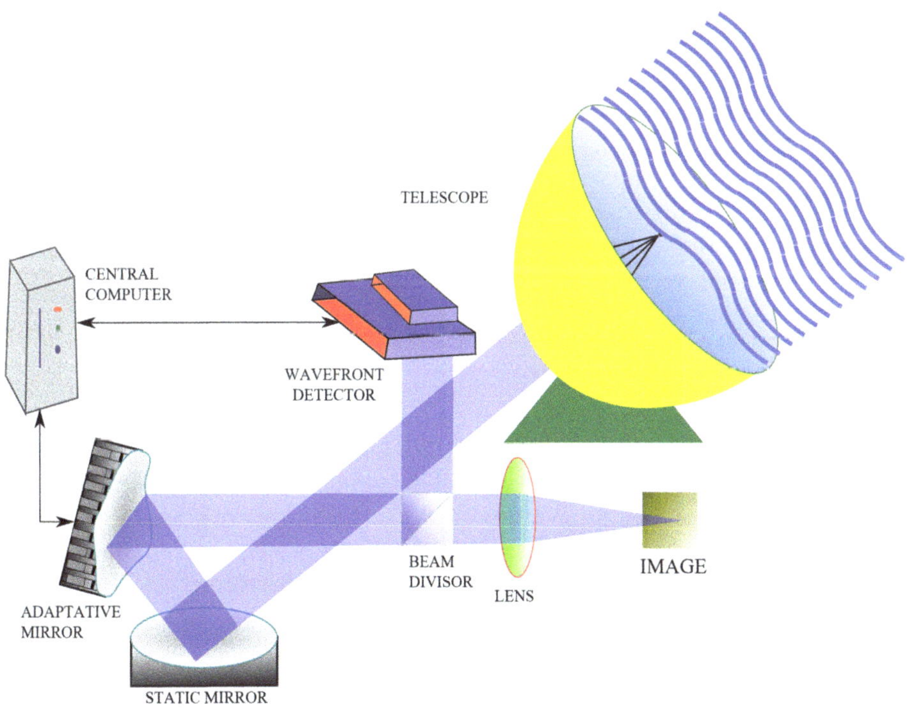

Figure 1.12. Schematic diagram of the basic configuration of a receiving system that uses technology to implement adaptive optics for the conformation of the received image. It can be seen that the adjustment processing is carried out by mirror deformation and feedback from the wavefront detector and actuators using a central processing unit.

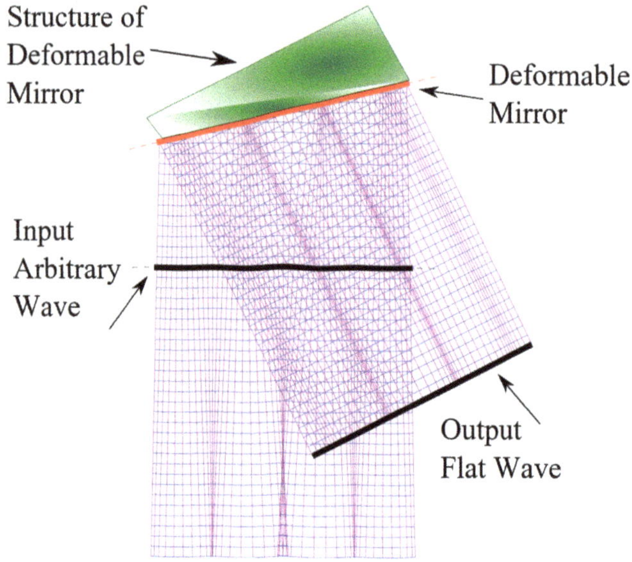

Figure 1.13. Example of an adaptive mirror; the wavefront input is shown in blue and the adaptive mirror, in red, is such that the output wavefront is flat.

1.9.2 The crux in adaptive optics: phase conjugation

Therefore the crux in adaptive optics is the surface, which when given an input wavefront cancels its phase, thereby obtaining a flat output wave. This procedure is also called phase conjugation.

Mathematically speaking, phase conjugation is fairly simple, but in practice, it is not. So far in this treatise we have not mentioned it, but the reader should know that light, as a propagating electric field, is an electromagnetic wave represented by an amplitude and a phase:

$$A(\rho, \theta) \exp\left[-i\Phi(\rho, \theta)\right]. \tag{1.52}$$

Phase conjugation is a procedure that flattens the phase of the electric field by multiplying it by $\exp[+i\Phi(\rho, \theta)]$ and then by the complex conjugate of the field. Physically, this means we need to add aberrations to the field at the right time using the deformable surface. Even though the concept of phase conjugation is based on the physical description of light in terms of electric fields, we are going to analytically solve it using only optical path theory and the Fermat principle, in chapters 2 and 6.

Figure 1.13 shows an example of phase conjugation performed by a deformable mirror.

1.9.3 Deformable mirrors

As we mentioned previously, deformable mirrors are used in large terrestrial telescopes to correct observations of celestial bodies. Deformable mirrors can be segmented, continuous, or edge-actuated mirrors. A segmented mirror has flat

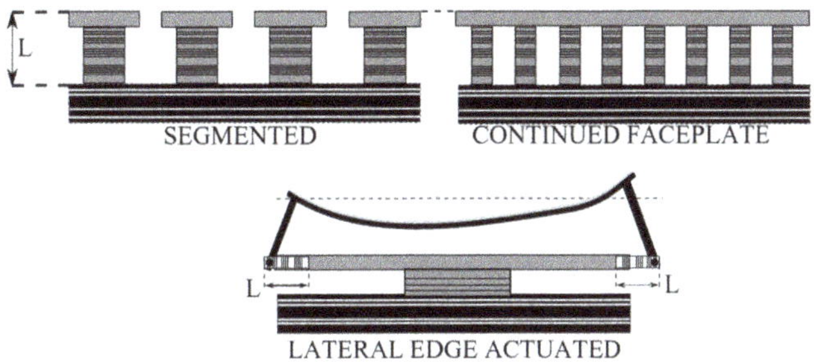

Figure 1.14. Different types of deformable mirror.

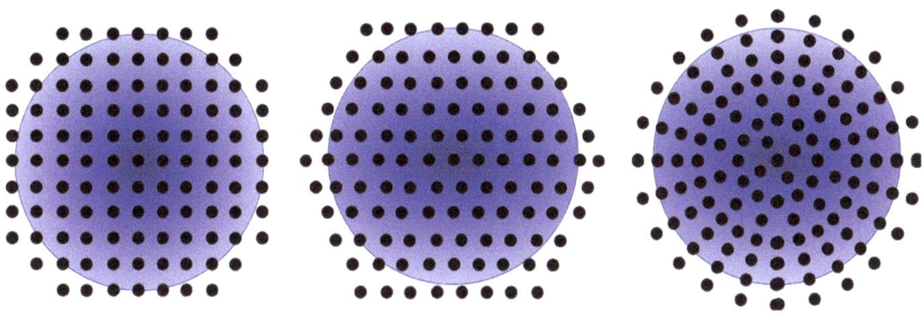

Figure 1.15. Different types of actuator array used in deformable mirrors.

segments that are moved either up and down by only one actuator or it may have three or more actuators that handle tip, tilt, and piston motions. Continuous mirrors work similarly to segmented mirrors, but the actuators deform a continuous mirror with an up-and-down movement. Finally, in edge-actuated mirrors, structural deformation of the surface takes place using mechanical modes.

Figure 1.14 shows schematics of segmented mirrors, continuous mirrors, and edge-actuated mirrors and figure 1.15 presents the different types of actuator array used in deformable mirrors.

1.9.4 Wavefront sensors

We mentioned that typical adaptive optical systems need a wavefront sensor, also called a wavefront analyser, to measure the wavefront of the light, so that the deformable mirror can subsequently perform phase conjuration. A wavefront analyser is an optical device used to analyze the shape of a wave surface, as the name suggests. The principle of the wavefront analyser is to break down a wavefront into elementary wavefronts and to determine the orientation of each elementary wavefront. When integrated, the measurements of these orientations allow the shape of the wavefront to be recovered.

Shack–Hartmann sensors are instruments used to measure the deformation of the wavefront (or phase) of an optical beam. Their use is quite widespread, and they are typically used in adaptive optical systems. In this section, we are going to study this type of sensor. The Hartmann–Shack wavefront analyser was first conceived by Johannes Franz Hartmann and later by American physicist Roland Shack.

In 1880, Johannes Franz Hartmann, a German astronomer, had the idea of placing a pierced plate in front of a Cassegrain-type telescope and of capturing the signal received from a star on a photographic plate in an intra- or extra-focal plane. Such a pierced plate samples the measured wavefront. If the plate has several holes then the behaviour of the wavefront is analysed in small portions, since the holes break down the elementary wavefront into wavefront portions. In Hartmann's time, measurements were made with a double decimetre directly onto the photographic plate. Later on, in 1970, Hartmann's principle was improved by the American physicist Roland Shack who replaced the holes with micro-lenses, making the measurement more precise.

A Shack–Hartmann sensor is made up of a matrix of micro-lenses, placed in front of a light-sensitive sensor, for example, a CCD (charge-coupled device) or a CMOS (complementary metal–oxide–semiconductor) camera. These two components are located in parallel planes. When a beam arrives, each micro-lens generates a focal point on the sensor relative to its reference position, which corresponds to an undistorted waveform. The position of the focal point varies, depending on the local deformation of the wavefront. See figure 1.16.

The image given by the sensor is thus a matrix of spots, the number of which corresponds to the number of micro-lenses. The deviation of each spot with respect to its reference gives the local derivative of the wavefront; the set of deviations directly gives the gradient of the wavefront. Algorithms then make it possible to recover the wavefront. If a wavefront is flat or spherical and thus has a point light source, the micro-lenses generate a matrix of spots whose spacing is perfectly regular. This spacing is directly related to the curvature of the wavefront, which is directly related to the distance from the source. The more it is curved, the greater the spacing between the spots. The sensor makes it possible to perform the wavefront measurement in real time. As the sensor is connected to a computer, a computer program searches the image for the position of the spots, calculates their deviation from their respective reference spot, then recovers the wavefront.

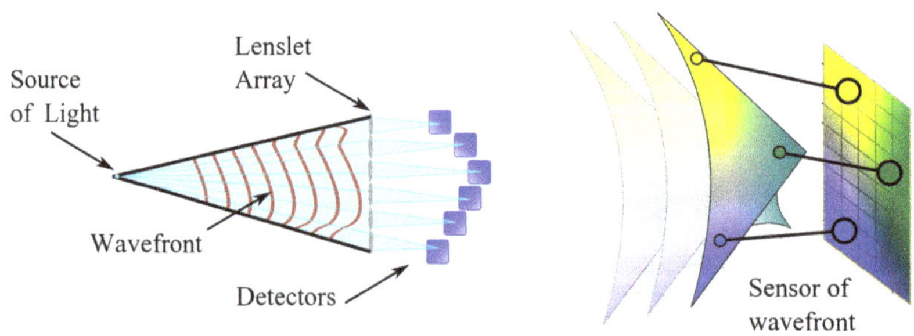

Figure 1.16. Diagram of the principle of the Shack–Hartmann sensor. The beam light is focused at a different point for each subaperture. The position of the centroid of the spot is proportional to the tilt of each subaperture.

1.10 Optical testing

In the previous section and its subsections, we studied adaptive optics because it is a direct application of the mathematical models that we are going to study in the next chapters. We now turn to optical testing for the same reason, which is that the mathematical models that we are going to study in the next chapters have direct applicability to optical testing.

We use ‘optical testing’ to denote the family of techniques used to test an optical element; normally, this means the surface of an optical element. As we have mentioned previously, if an optical element has defects on its surface, these change its performance because the laws of refraction and reflection directly depend on the normal vector of the surface. The history of optical testing starts in the seventeenth century. Galileo was the first astronomer to want to perform optical tests. He wanted to establish whether the surfaces of his lenses had the shapes he wanted, but he was not able to do it.

Today, interferometers are widely used in optical testing. The mathematical models that we are going to study in the next chapters, along with interferometers, can be very useful for optical testing. Thus, in the next subsection, we explore interferometers further.

1.10.1 Interferometry

The interferometry technique uses the superimposition of waves, typically electromagnetic waves, in order to obtain information. In optical testing, interferometry is used to extract information about the surface quality and surface irregularities of an optical element. Interferometry uses devices called interferometers to extract the abovementioned information, but interferometers are also widely used in science and industry to measure small displacements, changes in refractive index, etc.

In an interferometer, light from a single source is split into two beams that travel different optical paths and are then recombined to produce interference. The light is split into two identical beams by a beam splitter, e.g. a partially reflecting mirror. Each of these beams travels via a different route, called a path, and they are recombined before arriving at a detector. The resulting interference fringes at the detector provide information about the difference in the optical path lengths. The path difference is the difference between the distances travelled by the beams, which creates a phase difference between them. It is this introduced phase difference that creates the interference pattern between the initially identical waves.

If a single beam has been split along two paths, then the phase difference provides an indication of anything that changes the phase along the paths. This could be a physical change in the length of the path itself or a change in the refractive index along the path. Interferometry uses the principle of superposition to combine the waves so that the result of their combination has a significant property that is a diagnosis of the original state of the waves. This works because when two waves of the same frequency combine, the resulting intensity pattern is determined by the phase difference between the two waves: in-phase waves experience constructive interference, while out-of-phase waves experience destructive interference. Waves that are not completely in phase or completely out of phase will have an intermediate intensity pattern, which can be used to determine their relative phase difference.

Although there are several types of interferometer, we are just going to examine the one of the most used interferometers, the Fizeau interferometer. The Fizeau interferometer interferes with the reflections of two optical surfaces placed opposite each other, in order to control their quality, for example. See figure 1.17.

The Fizeau interferometer is relevant to our proposal because it is frequently used for optical testing. We will therefore explore it further.

The Fizeau interferometer is an optical assembly based on the principle of amplitude division interferometry, proposed in 1862 by the French physicist Hippolyte Fizeau. The Fizeau interferometer consists of a source, a beam splinter, a diverging lens, a test mirror, and the interferogram (where the interference is displayed).

The light source of the interferometer is typically a laser beam. The laser beam passes through the beam splitter and is divided into two beams. One of the beams goes directly to the interferogram. The other is refracted by the lens and then reflected by the mirror. There, it reverses, passes through the lens again and ends up at the interferogram. See figure 1.17.

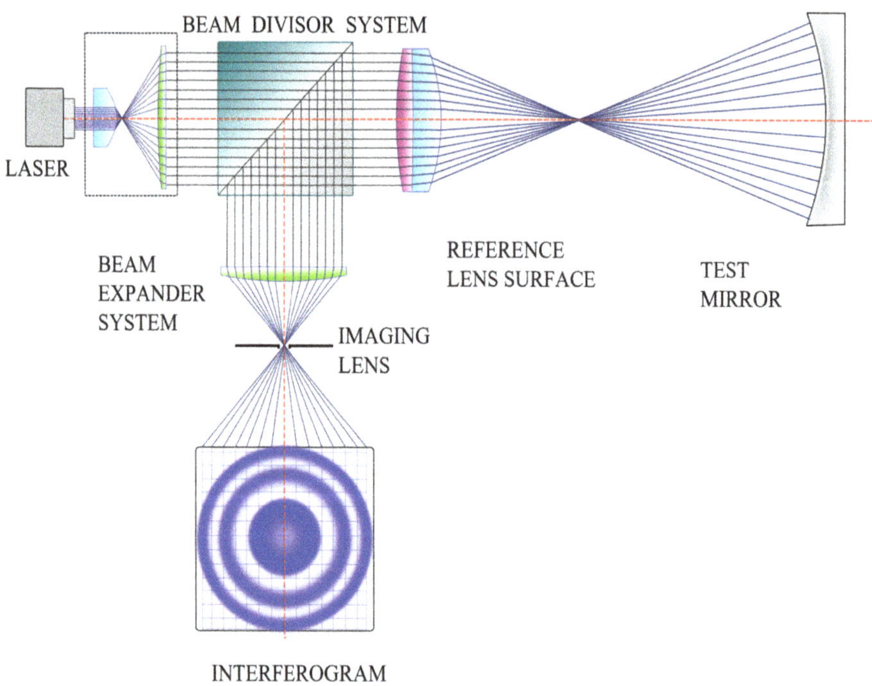

Figure 1.17. Diagram of a typical Fizeau interferometer used for optical testing. The source of light is expanded and then divided. One of the light beams goes to the reference surface and test mirror and the other goes directly to the imaging lens. Once both beams reach the imaging lens, the indifference patterns are shown in the interferogram.

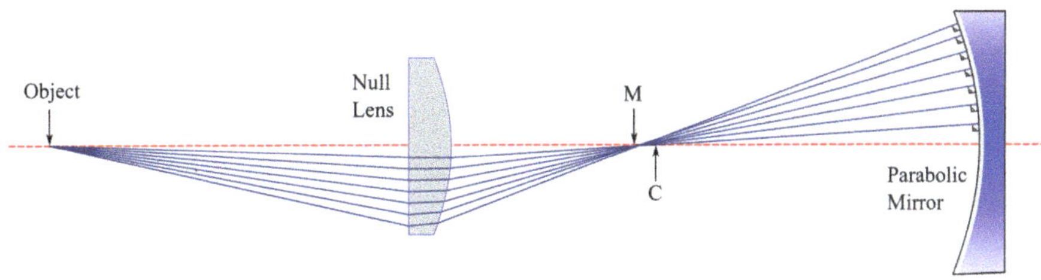

Figure 1.18. Diagram of a null lens with its respective null mirror.

If the lens and the mirror are such that the optical path lengths (OPLs) of all rays are the same then a uniform fringe pattern appears in the interferogram at the output of the Fizeau interferometer. When there is an optical path difference (OPD) between the rays that passed through the lens and the mirror, the Fizeau interferometer output contains nonuniform fringes.

The nonuniform fringes are caused by the OPD. Thus, Fizeau interferometers are very useful when lenses or mirrors are being tested to verify a certain shape. If a pair of mirrors or lenses has uniform fringes at the output, then the rays that passed through the lens and the mirror have a zero OPD. A defect on the surfaces of the lenses or mirrors can generate a non-zero OPD.

A null lens is a common name for these arrays. Null lenses are those lenses that, given a mirror, null the effect of the mirror in the Fizeau interferometer. Alternatively, if the shape of a mirror nulls the effect of the lens at the Fizeau interferometer, the mirror can be referred to as a null mirror. An example is shown in figure 1.18.

The null lens has an input spherical wavefront and the output wavefront is such that when the light strikes the mirror, the normal vectors of the mirror surface are parallel to the striking rays. This implies that the mirror is spheric.

1.11 End notes

In this chapter, we focused on the geometrical behaviour of light. From it, we successfully found Fermat's principle, Snell's law and the Malus–Dupin theorem.

We then demonstrated several essential concepts, such as stigmatism, and how the absence of stigmatism in a system forms aberrations.

Finally, we studied several techniques for which the equations of the following chapters have potential applications. These techniques are adaptive optics and optical testing.

Part II

Aspheric optical systems and the path of light

Chapter 2

General catoptric stigmatic surfaces

In this chapter, we are going to design a mirror such that given an input wavefront, it reflects it as a spherical wave or a plane wave. For reasons of clarity and pedagogical approach, we are going to design the mirror in two dimensions. During the design procedure, we do not use a numerical approach or approximation. The equations that describe the mirror are in analytic closed form and only based on Fermat's principle.

2.1 The crux of adaptive optics

Adaptive optics is an optical technique for counteracting the effects of the Earth's atmosphere on astronomical images in real time. In this technique, a deformable mirror supported by computer-controlled actuators is inserted into the optical path of the telescope to flatten the phase of the light and deliver a plane wave.

Deformable mirrors pose a difficult problem, so the adaptive optics paradigm is founded on polynomial approximations, such as Zernike polynomials, and numerically computing the coefficients. Since Zernike polynomials need to be truncated at some point, the solutions are not closed form. As a result, Zernike polynomials are an excellent method that yields excellent results, but in the end, they are approximations.

Our contribution in this chapter is a new approach for designing a mirror that reflects an arbitrary wavefront as a plane wave given an arbitrary wavefront. The approach we present is entirely analytical and closed form in two dimensions. We will study deformable mirrors in three dimensions in chapter 6. For reasons of clarity and pedagogical approach, it is good to first work on this problem and its solution in the two-dimensional case. The equation describing the abovementioned mirror can be obtained without using paraxial concepts or optimisations and in this chapter, we will find it.

doi:10.1088/978-0-7503-4705-1ch2

We have previously obtained closed-form solutions to optical problems that were not previously believed to have analytic solutions. The best example is the general equation for stigmatic lenses or lenses free of spherical aberration (see our books Analytical Lens Design and Stigmatic Optics in the references of this chapter). In the case of the general equation for stigmatic lenses, we rigorously applied Fermat's principle and only Fermat's principle. No other concept was implemented. In this chapter and book, we will again only use Fermat's principle without any additional concepts. Thus, all the results presented in this chapter are just natural implications of Fermat's principle.

2.2 General equation for deformable mirrors for images at a finite distance

In the introduction of this chapter, we set the goal of obtaining the general equation for a deformable mirror that can receive an aberrated wavefront and convert it into a plane wave. This is the main goal of the chapter, but first we are interested in studying the case in which the output wave is not a plane but a spherical wave. Thus, in this section, we explore the case of an aberrated input wavefront with a spherical output wavefront.

The diagram in figure 2.1 represents a generalized version of the problem we will be solving in this chapter. The origin of the Cartesian coordinate system is given by the vertex of the input wavefront w, which is a continuous function that depends only on one independent variable, r. The mirror's surface is described by (z_a, r_a), where z_a is the sagitta of the mirror or profile of the mirror and the parameter r_a is the radial component of the mirror's surface. We let z_o denote the distance from the input wavefront w to the mirror's surface. So, mathematically speaking, we want to find (z_a, r_a) such that the mirror reflects the input wave w as a spherical wavefront. See figure 2.1.

As the name of the book (optical path theory) suggests, we should develop a theory for the path taken by the light and study it. Therefore, we are interested in computing the optical path length (OPL) for the rays along the way. The path of any ray should start at w, then move towards (z_a, r_a), and be reflected as a ray of a spherical wavefront.

The optical path length of any ray can be described by

$$\text{OPL} = \sqrt{r_a^2 + (z_a - z_i - z_o)^2} + \sqrt{(r_a - r)^2 + (z_a - w)^2}, \tag{2.1}$$

and the OPL of the ray that goes along the axis is given by

$$\text{OPL} = z_o - z_i. \tag{2.2}$$

In the last expression the negative sign comes from the reflection. Both light paths should be the same such that we get an spherical wavefront from the reflection of w at (z_a, r_a). Thus

$$\boxed{\sqrt{r_a^2 + (z_a - z_i - z_o)^2} + \sqrt{(r_a - r)^2 + (z_a - w)^2} = z_o - z_i}. \tag{2.3}$$

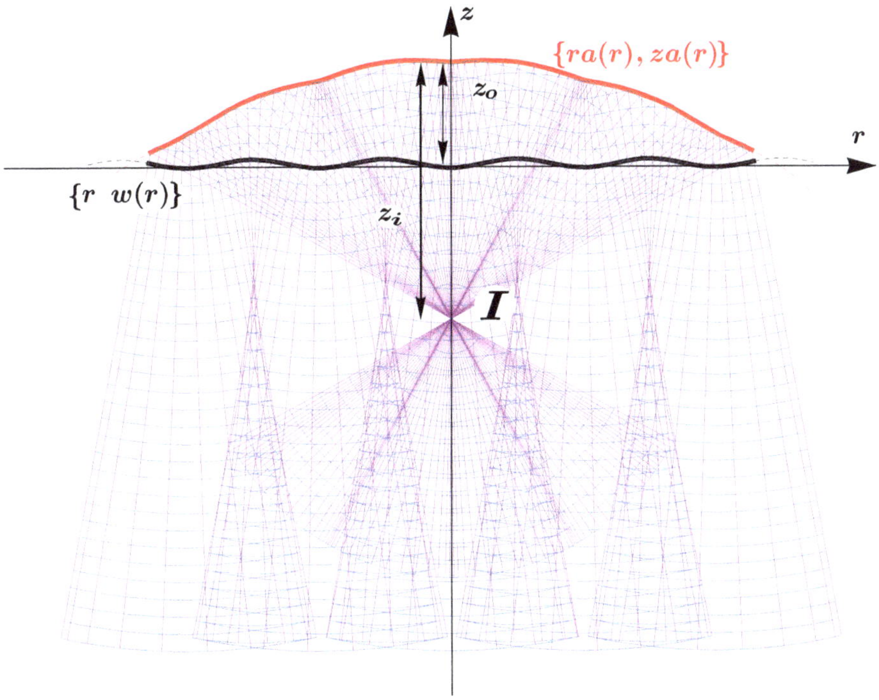

Figure 2.1. Diagram of an adaptive mirror system with a finite-distance image.

The next step is crucial; remember the Malus–Dupin theorem studied in the last chapter? It tells us that the rays are perpendicular to the wavefront, so the cosine directors of the rays of the wavefront w are the normal vectors of the wavefront w. These cosine directors are

$$w_r = -\frac{w'}{\sqrt{w'^2 + 1}}, \quad w_z = \frac{1}{\sqrt{w'^2 + 1}}, \tag{2.4}$$

where $w' \equiv dw/dr$ is the derivative of w to r, and w_r and w_z are the cosine directors in the r and z directions, respectively.

We mentioned that this step is crucial because we can express the surface mirrors in terms of the input rays, and for these rays, like the wavefront w and its cosine directors, the only independent variable is r. Thus, we can express the surface of the mirror as

$$r_a = r + \vartheta w_r, \quad z_a = w + \vartheta w_z, \tag{2.5}$$

using only the independent variable r, where ϑ is defined as,

$$\vartheta = \sqrt{(r_a - r)^2 + (z_a - w)^2}. \tag{2.6}$$

This parameter, ϑ, is determined by the distance a ray must travel from the wavefront to strike the mirror. We need to express ϑ in terms of the unknown variables, z_a and r_a. Thus, we substitute equations (2.6) and (2.5) into equation (2.3):

$$\sqrt{(w + \vartheta w_z - z_i - z_o)^2 + (\vartheta w_r + r)^2} + \vartheta = k, \tag{2.7}$$

where $k = z_i - z_o$ and we define the parameter τ as

$$\tau = w - z_i - z_o. \tag{2.8}$$

Replacing τ in equation (2.7) gives

$$\sqrt{(\vartheta w_r + r)^2 + (\vartheta w_z + \tau)^2} + \vartheta = k; \tag{2.9}$$

manipulating and squaring this yields

$$(\vartheta w_r + r)^2 + (\vartheta w_z + \tau)^2 = (k - \vartheta)^2, \tag{2.10}$$

and expanding and reddening this results in

$$r^2 + \vartheta^2 w_r^2 + 2r\vartheta w_r + \vartheta^2 w_z^2 + 2\tau\vartheta w_z + \tau^2 = k^2 - 2k\vartheta + \vartheta^2. \tag{2.11}$$

Finally, we can directly solve for ϑ in equation (2.11),

$$\vartheta = \frac{k^2 - r^2 - \tau^2}{2(k + rw_r + \tau w_z)}. \tag{2.12}$$

Now that we know the distance that each ray from the wavefront needs to travel to strike the mirror, we can substitute it into equation (2.5) to get the shape of the mirror of interest. The radial component of the mirror is

$$\boxed{r_a = r - \frac{w'[r^2 - 2w(z_i + z_o) + w^2 + 4z_i z_o]}{2\left(z_i\sqrt{w'^2 + 1} - z_o\sqrt{w'^2 + 1} + rw' - w + z_i + z_o\right)}.} \tag{2.13}$$

The sagittal component of the mirror is

$$\boxed{z_a = w + \frac{r^2 - 2w(z_i + z_o) + w^2 + 4z_i z_o}{2\left(z_i\sqrt{w'^2 + 1} - z_o\sqrt{w'^2 + 1} + rw' - w + z_i + z_o\right)}.} \tag{2.14}$$

Observe that we have not used any approximation to obtain equations (2.13) and (2.14). In fact, equations (2.13) and (2.14) are analytic closed-form solutions of the proposed problem. They are just a natural extension of Fermat's principle. We do not use any paraxial approximation, nor iteration, nor optimization. Observe that the OPL has presented the other necessary concepts in the solution: the orthogonality between the rays and waves, and the law of refraction. Both concepts are just natural implications of Fermat's principle.

Observe that, if Fermat's principle is the axiom, then the orthogonality between the rays and waves is a theorem and the law of refraction is another theorem. If Fermat's principle is true, what else is true? that is our main question in this treatise. So going back to our problem equations (2.13) and (2.14), they are true because they are natural implications of Fermat's principle, which is true. But also note that in this procedure, we obtained the only possible solution to this problem. Thus, given a set of input values z_o, z_i, and w, there is only one solution, which is given by equations (2.13) and (2.14). So, we have presented a uniqueness theorem.

As an additional remark before we start looking for the eikonal: since ϑ is determined by the distance a ray must travel from the wavefront to strike the mirror and we claim that equations (2.13) and (2.14) are the general solution to the problem presented, therefore $\vartheta \to z_o$ when $r \to 0$ for all possible input values z_o, z_i, and w. Note that $r \to 0$ is the axial ray, thus from equation (2.12) we compute the limit when $r \to 0$,

$$\lim_{r \to 0}(\vartheta) = -\frac{(z_o - z_i)^2 - (-z_i - z_o)^2}{4z_i} = z_o; \tag{2.15}$$

the result is as expected.

2.3 The eikonal, the wavefront, and ray tracing

To get the eikonal of the wavefront along the optical path, we need to recall another implication of Fermat's principle, i.e. Snell's law.

In particular, we consider Snell's law of reflection. This expression is the equation which states that information is reflected in the mirror given by (z_a, r_a). The vector form of Snell's law for the mirror (z_a, r_a) is given by

$$\boxed{\vec{v}_2 = [\vec{v}_1 - (\vec{n}_a \cdot \vec{v}_1)\vec{n}_a] - \vec{n}_a\sqrt{1 - (\vec{n}_a \times \vec{v}_1)^2} \quad \text{for} \quad \vec{v}_2, \vec{v}_1, \vec{n}_a \in \mathbb{R}^2,} \tag{2.16}$$

where $\vec{v}_1$ is the unit vector of the incident ray (the normal vector of the wavefront w), $\vec{v}_2$ is unit vector of the reflected ray, and finally $\vec{n}_a$ is the normal vector of the first surface. Please see figure 2.9.

$$\vec{v}_1 = \frac{-w'\vec{e}_1 + \vec{e}_2}{\sqrt{w'^2 + 1}}, \qquad \vec{v}_2 = \wp_r\vec{e}_1 + \wp_z\vec{e}_2, \qquad \vec{n}_a = \frac{z_a'\vec{e}_1 - r_a'\vec{e}_2}{\sqrt{r_a' + z_a'^2}}, \tag{2.17}$$

where $\vec{e}_1$ denotes the r direction and $\vec{e}_2$ denotes the z direction. The parameters $\wp_x$ and $\wp_y$ are the cosine directors of the reflected ray, thus they are the cosine directors of the rays of the spherical output wavefront. Finally, $r_a' \equiv dr_a/dr$ is the derivative of r_a to r, so we can obtain the derivative of equation (2.13):

$$r_a' = \frac{1}{2(w'^2+1)\left(z_i\sqrt{w'^2+1} - z_o\sqrt{w'^2+1} + rw' - w + z_i + z_o\right)^2}$$
$$\times \Big\{\Big[(z_i+z_o)w'^2 + (z_i - z_o)\sqrt{w'^2+1} - w(w'^2+1) + z_i + z_o\Big]$$
$$\times\Big[-w''(r^2+4z_iz_o) - w^2w'' + 2(z_i - z_o)\sqrt{w'^2+1}\,w'^2 \tag{2.18}$$
$$+2\Big((z_i+z_o)w'^2 + (z_i - z_o)\sqrt{w'^2+1} + rw' + z_i + z_o\Big)$$
$$+2rw'^3 + 2w([z_i+z_o]w'' - w'^2 - 1)\Big]\Big\},$$

and if $z_a' \equiv dz_a/dr$ is the derivative of z_a to r, we can obtain the derivative of equation (2.14):

$$z_a' = \frac{1}{2(w'^2+1)\left(z_i\sqrt{w'^2+1} - z_o\sqrt{w'^2+1} + rw' - w + z_i + z_o\right)^2}$$
$$\times\Big\{\Big([z_i - z_o]\sqrt{w'^2+1}\,w' + rw'^2 + r\Big)$$
$$\times\Big[-w''(r^2+4z_iz_o) - w^2w'' + 2(z_i - z_o)\sqrt{w'^2+1}\,w'^2 \tag{2.19}$$
$$+2\Big([z_i+z_o]w'^2 + [z_i - z_o]\sqrt{w'^2+1} + rw' + z_i + z_o\Big) + 2rw'^3$$
$$+2w\Big([z_i+z_o]w'' - w'^2 - 1\Big)\Big]\Big\}.$$

We are going to replace the unit vectors in equation (2.16). The procedure is long, so we will just present the main steps. We begin with $(\vec{\mathbf{n}}_a \times \vec{\mathbf{v}}_1)^2$:

$$(\vec{\mathbf{n}}_a \times \vec{\mathbf{v}}_1) = \frac{z_a' - r_a'w'}{\sqrt{w'^2+1}\sqrt{r_a'^2 + z_a'^2}}(\vec{\mathbf{e}}_1 \times \vec{\mathbf{e}}_2); \tag{2.20}$$

from basic vector algebra, we know $(\vec{\mathbf{e}}_1 \times \vec{\mathbf{e}}_2)^2 = \vec{\mathbf{e}}_3^{\,2} = 1$, so

$$(\vec{\mathbf{n}}_a \times \vec{\mathbf{v}}_1)^2 = \frac{(z_a' + r_a'w')^2}{(w'^2+1)(r_a'^2 + z_a'^2)}. \tag{2.21}$$

This implies that the squared root in equation (2.16) is given by

$$-\vec{\mathbf{n}}_a\sqrt{1 + (\vec{\mathbf{n}}_a \times \vec{\mathbf{v}}_1)^2} = -\frac{(z_a'\vec{\mathbf{e}}_1 - \vec{\mathbf{e}}_2)}{\sqrt{1 + z_a'^2}}\sqrt{1 - \frac{(z_a' - r_a'w')^2}{(w'^2+1)(r_a'^2 + z_a'^2)}}. \tag{2.22}$$

We now turn to the term $\vec{\mathbf{v}}_1 - (\vec{\mathbf{n}}_a \cdot \vec{\mathbf{v}}_1)\vec{\mathbf{n}}_a$ in Snell's law; this term is given by

$$\vec{\mathbf{v}}_1 - (\vec{\mathbf{n}}_a \cdot \vec{\mathbf{v}}_1)\vec{\mathbf{n}}_a = \frac{-w'\vec{\mathbf{e}}_1 + \vec{\mathbf{e}}_2}{\sqrt{w'^2 + 1}} - \left[-\frac{r_a' + w'z_a'}{\sqrt{w'^2 + 1}\sqrt{r_a'^2 + z_a'^2}}\right]\frac{(z_a'\vec{\mathbf{e}}_1 - r_a'\vec{\mathbf{e}}_2)}{\sqrt{r_a'^2 + z_a'^2}}. \quad (2.23)$$

By manipulating the last expression, we obtain

$$\vec{\mathbf{v}}_1 - (\vec{\mathbf{n}}_a \cdot \vec{\mathbf{v}}_1)\vec{\mathbf{n}}_a = \frac{r_a'(z_a' - r_a'w')}{\sqrt{w'^2 + 1}(r_a'^2 + z_a'^2)}\vec{\mathbf{e}}_1 + \frac{z_a'(z_a' - r_a'w')}{\sqrt{w'^2 + 1}(r_a'^2 + z_a'^2)}\vec{\mathbf{e}}_2. \quad (2.24)$$

Replacing equations (2.22) and (2.24) in (2.16) gives

$$\wp_r = \frac{r_a'(z_a' - r_a'w')}{\sqrt{w'^2 + 1}(r_a'^2 + z_a'^2)} + \frac{z_a'\sqrt{1 - \dfrac{(z_a' - r_a'w')^2}{(w'^2 + 1)(r_a'^2 + z_a'^2)}}}{\sqrt{r_a'^2 + z_a'^2}} \quad (2.25)$$

and

$$\wp_z = \frac{z_a'(z_a' - r_a'w')}{\sqrt{w'^2 + 1}(r_a'^2 + z_a'^2)} - \frac{r_a'\sqrt{1 - \dfrac{(z_a' - r_a'w')^2}{(w'^2 + 1)(r_a'^2 + z_a'^2)}}}{\sqrt{r_a'^2 + z_a'^2}}. \quad (2.26)$$

We now have expression for the cosine directors of each ray that depend only on r (here, we are talking about equations (2.4), (2.25), and (2.26)).

From the above, it is easy to see where the ray is located at time t. Using these expressions, i.e. (2.4), (2.25), and (2.26), we can obtain the eikonal for each ray at time t. The eikonal is easy to construct as a piecewise function given by

$$\boxed{H(t, r_a) = \begin{cases} [w + tw_z, r + tw_r], & t < \vartheta, \\ [z_a + \wp_z(t - \vartheta), r_a + \wp_r(t - \vartheta)], & t > \vartheta, \end{cases}} \quad (2.27)$$

where the condition $t < \vartheta$ means that the ray has not yet struck the mirror's surface, $t > \vartheta$ is the condition for the reflected rays, and z_a and r_a are given by (2.14) and (2.13).

Once again, observe that we have obtained $H(t, r_a)$ with no numerical approximation or assumptions. Thus, $H(t, r_a)$ can be seen as just another natural implication of Fermat's principle.

2.4 Mathematica code

Now that we have interesting equations, we can talk about equations (2.14), (2.13), and (2.27). Let us explore their performance using `Mathematica`.

The code presented in this section computes the mirrors (z_a, r_a) and the eikonal H (t, r_a) for a given set of input values.

In[1]:= $w[r\_]:=-\frac{r^2}{40}$

$wr[r\_]:=-\frac{w'[r]}{\sqrt{w'[r]^2+1}}$

$wz[r\_]:=\frac{1}{\sqrt{w'[r]^2+1}}$

$k=zo-zi;$

$\vartheta[r\_]:=\frac{k^2-r^2-\tau[r]^2}{2\ (k+r\ wr[r]+\tau[r]\ wz[r])}$

$ra[r\_]:=r-\frac{(r^2+4\ zi\ zo-2\ (zi+zo)\ w[r]+w[r]^2)\ w'[r]}{2\ (zi+zo-w[r]+r\ w'[r]+zi\ \sqrt{1+w'[r]^2}-zo\ \sqrt{1+w'[r]^2})}$

$za[r\_]:=w[r]+\frac{r^2+4\ zi\ zo-2\ (zi+zo)\ w[r]+w[r]^2}{2\ (zi+zo-w[r]+r\ w'[r]+zi\ \sqrt{1+w'[r]^2}-zo\ \sqrt{1+w'[r]^2})}$

$\tau[r\_]:=w[r]-zi-zo$

$\varrho r[r\_]:=\frac{no\ ra'[r]\ (-ra'[r]\ w'[r]+za'[r])}{-Sign[w[r]]ni\ \sqrt{1+w'[r]^2}\ (ra'[r]^2+za'[r]^2)}$

$$+\frac{za'[r]}{\sqrt{ra'[r]^2+za'[r]^2}}\sqrt{1-\frac{no^2\ (-ra'[r]\ w'[r]+za'[r])^2}{ni^2\ (1+w'[r]^2)\ (ra'[r]^2+za'[r]^2)}}$$

$$\varrho z[r\_]:=\frac{no\ za'[r]\ (-ra'[r]\ w'[r]+za'[r])}{-Sign[w[r]]ni\ \sqrt{1+w'[r]^2}\ (ra'[r]^2+za'[r]^2)}$$

$$-\frac{ra'[r]}{\sqrt{ra'[r]^2+za'[r]^2}}\sqrt{1-\frac{no^2\ (-ra'[r]\ w'[r]+za'[r])^2}{ni^2\ (1+w'[r]^2)\ (ra'[r]^2+za'[r]^2)}}$$

```
Homotopy[r_,tt_]:={{
  Piecewise[{
    {wr[r]tt+r,tt<ϑ[r]},
    {-Sign[w[r]]vϱr[r](tt-ϑ[r])+ra[r],tt>ϑ[r]}}],
  Piecewise[{
    {wz[r]tt+w[r],tt<ϑ[r]},
    {-Sign[w[r]]vϱz[r](tt-ϑ[r])+za[r],tt>ϑ[r]} }]
  }
}

c=1;
v=c/ni;

no=1;
ni=1;

zo=40;
zi =-20;

rmax=11;

Surface:=ParametricPlot[{{r,w[r]},{ra[r],za[r]},{3r,w[3r]},
{ra[1.75r],za[1.75r]}},{r,-rmax,rmax},AspectRatio→1/1,
AxesOrigin→{-50,-50}, PlotStyle→{{Black,Thickness[0.005]},
{Red,Thickness[0.005]},{Black,Dashed,Thin},{Red,Dashed,Thin}},
```

```
AxesLabel→{Style[r,FontSize→14,Bold],Style[z,FontSize→14,Bold]},
AxesStyle→Arrowheads[{0.05}]]

Waves:=Table[{ParametricPlot[Homotopy[r,tt],{r,-rmax,rmax},
PlotRange→All,PlotStyle→{Blue,Thin}]},{tt,-44,90,2}]

Rays:=Table[{ParametricPlot[Homotopy[r,tt],{tt,-44,90},
PlotRange→All,PlotStyle→{Purple,Thin}]},{r,-rmax,rmax}]

ShowOpt={PlotRange->All,ImageSize→{480,Automatic},
AspectRatio→Automatic};

Show[Surface,Rays,Waves,ShowOpt]
```

2.5 Examples

Example of mirrors and eikonals given by equations (2.13), (2.14), and (2.27) are presented in this section. The figures associated with this gallery are figures 2.2, 2.3, 2.4, 2.5, 2.6, 2.7, and 2.8.

Note that for all the examples the perpendicularity of the wavefronts with respect to the rays holds. All the rays have the same OPL. The output produced is a perfect spherical wave. The output spherical waves have their centres at z_i as predicted.

We should note a limitation of equations (2.13), (2.14), and (2.27) that must be taken into account. Mirrors can only overlap if two or more rays strike the same point on the surface. The reason for this is because all rays must share the same OPL, so from the derivation of equations (2.13), (2.14), and (2.27), two rays cannot touch the same point on the mirror's surface. Equation (2.3) does not take account of the case of two rays that strike the same point. So, in order that a non-axial ray and its neighbouring ray can have the same OPL, they should not strike the same point on the mirror's surface.

2.6 The general equation for deformable mirrors for images at infinity

In the previous section, we considered the case of a spherical output wavefront; in this section, we focus on the case of a planar output wavefront. To achieve our goal,

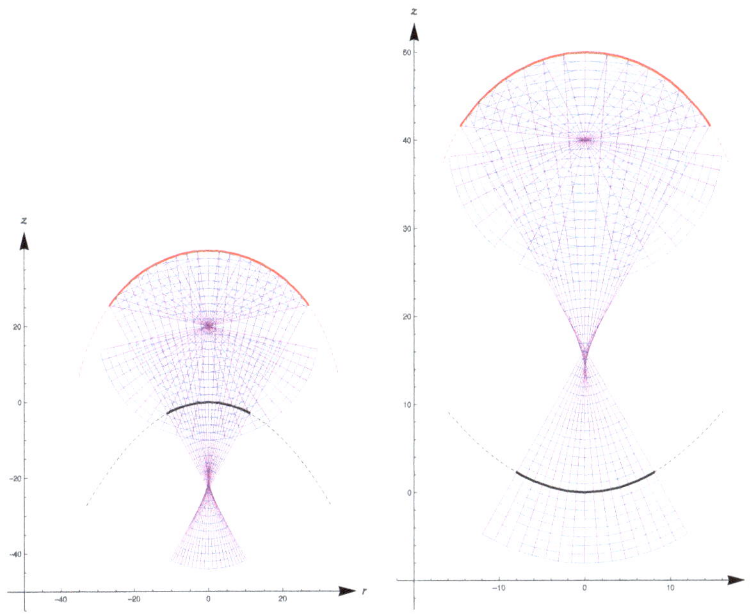

(a) A convex parabolic wavefront

(b) A concave parabolic wavefront

Figure 2.2. Design specifications for parabolic wavefronts: (a) the input wavefront is $-r^2 \div 40$, refractive index $n = 1.0$, object distance $z_o = 30$ mm, image distance $z_i = -20$ mm, r_{max} is 11 mm, (r_a, z_a) = equations (2.13), (2.14), and for the eikonal $H(t, r_a)$ = equation (2.27); (b) the input wavefront is $r^2 \div 28$, refractive index $n = 1.0$, object distance $z_o = 50$, image distance $z_i = -10$, r_{max} is 8 mm, (r_a, z_a) = equations (2.13), (2.14), and for the eikonal $H(t, r_a)$ = equation (2.27).

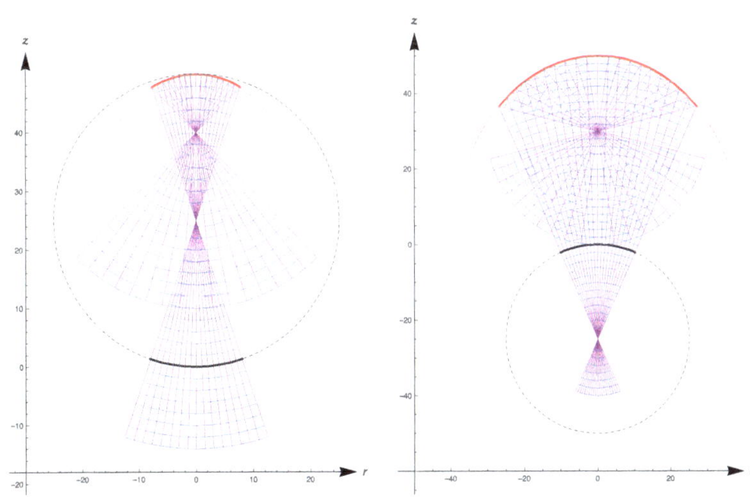

(a) A concave circular wavefront

(b) A convex circular wavefront

Figure 2.3. Design specifications for a circular wavefronts: (a) the input wavefront is $25 - \sqrt{25^2 - r^2}$, refractive index $n = 1.0$, object distance $z_o = 50$ mm, image distance $z_i = -10$ mm, r_{max} is 8 mm, (r_a, z_a) = equations (2.13), (2.14), and for the eikonal $H(t, r_a)$ = equation (2.27); (b) the input wavefront is $-25 + \sqrt{25^2 - r^2}$, refractive index $n = 1.0$, object distance $z_o = 50$ mm, image distance $z_i = -20$ mm, r_{max} is 10 mm, (r_a, z_a) = equations (2.13), (2.14), and for the eikonal $H(t, r_a)$ = equation (2.27).

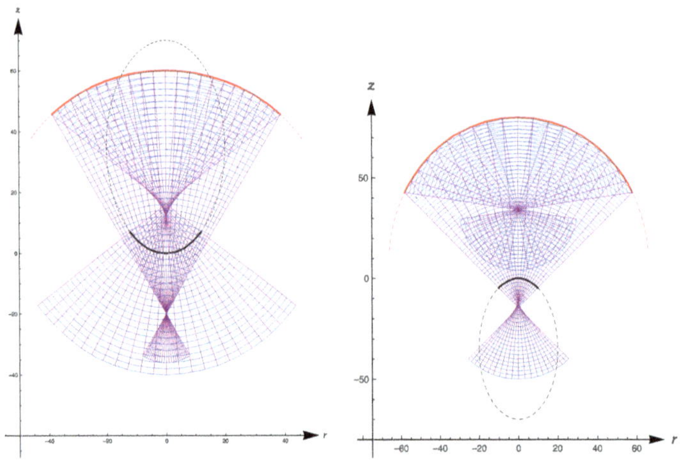

(a) A concave oblate ellipse wavefront (b) A convex oblate ellipse wavefront

Figure 2.4. Design specifications for oblate elliptical wavefronts: (a) the input wavefront is $b - \sqrt{(-a^2*b^2 + b^2r^2)/ - a^2}$, refractive index $n = 1.0$, object distance $z_o = 60$ image distance $z_i = -80$ mm, r_{max} is 12 mm, constant $a = 20$, constant $b = 35$, (r_a, z_a) = equations (2.13), (2.14), and for the eikonal $H(t, r_a)$ = equation (2.27); (b) the input wavefront is $-b + \sqrt{(-a^2*b^2 + b^2r^2)/ - a^2}$, refractive index $n = 1.0$, object distance $z_o = 80$ mm, image distance $z_i = -46$ mm, r_{max} is 10 mm, constant $a = 20$, constant $b = 35$, (r_a, z_a) = equations (2.13), (2.14), and for the eikonal $H(t, r_a)$ = equation (2.27).

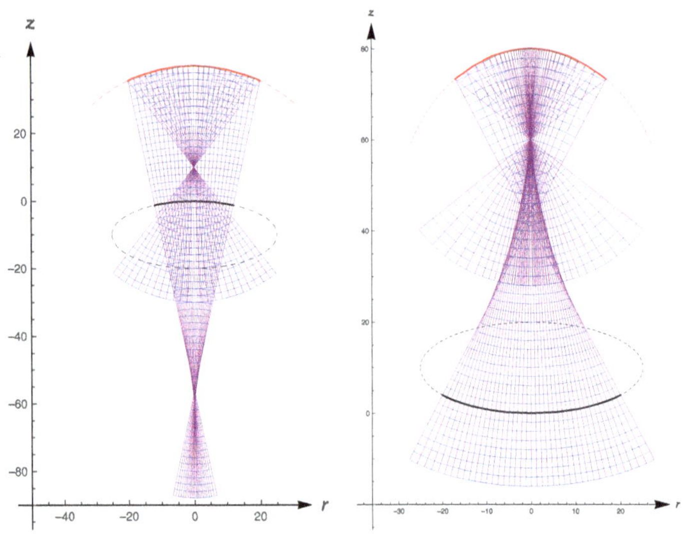

(a) A concave prolate ellipse wavefront (b) A convex prolate ellipse wavefront

Figure 2.5. Design specifications for prolate elliptical wavefronts: (a) the input wavefront is $-b + \sqrt{(-a^2*b^2 + b^2r^2)/ - a^2}$, refractive index $n = 1.0$, object distance $z_o = 40$ mm, image distance $z_i = -30$ mm, r_{max} is 12 mm, constant $a = 25$, constant $b = 10$, (r_a, z_a) = equations (2.13), (2.14), and for the eikonal $H(t, r_a)$ = equation (2.27); (b) the input wavefront is $b - \sqrt{(-a^2*b^2 + b^2r^2)/ - a^2}$, refractive index $n = 1.0$, object distance $z_o = 80$ mm, image distance $z_i = -20$ mm, r_{max} is 20 mm, constant $a = 25$, constant $b = 10$, (r_a, z_a) = equations (2.13), (2.14), and for the eikonal $H(t, r_a)$ = equation (2.27).

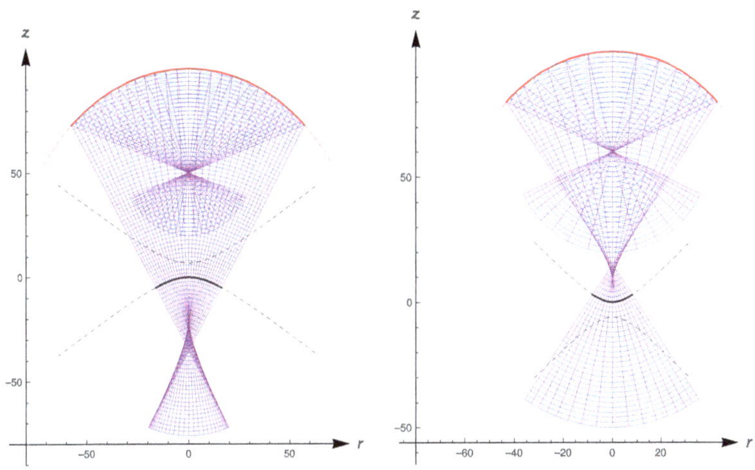

(a) A convex hyperbolic wavefront (b) A concave hyperbolic wavefront

Figure 2.6. Design specifications for hyperbolic wavefronts: (a) the input wavefront is $b - \sqrt{(-a^2 * b^2 - b^2 r^2)/ - a^2}$, refractive index $n = 1.0$, object distance $z_o = 100$ mm, image distance $z_i = -50$ mm, r_{max} is 16 mm, constant $a = 16$, constant $b = 12$, (r_a, z_a) = equations (2.13), (2.14), and for the eikonal $H(t, r_a)$ = equation (2.27); (b) the input wavefront is $-b + \sqrt{(-a^2 * b^2 - b^2 r^2)/ - a^2}$, refractive index $n = 1.0$, object distance $z_o = 100$ mm, image distance $z_i = -40$ mm, r_{max} is 8 mm, constant $a = 9$, constant $b = 9$, (r_a, z_a) = equations (2.13), (2.14), and for the eikonal $H(t, r_a)$ = equation (2.27).

we first consider that the mirror reflects the light of the wavefront to a single point image with height r_i and the horizontal distance between the mirror and the image point is z_i. So, r_i is the radial distance from the mirror vertex to the image and z_i is the sagittal distance from the mirror vertex to the image, so the point image is located at (r_i, z_i). See figure 2.9.

Based on this assumption, the OPLs can be set. So once again we recall Fermat's principle and its implications. We know from Fermat's principle that we can only obtain an image point at (r_i, z_i) if the OPLs of every ray are the same. By comparing the OPL of an axial ray with the OPL of a non-axial ray, we can say that they are equal. Thus, the OPL of the non-axial ray is given by the expression

$$\mathrm{OPL} = \sqrt{(r_a - r_i)^2 + (z_a - z_i - z_o)^2} + \sqrt{(r_a - r)^2 + (z_a - w)^2}, \tag{2.28}$$

and the OPL of the axial ray is

$$\mathrm{OPL} = \sqrt{r_i^2 + z_i^2} + z_o, \tag{2.29}$$

and in fact they are equal,

$$\sqrt{(r_a - r_i)^2 + (z_a - z_i - z_o)^2} + \sqrt{(r_a - r)^2 + (z_a - w)^2} = \sqrt{r_i^2 + z_i^2} + z_o. \tag{2.30}$$

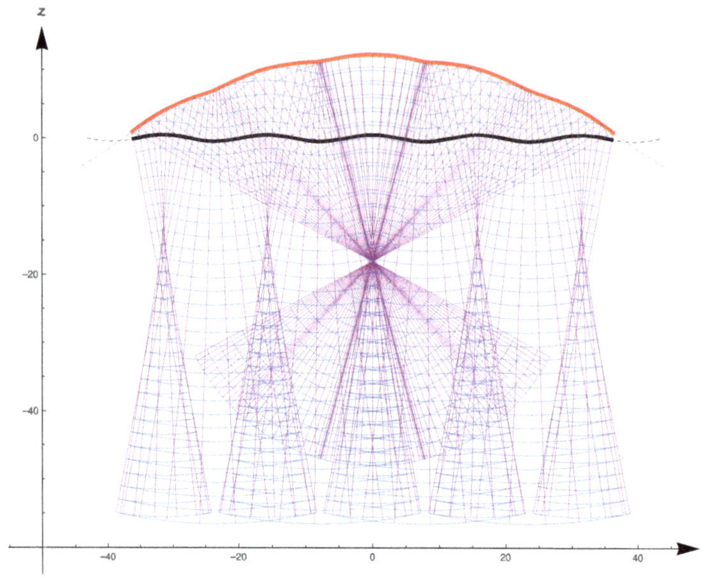

(a) A positive cosine wavefront

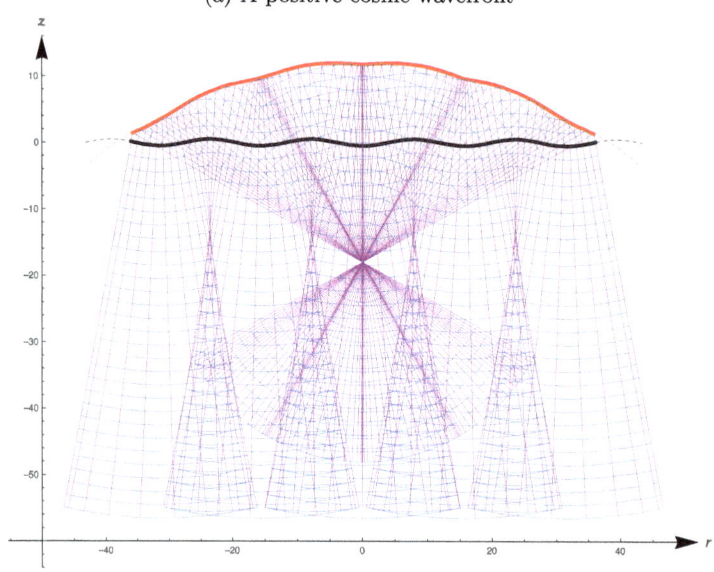

(b) A negative cosine wavefront

Figure 2.7. Design specifications for cosinoidal wavefronts: (a) the input wavefront is $\cos(2r/5)/2$, refractive index $n = 1.0$, object distance $z_o = 12$ mm, image distance $z_i = -30$ mm, $\mathrm{r}_{\max}$ is 36 mm, (r_a, z_a) = equations (2.13), (2.14), and for the eikonal $H(t, r_a)$ = equation (2.27); (b) the input wavefront is $-\cos(2r/5)/2$, refractive index $n = 1.0$, object distance $z_o = 12$ mm, image distance $z_i = -30$ mm, $\mathrm{r}_{\max}$ is 36 mm, (r_a, z_a) = equations (2.13), (2.14), and for the eikonal $H(t, r_a)$ = equation (2.27).

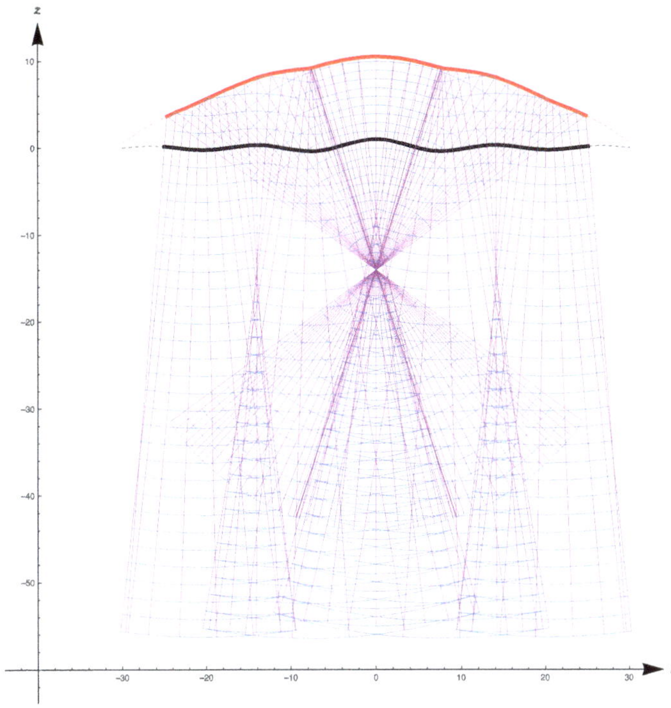

Figure 2.8. Design specifications for a Bessel zero-order wavefront. The input wavefront is $J_0(2/r)$, refractive index $n = 1.0$, object distance $z_o = 10$ mm, image distance $z_i = -24$ mm, r_{max} is 25 mm, $z_b =$ equations (2.13), (2.14), and for the eikonal $H(t, r_a) =$ equation (2.27).

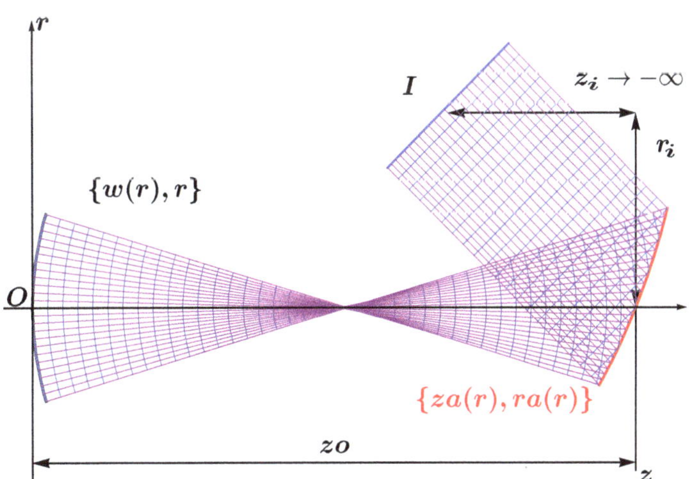

Figure 2.9. Diagram of an adaptive mirror system with the image at infinity.

But note that this OPL is for an image at a finite distance. We need to recover the parallelism inherent in a plane wave. Therefore, we compute the limit when $r_i \to pz_i$ over equation (2.30). Let $p \in \mathbb{R}$ be a real constant related to the inclination of the output wave plane, then

$$\begin{aligned}&\lim_{r_i \to pz_i}\left[\sqrt{(r_a - r_i)^2 + (z_a - z_i - z_o)^2} + \sqrt{(r_a - r)^2 + (z_a - w)^2} - \sqrt{r_i^2 + z_i^2} - z_o\right]\\&= -\sqrt{(p^2 + 1)z_i^2} + \sqrt{(r_a - pz_i)^2 + (-z_a + z_i + z_o)^2} + \sqrt{(r - r_a)^2 + (w - z_a)^2} - z_o.\end{aligned} \tag{2.31}$$

We now compute the limit when $z_i \to -\infty$ over the last expression; the computation leads us to

$$\begin{aligned}&\lim_{z_i \to -\infty}\left[-\sqrt{(p^2 + 1)z_i^2} + \sqrt{\left(r_a - pz_i\right)^2 + (-z_a + z_i + z_o)^2}\right.\\&\left.+ \sqrt{(r - r_a)^2 + (w - z_a)^2} - z_o\right]\\&= \frac{p \lim_{z_i \to -\infty}(r_a) + \lim_{z_i \to -\infty}(z_a) - z_o}{\sqrt{p^2 + 1}}\\&+ \sqrt{\left[r - \lim_{z_i \to -\infty}(r_a)\right]^2 + \left[w - \lim_{z_i \to -\infty}(z_a)\right]^2} - z_o.\end{aligned} \tag{2.32}$$

The last equation gives the optical path difference (OPD) between the axial ray and the non-axial ray when the image is far away. Note that we have previously expressed the first surface in terms of the input rays. Recalling equation (2.4),

$$w_r = -\frac{w'}{\sqrt{w'^2 + 1}}, \quad w_z = \frac{1}{\sqrt{w'^2 + 1}}, \tag{2.33}$$

we can also express the mirror surface when the object is far away with the following cosine directors:

$$\lim_{z_i \to -\infty}(r_a) = r + \lim_{z_i \to -\infty}(\vartheta)w_r, \quad \lim_{z_i \to -\infty}(z_a) = w + \lim_{z_i \to -\infty}(\vartheta)w_z, \tag{2.34}$$

where

$$\lim_{z_i \to -\infty}(\vartheta) = \sqrt{\left[\lim_{z_i \to -\infty}(r_a) - r\right]^2 + \left[\lim_{z_i \to -\infty}(z_a) - w\right]^2}. \tag{2.35}$$

Note that the limit does not affect the cosine directors of the rays of the input wavefront. Substituting equations (2.34) and (2.35) into the OPD, i.e. equation (2.32), yields

$$\frac{p\left[\lim_{z_i \to -\infty}(\vartheta)w_r + r\right] + \left[w + \lim_{z_i \to -\infty}(\vartheta)w_z\right] - z_o}{\sqrt{p^2 + 1}} - z_o + \lim_{z_i \to -\infty}(\vartheta) = 0. \tag{2.36}$$

Solving for $\lim_{z_i \to -\infty}(\vartheta)$ in the last expression,

$$\lim_{z_i \to \infty}(\vartheta) = \frac{\sqrt{p^2 + 1}\, z_o - pr - w + z_o}{\sqrt{p^2 + 1} + pw_r + w_z}. \tag{2.37}$$

Finally, by substituting equation (2.37) into equation (2.34), we get the radial component of the mirror,

$$\boxed{\lim_{z_i \to \infty}(r_a) = r + \frac{w'\left[-\left(\sqrt{p^2 + 1} + 1\right)z_o + pr + w\right]}{\sqrt{p^2 + 1}\sqrt{w'^2 + 1} - pw' + 1},} \tag{2.38}$$

and the sagittal component of the mirror,

$$\boxed{\lim_{z_i \to \infty}(z_a) = w + \frac{-\left(\sqrt{p^2 + 1} + 1\right)z_o + pr + w}{-\sqrt{p^2 + 1}\sqrt{w'^2 + 1} + pw' - 1}.} \tag{2.39}$$

Again, we use neither the paraxial concept, nor numerical approximations, nor optimisations to obtain equations (2.38) and (2.39). In fact, equations (2.38) and (2.39) are analytic closed-form solutions of the proposed problem. Thus, we can confirm that equations (2.38) and (2.39) are natural implications of Fermat's principle.

Observe that based on the procedure used to obtain equations (2.38) and (2.39), we can conclude that in the case where the image is far away, these are also uniqueness theorems. This means that for given input values, equations (2.38) and (2.39) compute a single mirror such that it performs the required task of transforming an input wavefront w into a perfect plane wave. Therefore, the set of mirrors designed with equations (2.38) and (2.39) contains all possible mirrors that give a plane wave from an input wavefront w.

Let us verify that the distance from the vertex of the input wavefront w to the vertex of the mirror is z_o. We compute the limit $r \to 0$ over equation (2.37):

$$\lim_{r \to 0}\left[\lim_{z_i \to \infty} \vartheta\right] = \frac{\sqrt{p^2 + 1}\, z_o + z_o}{\sqrt{p^2 + 1} + 1} = z_o. \tag{2.40}$$

The result is as expected. Remember that when we set $r \to 0$, we are taking the axial ray, i.e. the ray that starts from the vertex of w and is reflected at the vertex of the mirror.

2.7 The eikonal, the wavefront, and ray tracing

It is easy to obtain the eikonal of these mirrors when the image is far. We only need to compute the limit $z_i \to \infty$ over equations (2.13), and (2.14). But also we can get them by computing the derivative with respect to r over equation (2.38):

$$\begin{aligned}\lim_{z_i \to \infty}(r_a') = {} & \frac{1}{\sqrt{w'^2+1}\left(\sqrt{p^2+1}\sqrt{w'^2+1}-pw'+1\right)^2} \times \left\{\sqrt{p^2+1}\,w'^4\right.\\ & - pw'^3\left(\sqrt{p^2+1}+\sqrt{w'^2+1}\right)+w'^2\\ & \left(p^2\sqrt{w'^2+1}+3\sqrt{p^2+1}+2\sqrt{w'^2+1}\right)-pw'\\ & \left(\sqrt{p^2+1}+\sqrt{w'^2+1}\right)+p^2\sqrt{w'^2+1}+2\left(\sqrt{p^2+1}+\sqrt{w'^2+1}\right)\\ & + w''\left[-\left(\sqrt{p^2+1}+1\right)z_o\left(\sqrt{w'^2+1}+1\right)\right.\\ & \left.\left.+pr\left(\sqrt{p^2+1}+\sqrt{w'^2+1}\right)+w\left(\sqrt{p^2+1}+\sqrt{w'^2+1}\right)-p^2 z_o\right]\right\},\end{aligned} \tag{2.41}$$

and the derivative with respect to r over equation (2.39):

$$\begin{aligned}\lim_{z_i \to \infty}(z_a') = {} & \frac{1}{\sqrt{w'^2+1}\left(\sqrt{p^2+1}\sqrt{w'^2+1}-pw'+1\right)^2} \times\\ & \left\{-2p\sqrt{p^2+1}\,w'^4+w'^3\left(2p^2\sqrt{w'^2+1}+\sqrt{p^2+1}+\sqrt{w'^2+1}\right)\right.\\ & - pw'^2\left(3\sqrt{p^2+1}+\sqrt{w'^2+1}\right)+w'\\ & \left[w''\left(\sqrt{p^2+1}\,w+\sqrt{p^2+1}\,pr+p^2(-z_o)-\left(\sqrt{p^2+1}+1\right)z_o\right)+\right.\\ & \left.2p^2\sqrt{w'^2+1}+\sqrt{p^2+1}+\sqrt{w'^2+1}\right]+\\ & \left.p\sqrt{w'^2+1}\left(w''\left(\sqrt{p^2+1}\,z_o-pr-w+z_o\right)-1\right)-p\sqrt{p^2+1}\right\}.\end{aligned} \tag{2.42}$$

Using expressions (2.41) and (2.42) we can obtain the normal vector of the mirror when the image is far away. We can then apply the aforementioned limit to equations (2.25) and (2.26), the cosine directors of the output ray,

$$\lim_{z_i \to \infty}(\wp_r) = \frac{\lim_{z_i \to \infty}(r_a')\left[\lim_{z_i \to \infty}(z_a') - \lim_{z_i \to \infty}(r_a')w'\right]}{\sqrt{w'^2 + 1}\left[\lim_{z_i \to \infty}(r_a'^2) + \lim_{z_i \to \infty}(z_a'^2)\right]} + \frac{\lim_{z_i \to \infty}(z_a')\sqrt{1 - \frac{\left[\lim_{z_i \to \infty}(z_a') - \lim_{z_i \to \infty}(r_a')w'\right]^2}{(w'^2 + 1)\left[\lim_{z_i \to \infty}(r_a'^2) + \lim_{z_i \to \infty}(z_a'^2)\right]}}}{\sqrt{\lim_{z_i \to \infty}(r_a'^2) + \lim_{z_i \to \infty}(z_a'^2)}}, \tag{2.43}$$

and

$$\lim_{z_i \to \infty}(\wp_z) = \frac{\lim_{z_i \to \infty}(z_a')\left[\lim_{z_i \to \infty}(z_a') - \lim_{z_i \to \infty}(r_a')w'\right]}{\sqrt{w'^2 + 1}\left[\lim_{z_i \to \infty}(r_a'^2) + \lim_{z_i \to \infty}(z_a'^2)\right]} - \frac{\lim_{z_i \to \infty}(r_a')\sqrt{1 - \frac{\left[\lim_{z_i \to \infty}(z_a') - \lim_{z_i \to \infty}(r_a')w'\right]^2}{(w'^2 + 1)\left[\lim_{z_i \to \infty}(r_a'^2) + \lim_{z_i \to \infty}(z_a'^2)\right]}}}{\sqrt{\lim_{z_i \to \infty}(r_a'^2) + \lim_{z_i \to \infty}(z_a'^2)}}. \tag{2.44}$$

Finally, using these cosine directors, a direct construction of the eikonal $\lim_{z_i \to \infty} H(t, r_a)$ is given by

$$\boxed{\lim_{z_i \to \infty} H(t, r_a) = \begin{cases} [w + tw_z, r + tw_r], & t < \lim_{z_i \to \infty}(\vartheta), \\ \begin{bmatrix} \lim_{z_i \to \infty}(z_a) + \lim_{z_i \to \infty}(\wp_z)\left\{t - \lim_{z_i \to \infty}(\vartheta)\right\}, \\ \lim_{z_i \to \infty}(r_a) + \lim_{z_i \to \infty}(\wp_r)\left\{t - \lim_{z_i \to \infty}(\vartheta)\right\} \end{bmatrix}, & t > \lim_{z_i \to \infty}(\vartheta), \end{cases}} \tag{2.45}$$

where the piecewise condition seems very clear. It separates the input wavefront w from the plane wave at the output of the system.

2.8 Mathematica code

This section presents the `Mathematica` code for equations (2.38), (2.39), and (2.45); the outputs of the algorithm are presented in the following section.

```
In[2]:= ClearAll["Global`*"]

w[r_]:=16-√16^2-r^2

wr[r_]:=-w'[r]/√(w'[r]^2+1)

wz[r_]:=1/√(w'[r]^2+1)

ϑ[r_]:=(-r p+zo+√(1+p^2) zo-w[r])/(√(1+p^2)+p wr[r]+wz[r])

ra[r_]:=r+ϑ[r]wr[r]

za[r_]:=w[r]+ϑ[r]wz[r]

ϱr[r_]:=(no ra'[r] (-ra'[r] w'[r]+za'[r]))/(Sign[w[r]]ni √(1+w'[r]^2) (ra'[r]^2+za'[r]^2))
   +za'[r]/√(ra'[r]^2+za'[r]^2) √(1-(no^2 (-ra'[r] w'[r]+za'[r])^2)/(ni^2 (1+w'[r]^2) (ra'[r]^2+za'[r]^2)))

ϱz[r_]:=(no za'[r] (-ra'[r] w'[r]+za'[r]))/(Sign[w[r]]ni √(1+w'[r]^2) (ra'[r]^2+za'[r]^2))
   -ra'[r]/√(ra'[r]^2+za'[r]^2) √(1-(no^2 (-ra'[r] w'[r]+za'[r])^2)/(ni^2 (1+w'[r]^2) (ra'[r]^2+za'[r]^2)))

Homotopy={
{Piecewise[{
{wr[r]tt+r,tt<ϑ[r]},{-Sign[p]Sign[w[r]]v ϱr[r](tt-ϑ[r])+ra[r],tt>ϑ[r]}}],
Piecewise[{
{wz[r]tt+w[r],tt<ϑ[r]},{-Sign[p]Sign[w[r]] v ϱz[r](tt-ϑ[r])+za[r],tt>ϑ[r]}}]}};
```

```
c=1;
v=c/ni;
no=1;
ni=-1;
p=-4.6;
zo=60;
rmax=9;


Surface:=ParametricPlot[{{r,w[r]},{ra[r],za[r]},{5r,w[5r]},{ra[1.12r],za[1.12r]},
{5r,32-w[5r]}},{r,-rmax,rmax}, AxesOrigin→{-30,-26},
PlotStyle→{{Black,Thickness[0.005]},{Red,Thickness[0.005]},
{Black,Dashed,Thin},{Red,Dashed,Thin},{Black,Dashed,Thin}},
AxesLabel→{Style[r,FontSize→14,Bold],Style[z,FontSize→14,Bold]},
AxesStyle→Arrowheads[{0.05}]]

Waves:=Table[{ParametricPlot[Homotopy,{r,-rmax,rmax},
PlotRange→All,PlotStyle→{Blue,Thickness[0.0015]}]},{tt,-20,120,2}]

Rays:=Table[{ParametricPlot[Homotopy,{tt,-20,120},
PlotRange→All,PlotStyle→{Purple,Thickness[0.0015]}]},{r,-rmax,rmax}]

OptShow={ PlotRange->All};

Show[Surface, Rays,Waves,OptShow]
```

2.9 Examples

The examples computed in this section are presented in figures 2.10, 2.11, 2.12, 2.13, 2.14, 2.15, 2.16, 2.17, 2.18, 2.19, 2.20, 2.21, 2.22, 2.23, 2.24, 2.25, 2.26, and 2.27,

From all the examples computed, we can confirm that equations (2.38), (2.39), and (2.45) behaved as expected. In all the cases the output wavefront is a plane, as predicted.

For equations (2.38), (2.39), and (2.45) an overlap may occur if two rays strike at the same point of the mirror, for reasons previously explained in the previous sections of this chapter.

We have obtained equations (2.38), (2.39), and (2.45), and they accurately compute the mirrors that we wanted, but where is the adaptive optics? It is important to remark that if the wavefront w depends on time t, we can still use equations (2.38), (2.39), and (2.45) to simulate the design of an adaptive optics mirror. The mechanics of how the actuators should deform the mirror so that it has a given surface is a well known task that can be found in the adaptive optics books presented in the bibliography of this chapter. So, it is not a major challenge to set the actuators to get the mirror shapes described by equations (2.38), (2.39), and (2.45).

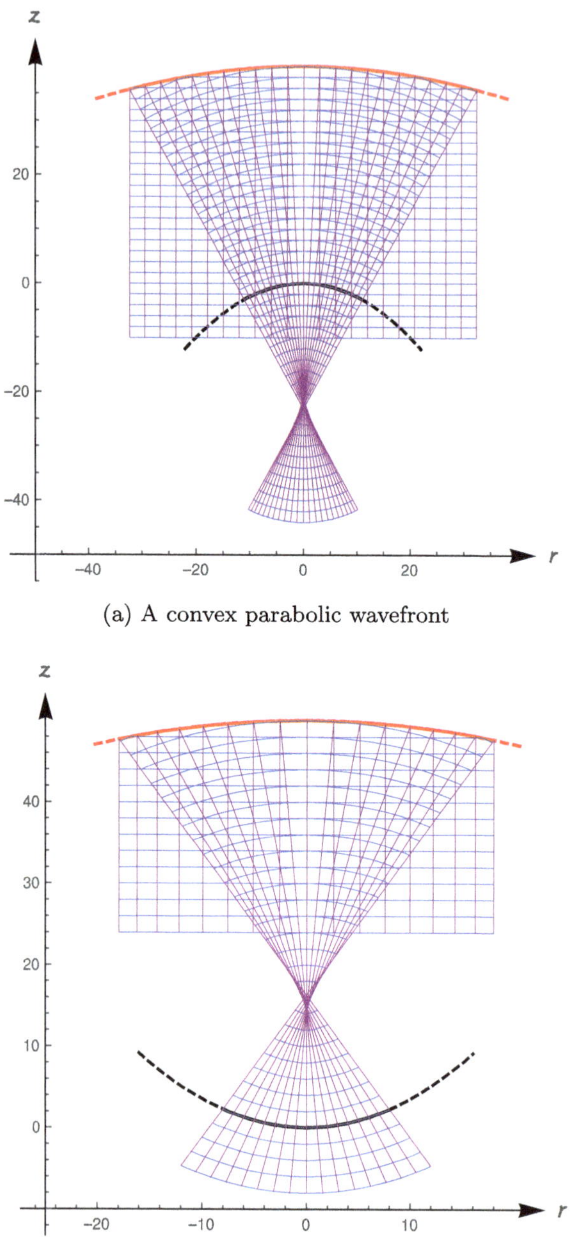

(a) A convex parabolic wavefront

(b) A concave parabolic wavefront

Figure 2.10. Design specifications for parabolic wavefronts: (a) the input wavefront is $r^2 \div 28$, refractive index $n = -1.0$, object distance $z_o = 50$ mm, r_{max} is 8 mm, constant $p = 0$, (r_a, z_a) = equations (2.38), (2.39), and for the eikonal $H(t, r_a)$ = equation (2.45); (b) the input wavefront is $-r^2 \div 40$, refractive index $n = -1.0$, object distance $z_o = 40$ mm, r_{max} is 13 mm, constant $p = -1000$, (r_a, z_a) = equations (2.38), (2.39), and for the eikonal $H(t, r_a)$ = equation (2.45).

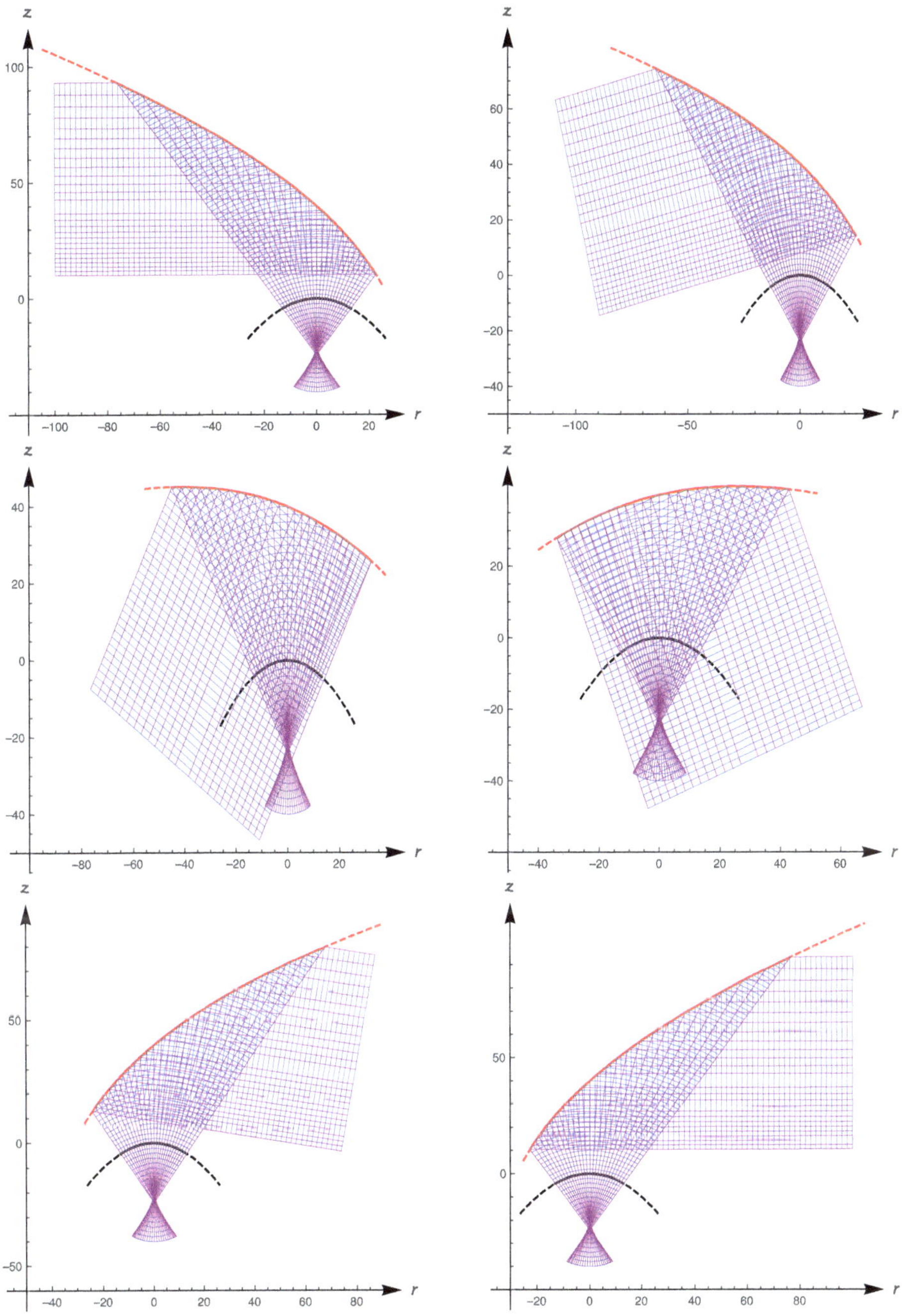

Figure 2.11. Design specifications for a convex parabolic wavefront input. The input wavefront is $-r^2 \div 40$, refractive index $n = -1.0$, object distance $z_o = 40\,\text{mm}$, $r_{\max}$ is 13 mm, constant $p = \{-1000, -4, -0.4, 0.6, 4, 1000\}$, (r_a, z_a) = equations (2.38), (2.39), and for the eikonal $H(t, r_a)$ = equation (2.45).

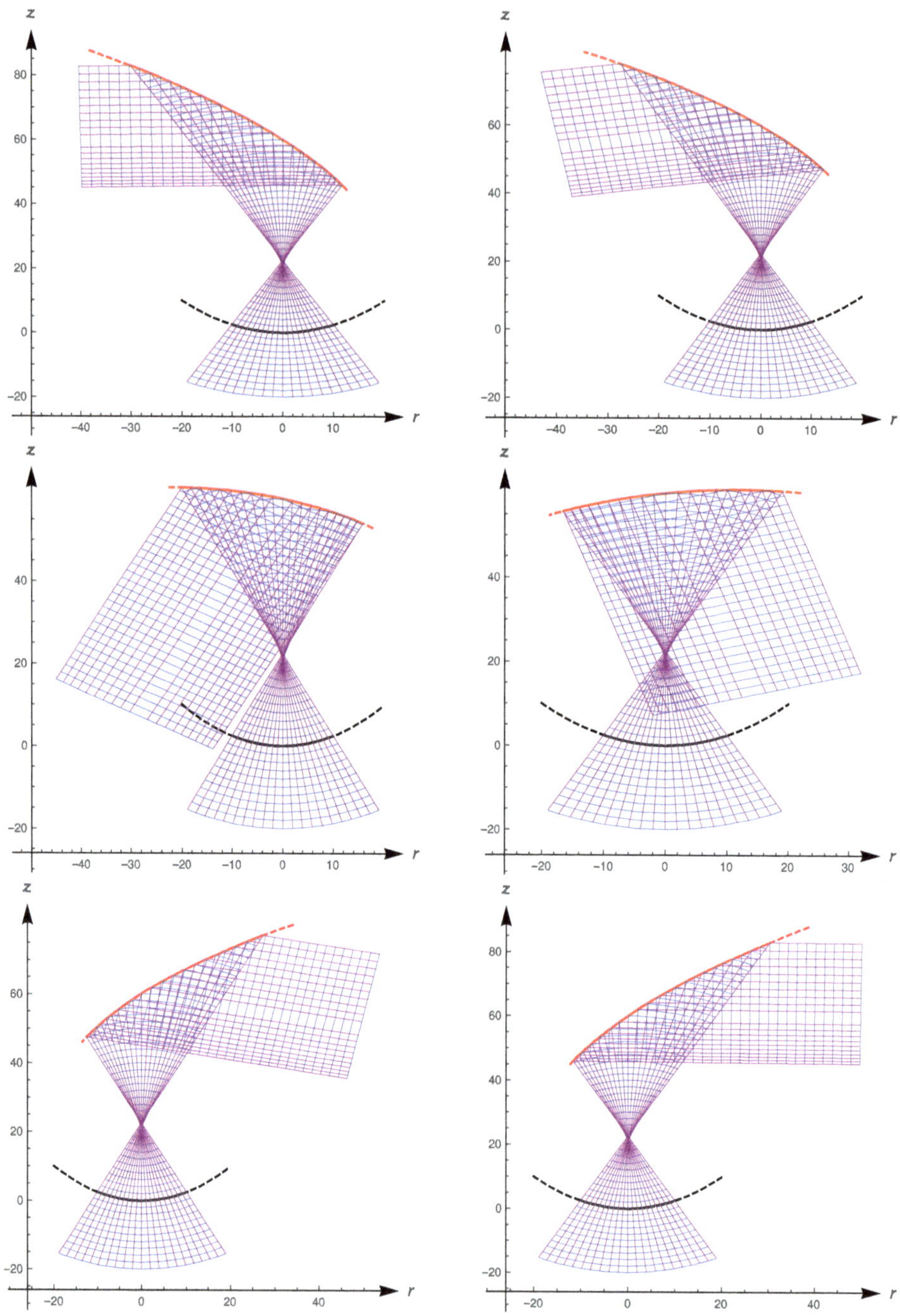

Figure 2.12. Design specifications for a concave parabolic wavefront input. The input wavefront is $r^2 \div 40$, refractive index $n = -1.0$, object distance $z_o = 60\,\text{mm}$, r_{max} is 10 mm, constant $p = \{-1000, -4, -0.4, 0.54, 6, 60\}$, (r_a, z_a) = equations (2.38), (2.39), and for the eikonal $H(t, r_a)$ = equation (2.45).

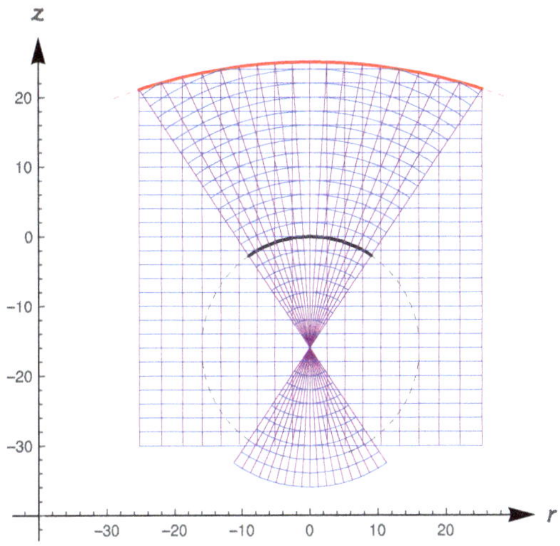

(a) A convex circular wavefront

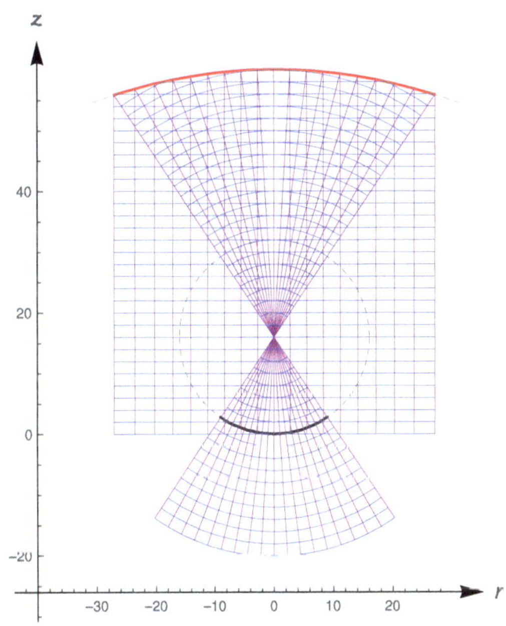

(b) A concave circular wavefront

Figure 2.13. Design specifications for circular wavefronts: (a) the input wavefront is $-16 + \sqrt{16^2 - r^2}$, refractive index $n = -1.0$, object distance $z_o = 25$ mm, r_{max} is 9 mm, constant $p = 0$, (r_a, z_a) = equations (2.38), (2.39), and for the eikonal $H(t, r_a)$ = equation (2.45); (b) the input wavefront is $16 - \sqrt{16^2 - r^2}$, refractive index $n = -1.0$, object distance $z_o = 60$ mm, r_{max} is 9 mm, constant $p = -0$, (r_a, z_a) = equations (2.38), (2.39), and for the eikonal $H(t, r_a)$ = equation (2.45).

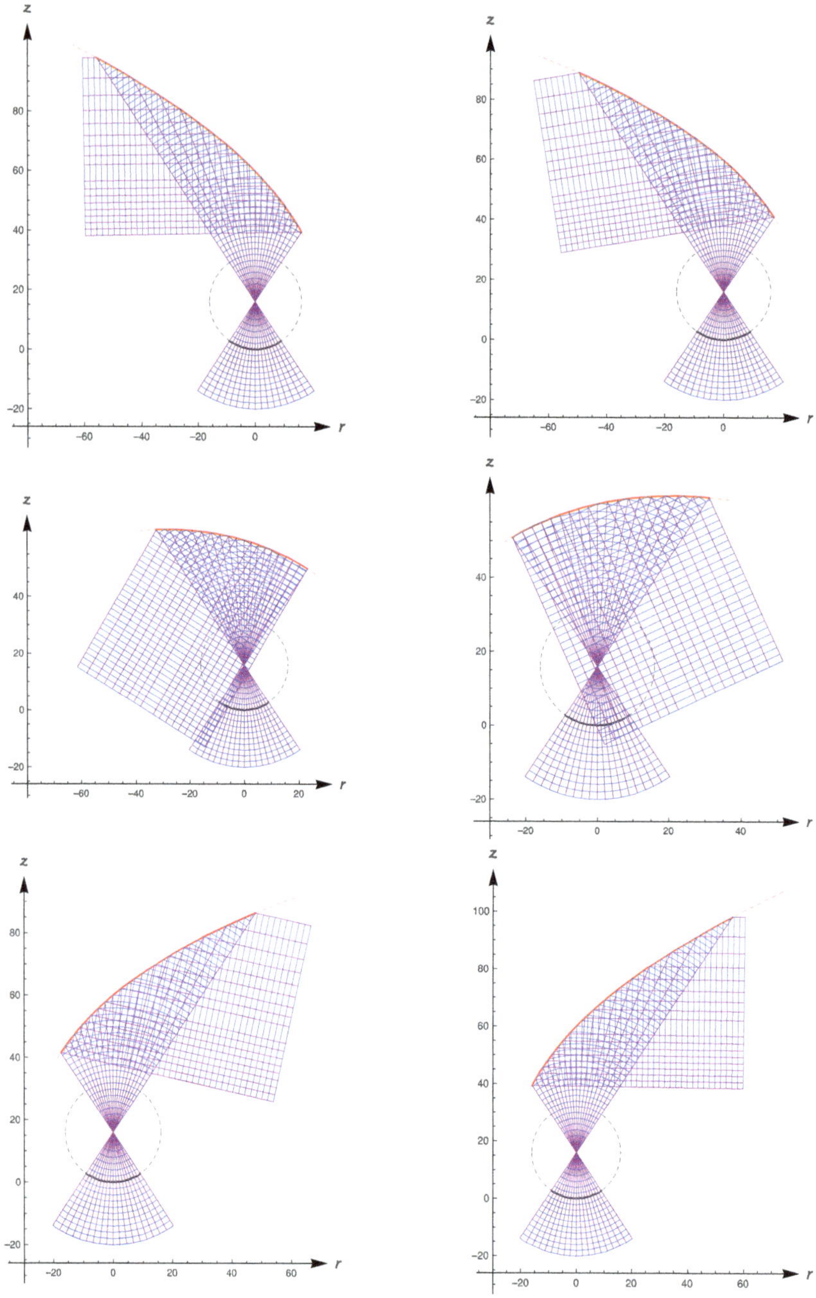

Figure 2.14. Design specifications for a concave circular wavefront input. The input wavefront is $16 - \sqrt{16^2 - r^2}$, refractive index $n = -1.0$, object distance $z_o = 60$ mm, $r_{\max}$ is 9 mm, constant $p = \{-80, -4.6, -0.46, 0.6, 6, 60\}$, (r_a, z_a) = equations (2.38), (2.39), and for the eikonal $H(t, r_a)$ = equation (2.45).

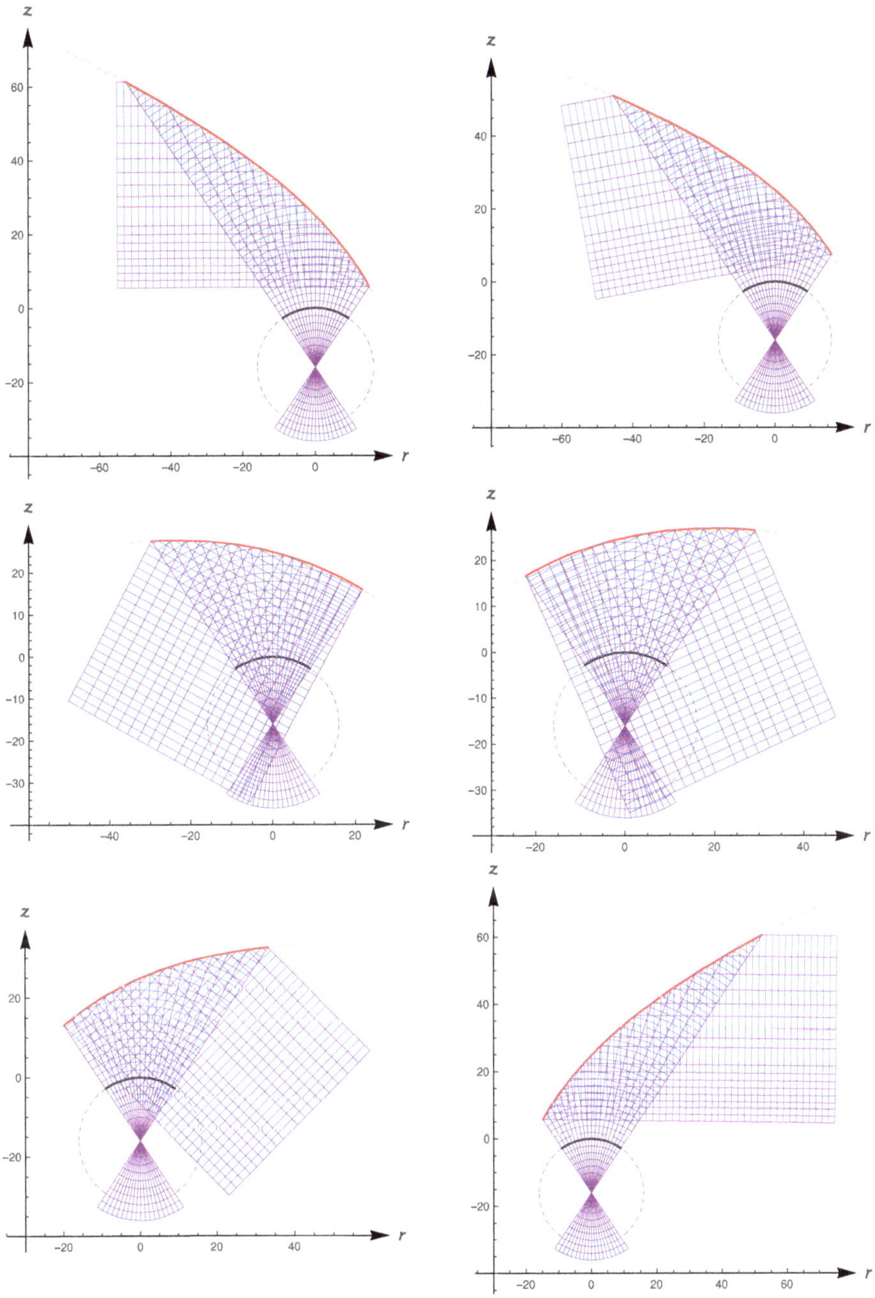

Figure 2.15. Design specifications for a convex circular wavefront input. The input wavefront is $-16 + \sqrt{16^2 - r^2}$, refractive index $n = -1.0$, object distance $z_o = 25\,\text{mm}$, r_{max} is 9 mm, constant $p = \{-80, -1, -0.44, 0.54, 5.4, 450\}$, (r_a, z_a) = equations (2.38), (2.39), and for the eikonal $H(t, r_a)$ = equation (2.45).

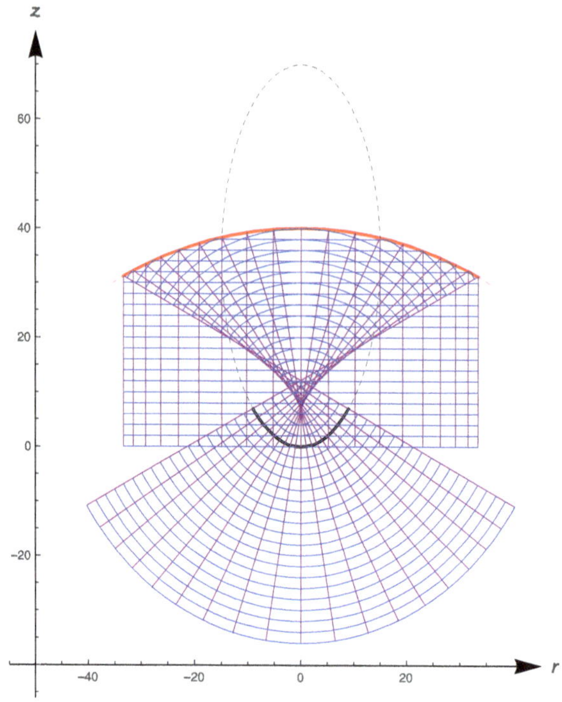

(a) A concave oblate ellipse wavefront

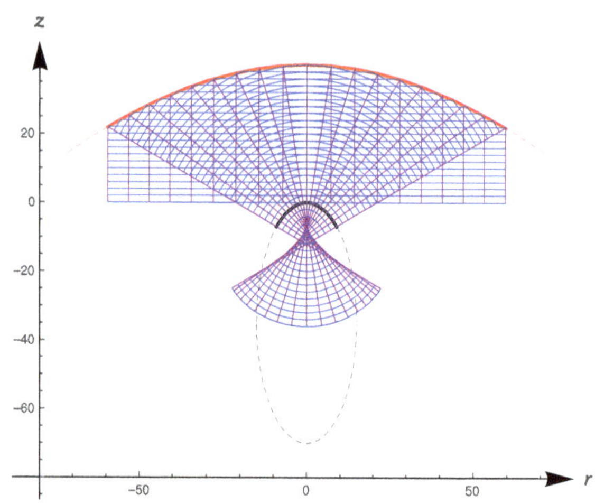

(b) A convex oblate ellipse wavefront

Figure 2.16. Design specifications for oblate elliptical wavefronts: (a) the input wavefront is$b - \sqrt{(-a^2*b^2 + b^2r^2)/ - a^2}$, refractive index $n = -1.0$, object distance $z_o = 40$ mm, r_{max} is 9 mm, constant $a = 15$, constant $b = 35$, constant $p = 0$, (r_a, z_a) = equations (2.38), (2.39), and for the eikonal $H(t, r_a)$ = equation (2.45); (b) the input wavefront is $-b + \sqrt{(-a^2*b^2 + b^2r^2)/ - a^2}$, refractive index $n = -1.0$, object distance $z_o = 40$ mm, r_{max} is 9 mm, constant $a = 15$, constant $b = 35$, constant $p = -0$, (r_a, z_a) = equations (2.38), (2.39), and for the eikonal $H(t, r_a)$ = equation (2.45).

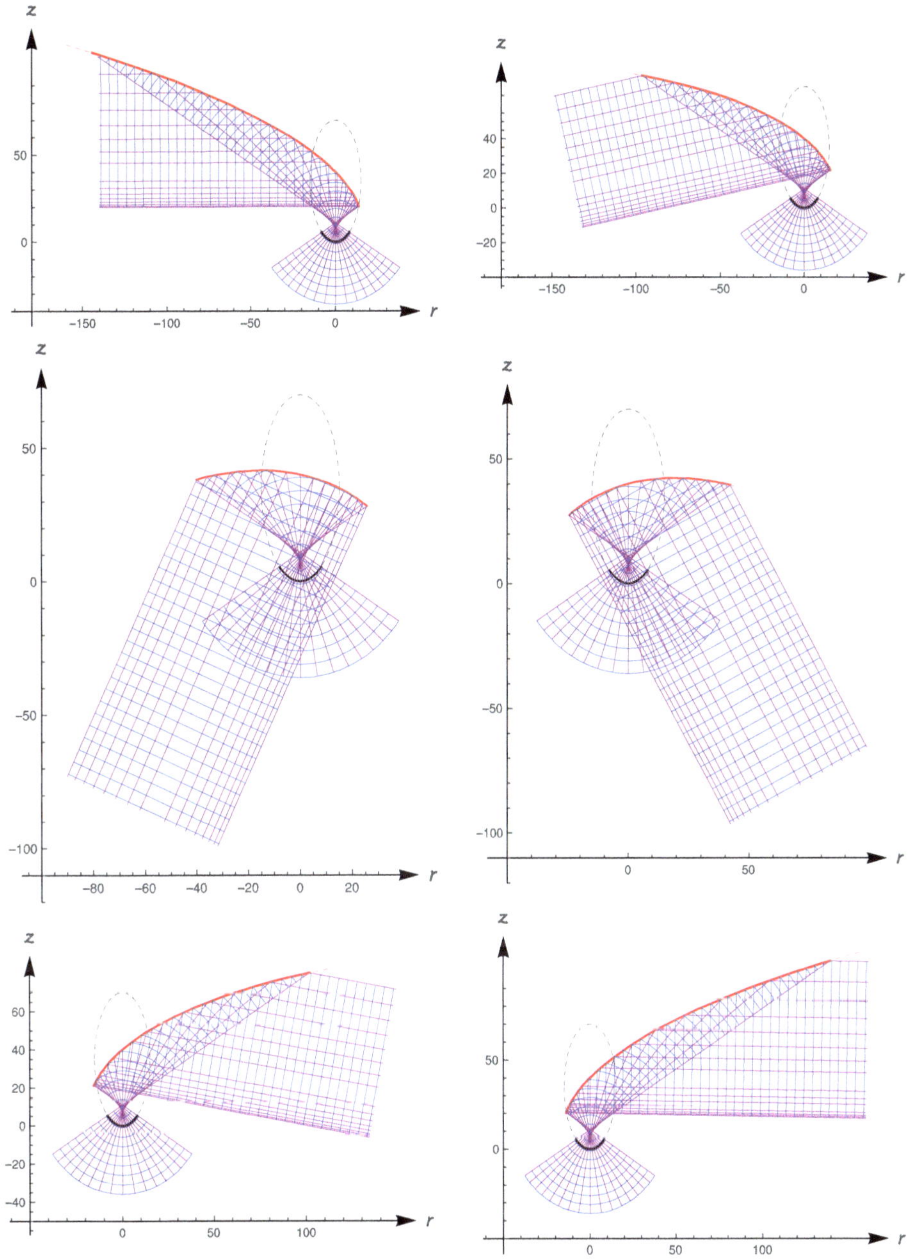

Figure 2.17. Design specifications for a concave oblate elliptical wavefront input. The input wavefront is $b - \sqrt{(-a^2 * b^2 + b^2 r^2)/ - a^2}$, refractive index $n = -1.0$, object distance $z_o = 40$ mm, r_{max} is 9 mm, constant $a = 15$, constant $b = 35$, constant $p = \{450, 4.5, 0.45 - 0.54, -5.4, -54\}$, (r_a, z_a) = equations (2.38), (2.39), and for the eikonal $H(t, r_a)$ = equation (2.45).

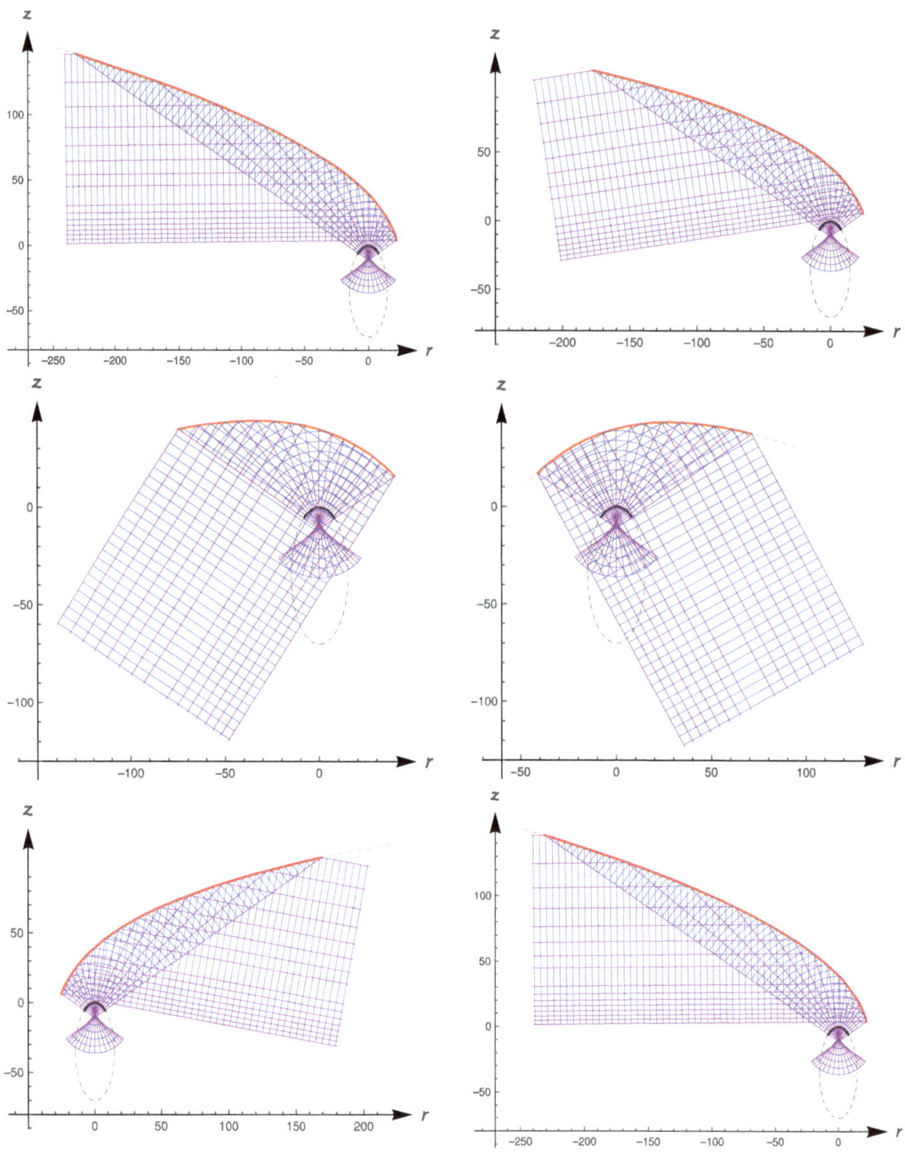

Figure 2.18. Design specifications for a convex oblate elliptical wavefront input. The input wavefront is $-b + \sqrt{(-a^2*b^2 + b^2r^2)/ - a^2}$, refractive index $n = -1.0$, object distance $z_o = 40$ mm, r_{max} is 9 mm, constant $a = 15$, constant $b = 35$, constant $p = \{85, 6.5, 0.65 - 0.55, -5.5, -88\}$, (r_a, z_a) = equations (2.38), (2.39), and for the eikonal $H(t, r_a)$ = equation (2.45).

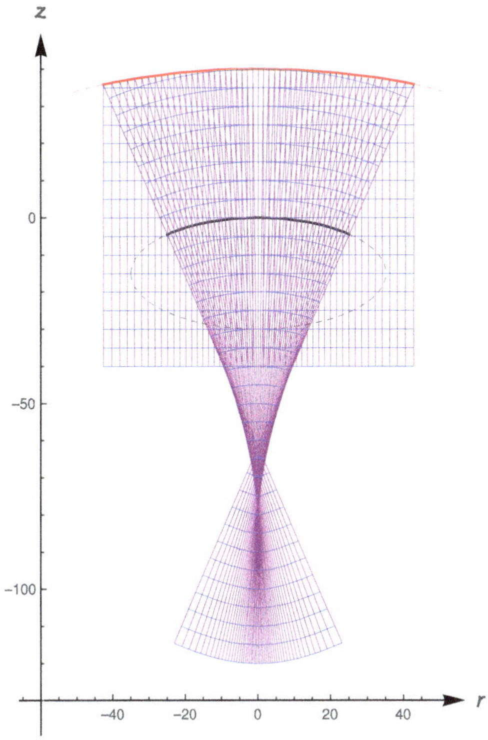

(a) A convex prolate ellipse wavefront

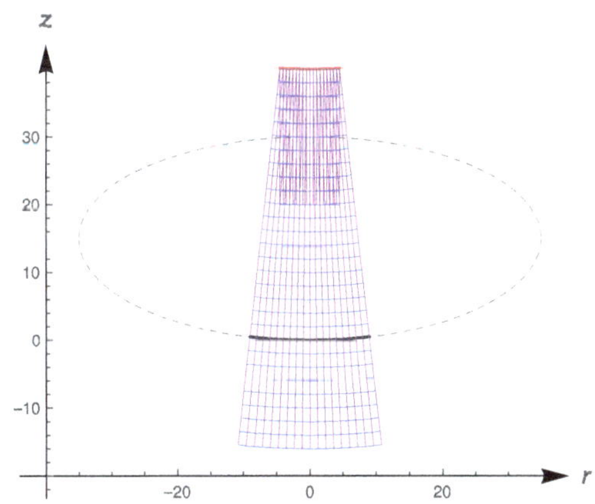

(b) A concave proolate ellipse wavefront

Figure 2.19. Design specifications for prolate elliptical wavefronts: (a) the input wavefront is$-b + \sqrt{(-a^2 * b^2 + b^2 r^2)/ - a^2}$, refractive index $n = -1.0$, object distance $z_o = 40$ mm, $r_{\max}$ is 25 mm, constant $a = 35$, constant $b = 15$, constant $p = 0$, (r_a, z_a) = equations 2.38, 2.39 (2.38), (2.39), and for the eikonal $H(t, r_a)$ − equation (2.45); (b) the input wavefront is $b - \sqrt{(-a^2 * b^2 + b^2 r^2)/ - a^2}$, refractive index $n = -1.0$, object distance $z_o = 40$ mm, $r_{\max}$ is 9 mm, constant $a = 35$, constant $b = 15$, constant $p = -0$, (r_a, z_a) = equations (2.38), (2.39), and for the eikonal $H(t, r_a)$ = equation (2.45).

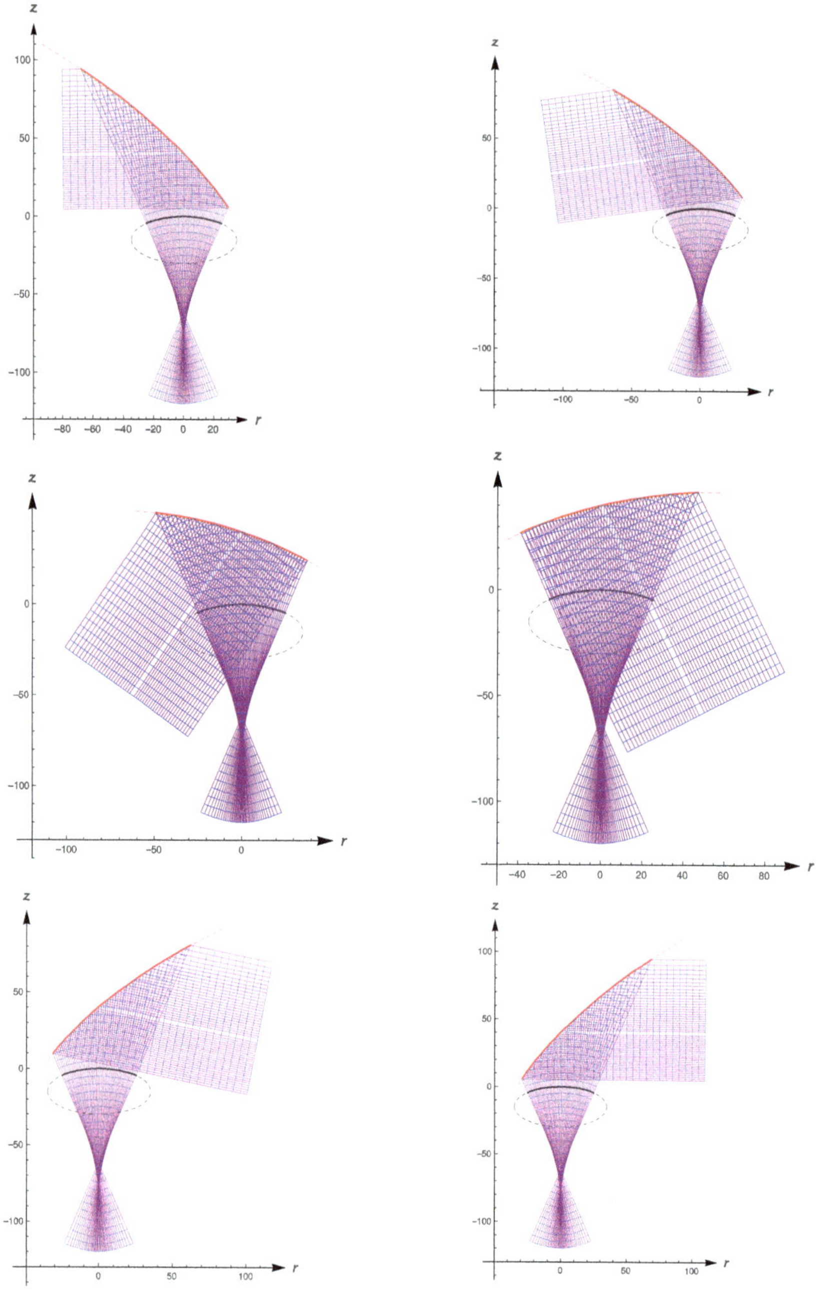

Figure 2.20. Design specifications for a convex prolate elliptical wavefront. The input wavefront is$-b + \sqrt{(-a^2*b^2 + b^2r^2)/ - a^2}$, refractive index $n = -1.0$, object distance $z_o = 40$ mm, r_{max} is 25 mm, constant $a = 35$, constant $b = 15$, constant $p = \{70, 7, 0.7, -0.5, -5, -150\}$, (r_a, z_a) = equations (2.38), (2.39), and for the eikonal $H(t, r_a)$ = equation (2.45).

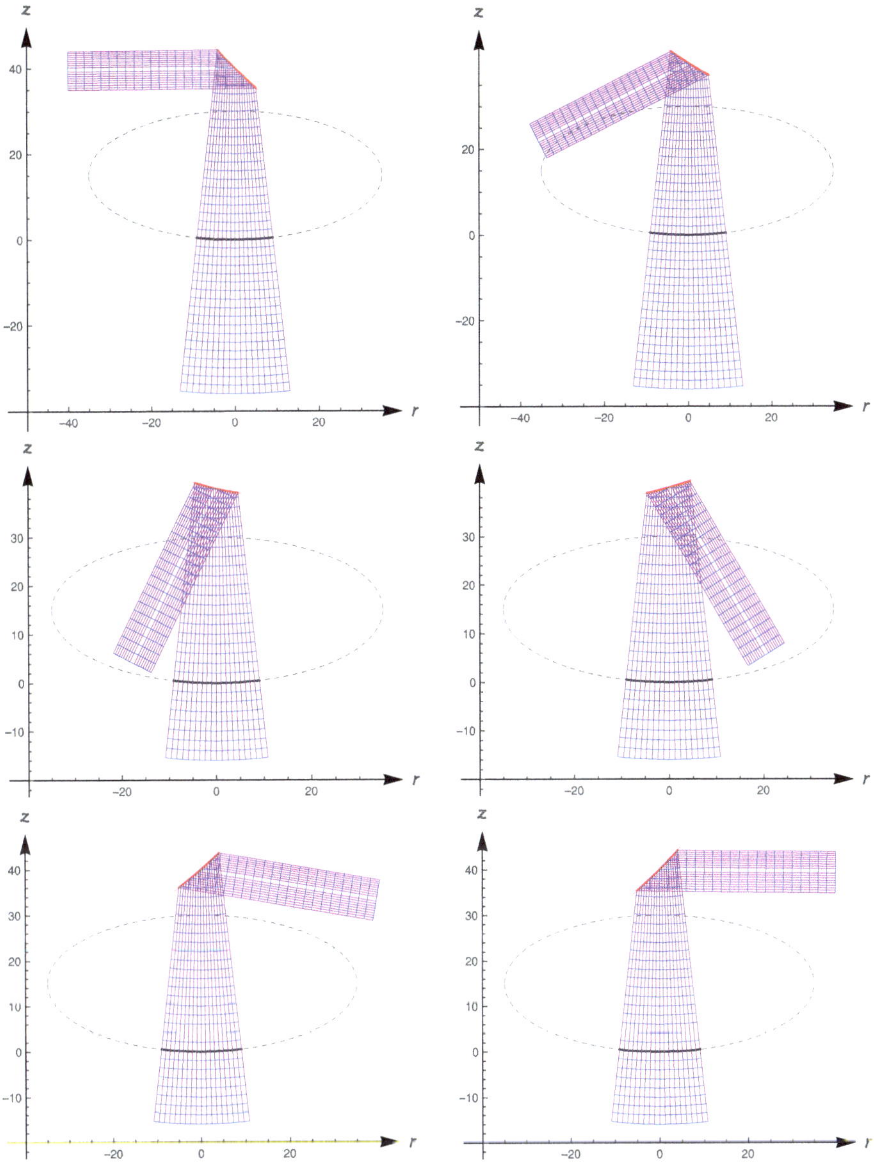

Figure 2.21. Design specifications for a concave prolate elliptical wavefront. The input wavefront is $b - \sqrt{(-a^2*b^2 + b^2r^2)/ - a^2}$, refractive index $n = -1.0$, object distance $z_o = 40$ mm, r_{max} is 9 mm, constant $a = 35$, constant $b = 15$, constant $p = \{150, 2, 0.5, -0.6, -6, -100\}$, (r_a, z_a) = equations (2.38), (2.39), and for the eikonal $H(t, r_a)$ = equation (2.45).

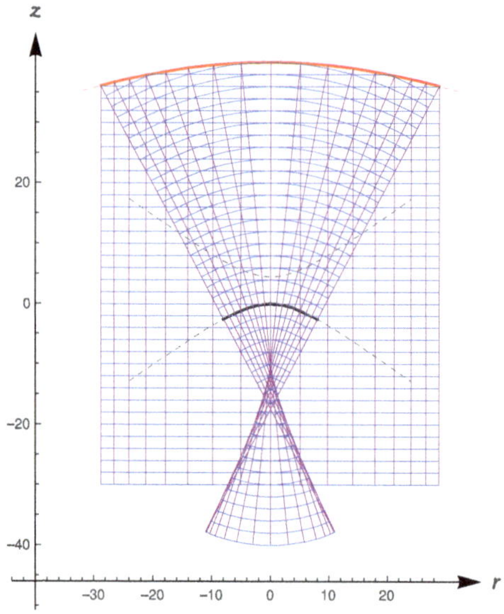

(a) A convex hyperbolic wavefront

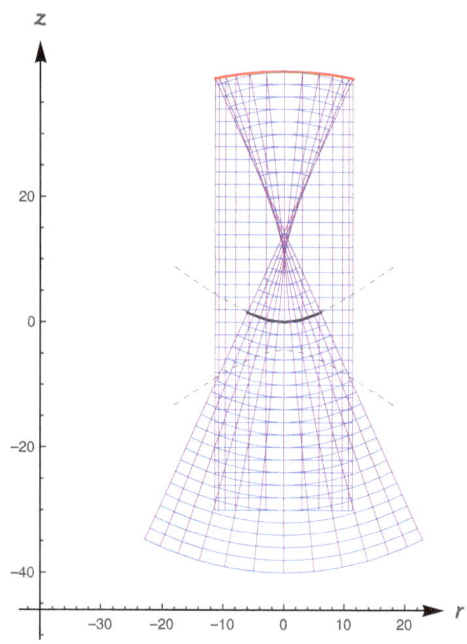

(b) A concave hyperbolic wavefront

Figure 2.22. Design specifications for hyperbolic wavefronts: (a) the input wavefront is $b - \sqrt{(-a^2{*}b^2 - b^2r^2)/ - a^2}$, refractive index $n = -1.0$, object distance $z_o = 40$ mm, r_{max} is 8 mm, constant $a = 7$, constant $b = 5$, constant $p = 0$, (r_a, z_a) = equations (2.38), (2.39), and for the eikonal $H(t, r_a)$ = equation (2.45); (b) the input wavefront is $-b + \sqrt{(-a^2{*}b^2 + b^2r^2)/ - a^2}$, refractive index $n = -1.0$, object distance $z_o = 40$ mm, r_{max} is 8 mm, constant $a = 7$, constant $b = 5$, constant $p = 0$, (r_a, z_a) = equations (2.38), (2.39), and for the eikonal $H(t, r_a)$ = equation (2.45).

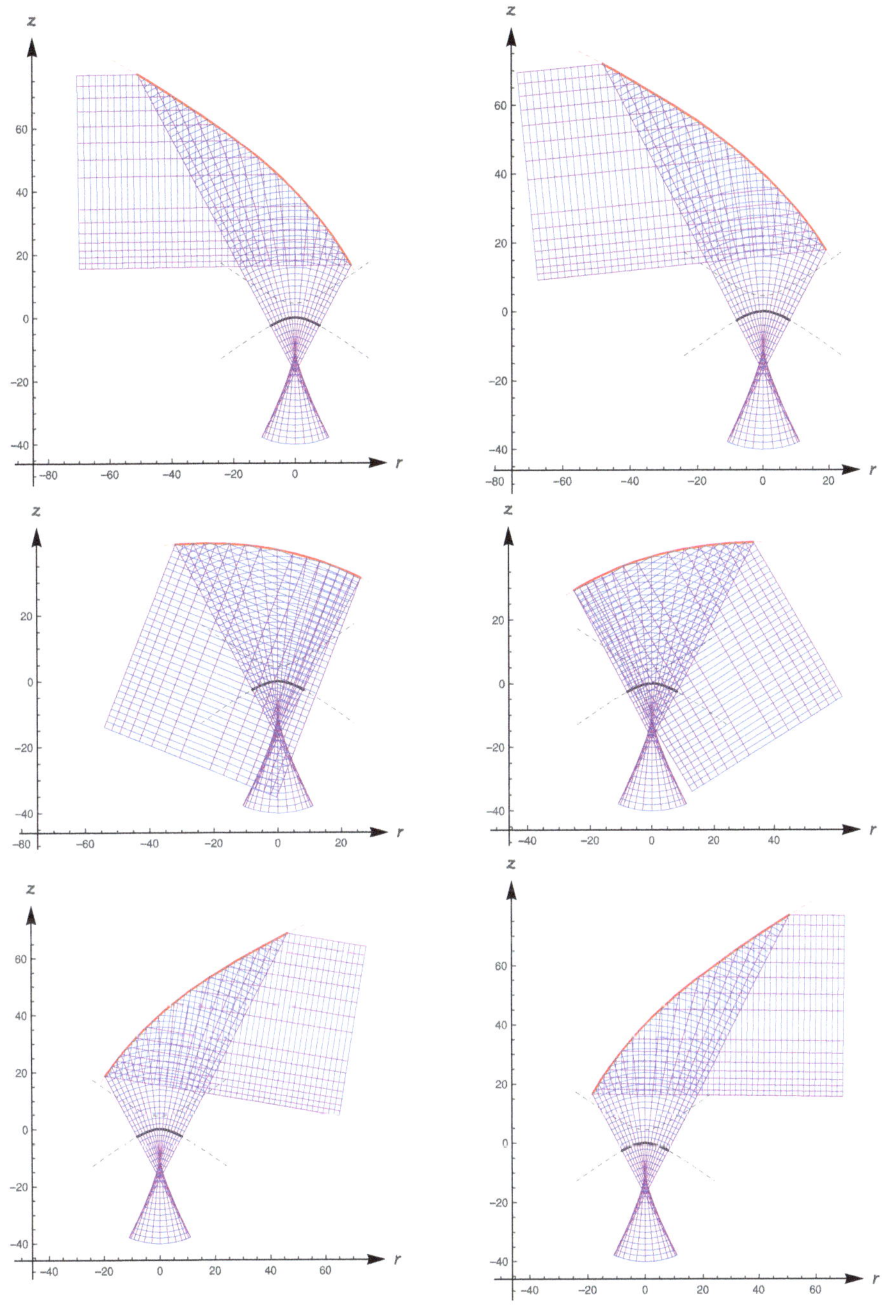

Figure 2.23. Design specifications for a convex hyperbolic wavefront. The input wavefront is $b - \sqrt{(-a^2*b^2 - b^2r^2)/ - a^2}$, refractive index $n = -1.0$, object distance $z_o = 40$ mm, r_{max} is 8 mm, constant $a = 7$, constant $b = 5$, constant $p = \{200, 1, 0.3, -0.43, -4.3, -400\}$, (r_a, z_a) = equations (2.38), (2.39), and for the eikonal $H(t, r_a)$ = equation (2.45).

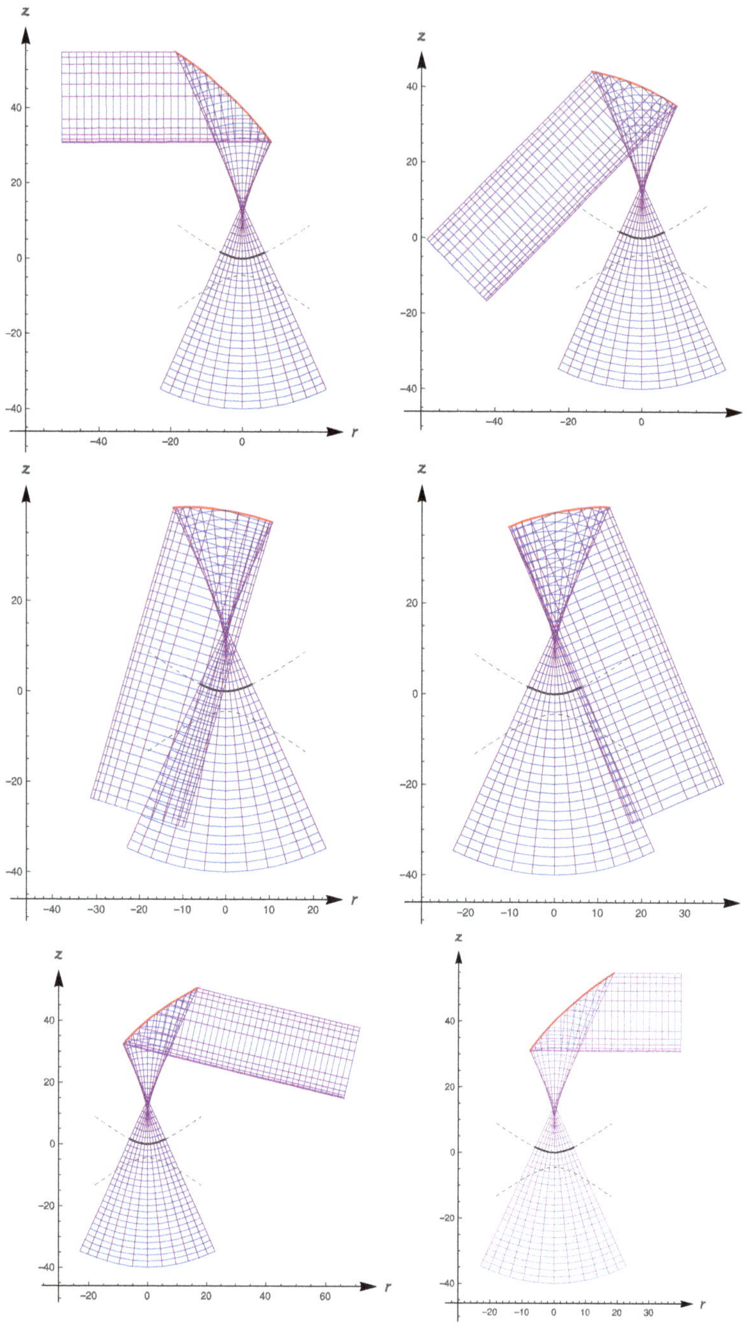

Figure 2.24. Design specifications for a concave hyperbolic wavefront. The input wavefront is$-b + \sqrt{(-a^2*b^2 + b^2r^2)/ - a^2}$, refractive index $n = -1.0$, object distance $z_o = 40$ mm, r_{max} is 8 mm, constant $a = 7$, constant $b = 5$, constant $p = \{200, 1, 0.3, -0.43, -4.3, -400\}$, (r_a, z_a) = equations (2.38), (2.39), and for the eikonal $H(t, r_a)$ = equation (2.45).

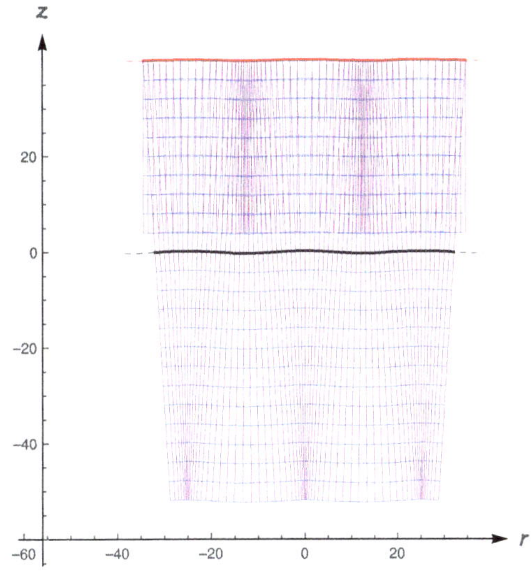

(a) A positive cosine wavefront

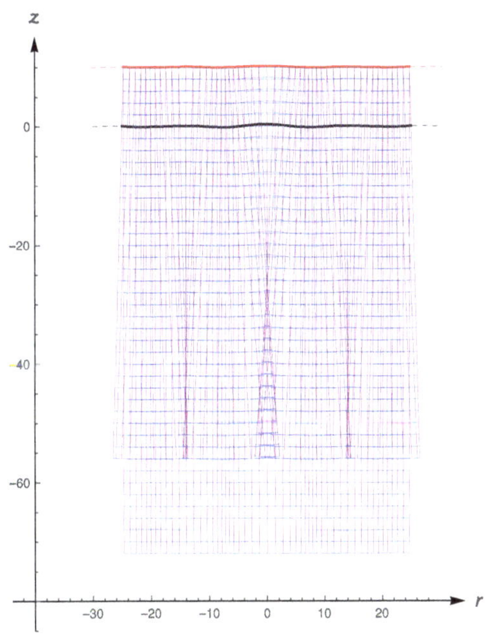

(b) A negative cosine wavefront

Figure 2.25. Design specifications for cosine wavefronts: (a) the input wavefront is $\cos(r/4)/4$, refractive index $n = -1.0$, object distance $z_o = 40$ mm, r_{max} is 32 mm, constant $p = 0$, (r_a, z_a) = equations (2.38), (2.39), and for the eikonal $H(t, r_a)$ = equation (2.45); (b) the input wavefront is $-\cos(r/4)/4$, refractive index $n = -1.0$, object distance $z_o = 40$ mm, r_{max} is 13 mm, constant $p = -1000$, (r_a, z_a) = equations (2.38), (2.39), and for the eikonal $H(t, r_a)$ = equation (2.45).

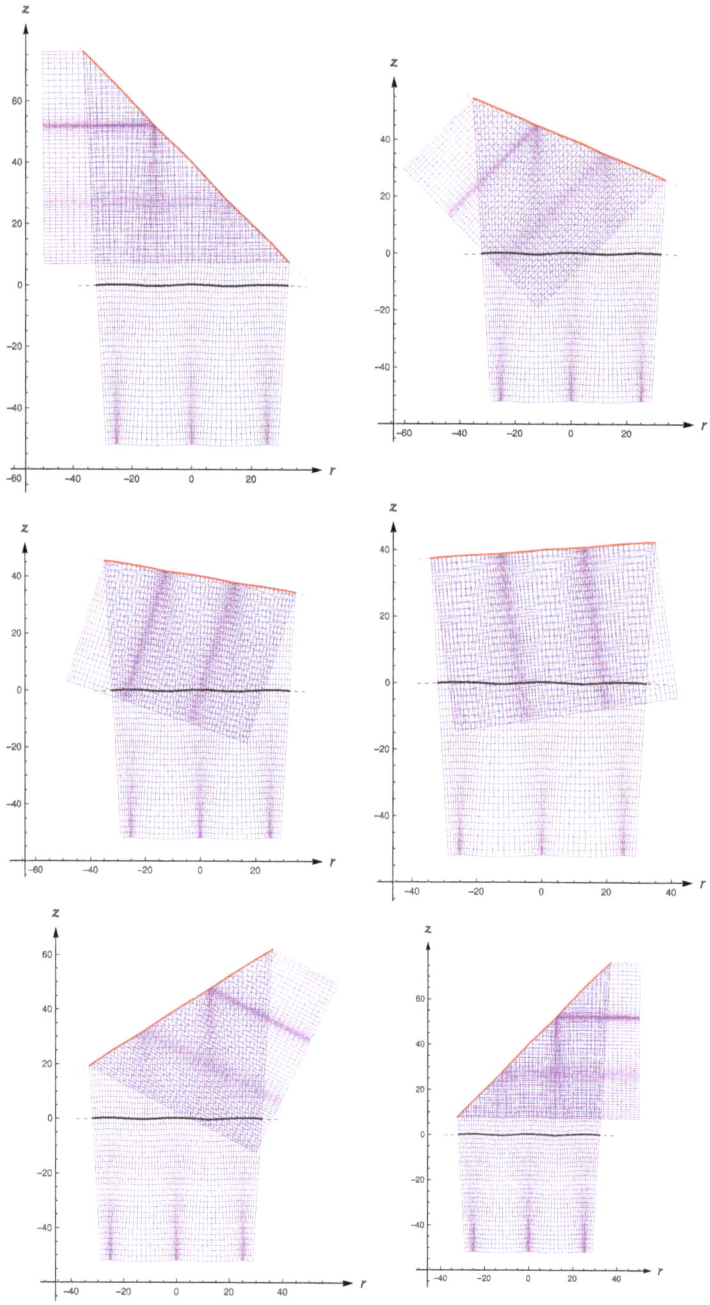

Figure 2.26. Design specifications for a positive cosine wavefront. The input wavefront is $\cos(r/4)/4$, refractive index $n = -1.0$, object distance $z_o = 40$ mm, r_{max} is 32 mm, constant $p = \{100, 1, 0.33, -0.15, -2, -200\}$, (r_a, z_a) = equations (2.38), (2.39), and for the eikonal $H(t, r_a)$ = equation (2.45).

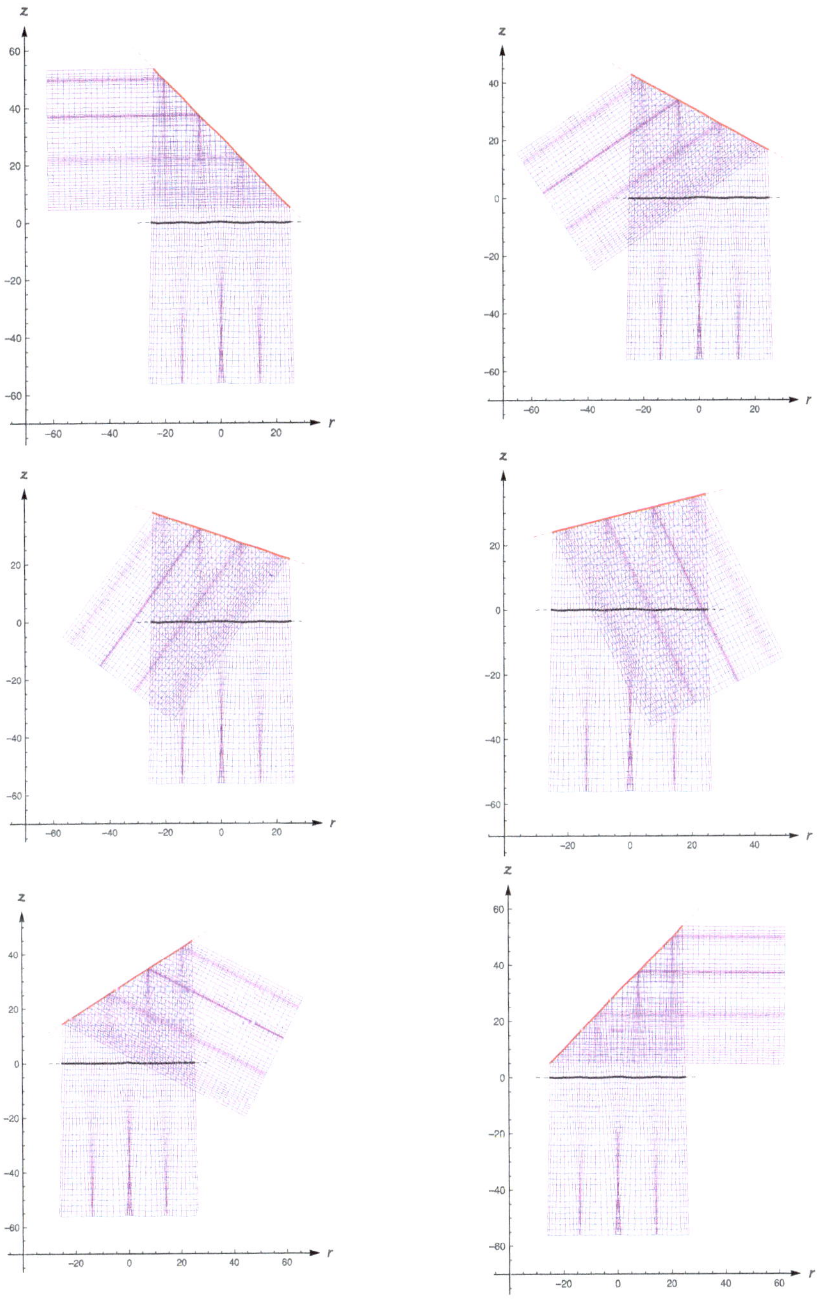

Figure 2.27. Design specifications for a negative cosine wavefront. The input wavefront is $-\cos(r/4)/4$, refractive index $n = -1.0$, object distance $z_o = 40$ mm, r_{max} is 32 mm, constant $p = \{100, 1, 0.33, -0.15, -2, -200\}$, (r_a, z_a) = equations (2.38), (2.39), and for the eikonal $H(t, r_a)$ = equation (2.45).

Also note that we have expressed everything in Cartesian coordinates and we have not used any expressions in cylindrical coordinates, such as the Zernike polynomials, the polynomials usually used to described wavefronts in adaptive optics. We do not use those coordinates because expressing the path of each ray in Cartesian coordinates is simpler than in other coordinate systems for the given problem. This gives us the general solution to the problem of designing a mirror such that it reflects the light of a given wavefront w as a plane wave. The derivation of equations (2.38), (2.39), and (2.45) shows that equations (2.38), (2.39), and (2.45) are not only the general solutions to the mentioned problem, but are also implicit uniqueness theorems. Thus, for any given input value, there is only one mirror that can accomplish the task described in two dimensions.

2.10 End notes

In this chapter we obtained a dimensional mirror that converts a given wavefront w by turning it into a spherical wavefront or a plane wave. The computations in `Mathematica` validated the equations that describe this mirror; equations (2.13), (2.14), and (2.27) are used for input spherical wavefronts and equations (2.38), (2.39), and (2.45) are used for plane output wavefronts.

Moreover, all the procedures presented in this chapter are just natural extensions of Fermat's principle, since we did not use any numerical approach or approximation and they are analytic closed-form solutions of the proposed problem. In chapter 6 we will return to the solution of the problem presented in this chapter, but this time in three dimensions, and see why we have solved the crux of adaptive optics.

Chapter 3

General dioptric stigmatic surfaces

In this chapter, we are going to design a refractive surface such that given an input wavefront, it refracts it as a spherical wave or a plane wave. The input wavefront can be any aspherical continuous function, so it is not restricted to being a spheric or plane wavefront. Thus, we call these refractive surfaces general dioptric stigmatic surfaces. During the design procedure, we do not use any numerical approach or approximation. The equations that describe the general dioptric stigmatic surfaces are analytic closed-form equations and are only based on Fermat's principle.

3.1 A more general solution than Cartesian ovals

Images produced by an optical system are influenced by the aberrations of the system. Stigmatized pairs are made up of an object's rays and an image point. When two rays from the object are convergent, they yield a single point free of aberrations; the pair is not the same as a traditional image. Stigmatic pairs only exist sporadically in optical systems, for example in conical mirrors, Cartesian ovals, stigmatic lenses, and other special stigmatic optical systems. A stigmatized image results from rays that all have same optical path length (OPL).

Cartesian ovals are stigmatic refractive surfaces. In other words, Cartesian ovals have a refractive surface shape that does not introduce spherical aberration when rays (beams) pass through the interface. Cartesian ovals have been studied for over a thousand years. While they are named after the French mathematician René Descartes, they were actually conceived in 984 by Ibn Sahl in what is now Basra, Iraq. It is well known that Ibn Sahl possessed a deep understanding of ancient Greek optics, but he explored a far wider range of topics, including refraction. His treatise, *On the Burning Instruments*, is magnificent for its time and it shows that Ibn Sahl was the first mathematician to have studied and understood stigmatic lens design (the design of lenses which are stigmatic). Around his time, i.e. the tenth century, the study of optics was primarily focused on dioptrics. Ibn Sahl studied 'burning mirrors,' both parabolic and ellipsoidal; he considered hyperbolic plano-convex lenses and hyperbolic biconvex lenses, all of which were presented in *On the Burning Instruments*. His most notable achievement was the development of Snell's law of refraction long before Snell himself.

As a result of his work, Sahl discovered the stigmatic refractive surface, now termed the Cartesian oval. Using two Cartesian ovals as stigmatic lenses, Sahl proposed that the rays they refracted were collimated along the optical axis. However, Cartesian ovals cannot be defined by any general equation proposed by Sahl. Later, many authors became interested in Sahl's works and in Cartesian ovals in general. We extensively studied Cartesian ovals in our last treatise, *Stigmatic Optics*, but here we are going to explore more general surfaces than Cartesian ovals. If the input wavefront is not spherical, but arbitrary, what is the shape of a refractive surface such that it converges this input wavefront to a single point image using just refraction? That is the problem presented in this chapter, which in fact is an extension of the problem presented in chapter 2.

3.2 General equation for stigmatic surfaces for images at finite distances

An overview of the problem geometry can be found in figure 3.1. The origin of the coordinate system is placed at the wavefront's vertex, and it is described by the function w, where r is the radial variable, just as in chapter 2. The distance between the wavefront vertex (the origin of the Cartesian coordinate system) and the refractive surface (z_a, r_a) is z_o. The distance from the vertex of the refractive surface to the image point is z_i. Thus the image is placed at $z_o + z_i$ along the optical axis. Let n_o and n_i denote the refractive indices of the object and image spaces, respectively. See figure 3.1.

So, mathematically speaking, the problem can be stated as follows: given a wavefront w, distances z_o and z_i, and the refractive indices of the object and image spaces, what is the refractive surface (z_a, r_a) such that the rays of the wavefront converge into a single point image located at $z_o + z_i$? This problem is a generalization of the classical problem of the stigmatic Cartesian ovals that only considers an axial object point and an axial image point. The wavefront w is a continuous function that is not necessarily a plane wavefront or a spherical wavefront.

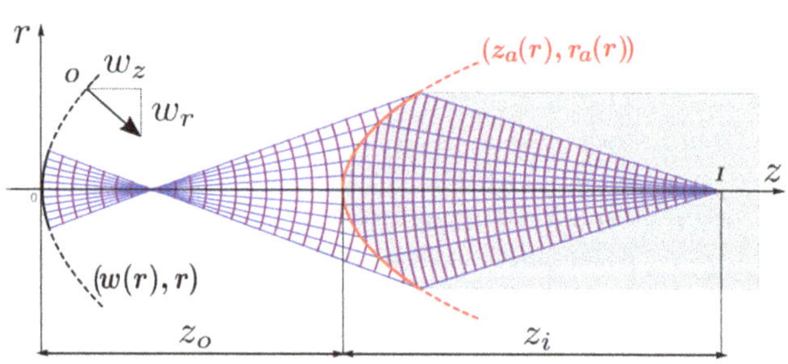

Figure 3.1. Basic configuration of a stigmatic refractive surface for a finite-distance image.

The only axiom of optical path theory is Fermat's principle, which says that the light traverses the minimum path. This principle implies that to obtain the stigmatic condition between the wavefront w and the image point, all the rays that emerge from the wavefront must have the same optical path length (OPL). The OPL of an axial ray is given by

$$n_i z_i + n_o z_o \equiv k, \tag{3.1}$$

where k is the constant value of the OPL. The OPL of a non-axial ray is given by

$$n_i\sqrt{(z_a - z_i - z_o)^2 + r_a^2} + n_o\sqrt{(r_a - r)^2 + (z_a - w)^2} \equiv k. \tag{3.2}$$

Both paths are equal, so

$$\boxed{n_i\sqrt{(z_a - z_i - z_o)^2 + r_a^2} + n_o\sqrt{(r_a - r)^2 + (z_a - w)^2} = n_i z_i + n_o z_o \equiv k.} \tag{3.3}$$

The right-hand side of equation (3.3) (the sum of the two square roots) is the OPL of a non-axial ray, and $n_i z_i + n_o z_o$ is the OPL of an axial ray (note that they must be equal).

Let us now focus on another implication of the optical path theory and Fermat's principle, which is the Malus–Dupin theorem studied in the first chapter of this treatise and successfully applied in chapter 2. The Malus–Dupin theorem will again be used here, with the same goal as in chapter 1. We express the first surface in terms of the input rays of the wavefront w. From this theorem, we know that the rays of the wavefront w are perpendicular to the wavefront. Therefore, the rays propagate along the normal vector of w. In this context, the cosine directors of the input rays are given by

$$w_r = -\frac{w'}{\sqrt{w'^2 + 1}}, \qquad w_z = \frac{1}{\sqrt{w'^2 + 1}}, \tag{3.4}$$

where $w' \equiv dw/dr$ is the derivative of w with respect to r, and w_r and w_z are the cosine directors in the r and z directions, respectively. So, we can propagate this cosine directors until they reach the refractive surface, as expressed by

$$r_a = r + \vartheta w_r, \qquad z_a = w + \vartheta w_z, \tag{3.5}$$

where

$$\vartheta = \sqrt{(r_a - r)^2 + (z_a - w)^2}. \tag{3.6}$$

Here, ϑ is the distance between the wavefront w and the refractive surface (z_a, r_a).

Using equations (3.4) and (3.6) we can express the refractive surface in parametric form, as follows:

$$r_a = r + \vartheta w_r, \quad z_a = w + \vartheta w_z. \tag{3.7}$$

We have now the structure of the solution. We only need to express ϑ in terms of the known parameters. To achieve this we substitute equations (3.5) and (3.6) into (3.3):

$$n_i\sqrt{(-z_i - z_o + \vartheta w_z + w)^2 + (\vartheta w_r + r)^2} + \vartheta n_o = k, \tag{3.8}$$

squaring both sides,

$$n_i^2[(-z_i - z_o + \vartheta w_z + w)^2 + (\vartheta w_r + r)^2] = (k - \vartheta n_o)^2, \tag{3.9}$$

and we define the parameter z_τ in order to reduce the size of equation (3.9):

$$z_\tau \equiv -w + z_i + z_o. \tag{3.10}$$

We now replace set z_τ in equation (3.9):

$$n_i^2[(\vartheta w_z - z_\tau)^2 + (\vartheta w_r + r)^2] = (k - \vartheta n_o)^2, \tag{3.11}$$

and expand it to obtain

$$n_i^2[(\vartheta^2 w_z^2 - 2\vartheta w_z z_\tau + z_\tau^2) + (\vartheta^2 w_r^2 + 2\vartheta w_r r + r^2)] = k^2 - 2k\vartheta n_o + \vartheta^2 n_o^2. \tag{3.12}$$

Note that in the last equation, we can reduce $\vartheta(w_x^2 + w_y^2) \to \vartheta$, thus:

$$n_i^2[\vartheta^2 - 2\vartheta w_z z_\tau + z_\tau^2 + 2\vartheta w_r r + r^2] = k^2 - 2k\vartheta n_o + \vartheta^2 n_o^2, \tag{3.13}$$

and by reordering the terms, we obtain

$$\vartheta^2(n_i^2 - n_o^2) + 2\vartheta[-n_i^2(w_z z_\tau) + n_i^2(w_r r) + kn_o] + [-k^2 + n_i^2(z_\tau^2 + r^2)] = 0. \tag{3.14}$$

We multiply both sides by $4(n_i^2 - n_o^2)$, which yields

$$\begin{aligned} &4\vartheta^2(n_i^2 - n_o^2)^2 + 8\vartheta(n_i^2 - n_o^2)[-n_i^2(w_z z_\tau) + n_i^2(w_r r) + kn_o] \\ &= -4(n_i^2 - n_o^2)[-k^2 + n_i^2(z_\tau^2 + r^2)]. \end{aligned} \tag{3.15}$$

We sum both sides with $+[-n_i^2(w_z z_\tau) + n_i^2(w_r r) + kn_o]^2$, giving

$$
\begin{aligned}
&4\vartheta^2(n_i^2 - n_o^2)^2 + 8\vartheta\left(n_i^2 - n_o^2\right)\left[-n_i^2\left(w_z z_\tau\right) + n_i^2\left(w_r r\right) + kn_o\right] \\
&\quad + 4[-n_i^2(w_z z_\tau) + n_i^2(w_r r) + kn_o]^2 \\
&= [-n_i^2(w_z z_\tau) + n_i^2(w_r r) + kn_o]^2 - 4\left(n_i^2 - n_o^2\right)\left[-k^2 + n_i^2\left(z_\tau^2 + r^2\right)\right].
\end{aligned}
\tag{3.16}
$$

By regrouping the left-hand side of equation (3.16), we obtain

$$
\begin{aligned}
&\left\{2\left(n_i^2 - n_o^2\right)\vartheta + 2\left[-n_i^2\left(w_z z_\tau\right) + n_i^2\left(w_r r\right) + kn_o\right]\right\}^2 \\
&= 4[-n_i^2(w_z z_\tau) + n_i^2(w_r r) + kn_o]^2 - 4\left(n_i^2 - n_o^2\right)\left[-k^2 + n_i^2\left(z_\tau^2 + r^2\right)\right],
\end{aligned}
\tag{3.17}
$$

and by taking the square root of both sides, we obtain

$$
\begin{aligned}
&(n_i^2 - n_o^2)\vartheta + [-n_i^2(w_z z_\tau) + n_i^2(w_r r) + kn_o] \\
&= \sqrt{[-n_i^2(w_z z_\tau) + n_i^2(w_r r) + kn_o]^2 - (n_i^2 - n_o^2)[-k^2 + n_i^2(z_\tau^2 + r^2)]}.
\end{aligned}
\tag{3.18}
$$

Finally, solving for ϑ, we get,

$$
\begin{aligned}
\vartheta = \frac{n_i^2(rw_r - w_z z_\tau) + kn_o}{n_o^2 - n_i^2} &+ \mathrm{sgn}(\mathrm{z}_i)\mathrm{sgn}(\mathrm{n}_i - \mathrm{n}_o) \\
&\frac{\sqrt{(\mathrm{n}_o^2 - \mathrm{n}_i^2)(\mathrm{n}_i^2(\mathrm{r}^2 + \mathrm{z}_\tau^2) - \mathrm{k}^2) + (\mathrm{n}_i^2(\mathrm{rw}_r - \mathrm{w}_z\mathrm{z}_\tau) + \mathrm{kn}_o)^2}}{\mathrm{n}_o^2 - \mathrm{n}_i^2}.
\end{aligned}
\tag{3.19}
$$

Note that $\mathrm{sgn}(\mathrm{z}_i)\mathrm{sgn}$ implies that the image can be real or virtual. The expression $\mathrm{sgn}(\mathrm{z}_i)\mathrm{sgn}$ has been obtained via trial and error during the computation of the shapes of refractive surfaces; it restricts the solution to valid physical cases only. The other cases are mathematical solutions with no physical interpretation, since they do not follow $z_a(r = 0) = z_o$. The parameter $\mathrm{sgn}(\mathrm{z}_i)$ originates from the fact that the image can be real or virtual; $\mathrm{sgn}(\mathrm{n}_i - \mathrm{n}_o)$ originates from the case of $n_o > n_i$ or the case of $n_o < n_i$.

Once we have obtained ϑ, we can properly express r_a and z_a by substituting equation (3.19) into equation (3.5); for r_a we obtain

$$
\begin{aligned}
r_a = r - &\frac{w'}{\sqrt{w'^2+1}}\\
&\times\left\{\frac{n_i^2\left[r\left(-\frac{w'}{\sqrt{w'^2+1}}\right)-\left(\frac{1}{\sqrt{w'^2+1}}\right)z_\tau\right]+kn_o}{n_o^2-n_i^2}+\mathrm{sgn}(z_i)\mathrm{sgn}(n_i-n_o)\right.\\
&\left.\times\frac{\sqrt{(n_o^2-n_i^2)[n_i^2(r^2+z_\tau^2)-k^2]+\left\{n_i^2\left[r\left(-\frac{w'}{\sqrt{w'^2+1}}\right)-\left(\frac{1}{\sqrt{w'^2+1}}\right)z_\tau\right]+kn_o\right\}^2}}{n_o^2-n_i^2}\right\},
\end{aligned}
\tag{3.20}
$$

and for z_a we have

$$
\begin{aligned}
z_a = w + &\frac{1}{\sqrt{w'^2+1}}\\
&\times\left\{\frac{n_i^2\left[r\left(-\frac{w'}{\sqrt{w'^2+1}}\right)-\left(\frac{1}{\sqrt{w'^2+1}}\right)z_\tau\right]+kn_o}{n_o^2-n_i^2}+\mathrm{sgn}(z_i)\mathrm{sgn}(n_i-n_o)\right.\\
&\left.\times\frac{\sqrt{(n_o^2-n_i^2)[n_i^2(r^2+z_\tau^2)-k^2]+\left\{n_i^2\left[r\left(-\frac{w'}{\sqrt{w'^2+1}}\right)-\left(\frac{1}{\sqrt{w'^2+1}}\right)z_\tau\right]+kn_o\right\}^2}}{n_o^2-n_i^2}\right\}.
\end{aligned}
\tag{3.21}
$$

Equations (3.20) and (3.21) are the most important result in this chapter. They describe, using a mathematical closed form, the refractive surface required so that the rays coming from a given wavefront geometrically converge into a single image point. The equations were deduced without any approximation, either numerical or paraxial.

Here, we present a completely analytical procedure. A numerical approximation or a series of iterations is never used; this result is strictly based on Fermat's principle. Thus equations (3.20) and (3.21) are natural implications of Fermat's principle. This is evident if the step-by-step deduction of (3.20) and (3.21) is followed. Only algebraic steps are used in the derivations mentioned. The key expression is equation (3.3), because it requires all rays to share the same OPL.

Also note that in this procedure we obtained the only possible solution of the proposed problem. So, given a set of input values, i.e. the refraction indices n_o, n_i, the distances z_o, z_i, and the wavefront w, there is only one solution given by equations (3.20) and (3.21). So, we have again presented uniqueness theorems.

3.3 The wavefronts of images at finite distances

To obtain the eikonal of the wavefront along the optical path, we again need to recall another implication of Fermat's principle, Snell's law. But this time, we are considering Snell's law of refraction, which is the equation that yields information for the case of light crossing from one medium to another. The vector form of Snell's law of refraction for this surface is given by the following equation, equation (3.22):

$$\boxed{\vec{\mathbf{v}}_2 = \frac{n_o}{n_i}[\vec{\mathbf{v}}_1 - (\vec{\mathbf{n}}_o \cdot \vec{\mathbf{v}}_1)\vec{\mathbf{n}}_o] - \vec{\mathbf{n}}_o\sqrt{1 - \frac{n_o^2}{n_i^2}(\vec{\mathbf{n}}_o \times \vec{\mathbf{v}}_1)^2} \quad \text{for } \vec{\mathbf{v}}_2, \vec{\mathbf{v}}_1, \vec{\mathbf{n}}_o \in \mathbb{R}^2,} \tag{3.22}$$

where $\vec{\mathbf{v}}_1$ is the unit vector of the incident ray, $\vec{\mathbf{v}}_2$ is the unit vector of the refracted ray, and $\vec{\mathbf{n}}_o$ is the normal vector of the first surface (see figure 3.14). The unit vectors mentioned are given by

$$\vec{\mathbf{v}}_1 = \frac{-w'\vec{\mathbf{e}}_1 + \vec{\mathbf{e}}_2}{\sqrt{w'^2 + 1}}, \qquad \vec{\mathbf{v}}_2 = \wp_r\vec{\mathbf{e}}_1 + \wp_z\vec{\mathbf{e}}_2, \qquad \vec{\mathbf{n}}_o = \frac{z_a'\vec{\mathbf{e}}_1 - r_a'\vec{\mathbf{e}}_2}{\sqrt{r_a' + z_a'^2}}, \tag{3.23}$$

where $\vec{\mathbf{e}}_1$ denotes the r direction, $\vec{\mathbf{e}}_2$ denotes the z direction, $\wp_r$ and $\wp_z$ are the cosine directors of the refracted ray in the r and z directions, $r_a' \equiv dr_a/dr$ is the derivative of r_a to r, and $z_a' \equiv dr_a/dr$ is the derivative of z_a to r. We do not formulas for r_a' and z_a' because they are too large to present in book format, but they can easily be computed using `Mathematica`.

We substitute the unit vectors of equation (3.23) into equation (3.22), starting with $(\vec{\mathbf{n}}_o \times \vec{\mathbf{v}}_1)^2$, which is inside the square root of equation (3.22):

$$(\vec{\mathbf{n}}_o \times \vec{\mathbf{v}}_1) = \frac{z_a' + r_a'w'}{\sqrt{w'^2 + 1}\sqrt{r_a'^2 + z_a'^2}}(\vec{\mathbf{e}}_1 \times \vec{\mathbf{e}}_2); \tag{3.24}$$

squaring this yields

$$(\vec{\mathbf{n}}_o \times \vec{\mathbf{v}}_1)^2 = \frac{(z_a' + r_a'w')^2}{(w'^2 + 1)(r_a'^2 + z_a'^2)}. \tag{3.25}$$

Thus, the square root in Snell's law at the refractive surface is

$$-\vec{\mathbf{n}}_o\sqrt{1 + \frac{n_o^2}{n_i^2}(\vec{\mathbf{n}}_o \times \vec{\mathbf{v}}_1)^2} = -\frac{(z_a'\vec{\mathbf{e}}_1 - \vec{\mathbf{e}}_2)}{\sqrt{1 + z_a'^2}}\sqrt{1 - \frac{n_o^2}{n_i^2}\frac{(z_a' - r_a'w')^2}{(w'^2 + 1)(r_a'^2 + z_a'^2)}}. \tag{3.26}$$

We now focus on the term $\vec{v}_1 - (\vec{n}_o \cdot \vec{v}_1)\vec{n}_o$:

$$\vec{v}_1 - (\vec{n}_o \cdot \vec{v}_1)\vec{n}_o = \frac{-w'\vec{e}_1 + \vec{e}_2}{\sqrt{w'^2 + 1}} - \left[-\frac{r_a' + w'z_a'}{\sqrt{w'^2 + 1}\sqrt{r_a'^2 + z_a'^2}} \right] \frac{(z_a'\vec{e}_1 - r_a'\vec{e}_2)}{\sqrt{r_a'^2 + z_a'^2}}. \quad (3.27)$$

Simplifying and multiplying by $\frac{n_o}{n_i}$, we obtain

$$\frac{n_o}{n_i}[\vec{v}_1 - (\vec{n}_o \cdot \vec{v}_1)\vec{n}_o] = \frac{n_o}{n_i} \frac{r_a'(z_a' - r_a'w')}{\sqrt{w'^2 + 1}\,(r_a'^2 + z_a'^2)} \vec{e}_1 + \frac{n_o}{n_i} \frac{z_a'(z_a' - r_a'w')}{\sqrt{w'^2 + 1}\,(r_a'^2 + z_a'^2)} \vec{e}_2. \quad (3.28)$$

Substituting equations (3.26) and (3.28) into (3.22) and separating components gives

$$\wp_r = \frac{n_o r_a'(z_a' - r_a'w')}{n_i\sqrt{w'^2 + 1}\,(r_a'^2 + z_a'^2)} + \frac{z_a'\sqrt{1 - \dfrac{n_o^2(z_a' - r_a'w')^2}{n_i^2(w'^2 + 1)(r_a'^2 + z_a'^2)}}}{\sqrt{r_a'^2 + z_a'^2}}, \quad (3.29)$$

and

$$\wp_z = \frac{n_o z_a'(z_a' - r_a'w')}{n_i\sqrt{w'^2 + 1}\,(r_a'^2 + z_a'^2)} - \frac{r_a'\sqrt{1 - \dfrac{n_o^2(z_a' - r_a'w')^2}{n_i^2(w'^2 + 1)(r_a'^2 + z_a'^2)}}}{\sqrt{r_a'^2 + z_a'^2}}. \quad (3.30)$$

We have an expression for the cosine directors of each ray at the input in equation (3.4), we have expressions of the cosine directors of each ray at the output in equations (3.29) and (3.30), and we know the separating condition ϑ from equation (3.19). So, we can easily construct the eikonal of this optical system, which is given by the following piecewise function,

$$\boxed{H(t, r_a) = \begin{cases} [w + tw_z, r + tw_r], & t < \vartheta, \\ [z_a + \wp_z(t - \vartheta), r_a + \wp_r(t - \vartheta)], & t > \vartheta, \end{cases}} \quad (3.31)$$

where the condition $t < \vartheta$ means that the ray has not yet struck the refractive surface, and $t > \vartheta$ is the condition for the refracted rays.

Note $H(t, r_a)$ was constructed not only as a piecewise function that separates the input and outputs of the system but it was also constructed without using numerical approximation or assumptions. Therefore, we can confirm that $H(t, r_a)$ is just a natural implication of Fermat's principle, like equations (3.20) and (3.21).

Observe that $H(t, r_a)$ is an implicit uniqueness theorem, since it is constructed from equations (3.20) and (3.21) which are themselves implicit uniqueness theorems.

3.4 Mathematica code

The Mathematica code for equations (3.20), (3.21), and (3.31) is provided in this section. The output of the algorithm is presented in figures 3.2, 3.3, 3.4, 3.5, 3.6, 3.7, 3.8, 3.9, 3.10, 3.11, 3.12, and 3.13.

```
In[3]:=  ClearAll["Global`*"]
```

$$w[r\_]:=r^2/28$$

$$wr[r\_]:=-\frac{w'[r]}{\sqrt{w'[r]^2+1}}$$

$$wz[r\_]:=\frac{1}{\sqrt{w'[\ ]^2+1}}$$

$$z\tau = (zi\overset{r}{+}zo-w[r]);$$

$$\vartheta[r\_]:=\frac{1}{-ni^2+no^2}\left(k\ no+ni^2\ (r\ wr[r]-z\tau\ wz[r])+\frac{1}{2}\sqrt{\left(-4\ (ni^2-no^2)\times(-k^2+ni^2\ (r^2+(zi+zo)^2)+ni^2\ w[r]\ (-2\ (zi+zo)+w[r]))+4\ (k\ no+ni^2\ (r\ wr[r]-(zi+zo-w[r])\ wz[r]))^2\right)}\right)$$

$$ra[r\_]:=r+\vartheta[r]wr[r]$$

$$za[r\_]:=w[r]+\vartheta[r]wz[r]$$

$$\varrho r[r\_]:=\frac{no\ ra'[r]\ (-ra'[r]\ w'[r]+za'[r])}{ni\ \sqrt{1+w'[r]^2}\ (ra'[r]^2+za'[r]^2)}+\frac{za'[r]}{\sqrt{ra'[r]^2+za'[r]^2}}\sqrt{}(1-\frac{(no^2\ (-ra'[r]\ w'[r]+za'[r])^2)}{(ni^2\ (1+w'[r]^2)\ (ra'[r]^2+za'[r]^2))})$$

$$\varrho z[r\_]:=\frac{no\ za'[r]\ (-ra'[r]\ w'[r]+za'[r])}{ni\ \sqrt{1+w'[r]^2}\ (ra'[r]^2+za'[r]^2)}-\frac{ra'[r]}{\sqrt{ra'[r]^2+za'[r]^2}}\sqrt{\frac{(no^2\ (-ra'[r]\ w'[r]+za'[r])^2)}{(ni^2\ (1+w'[r]^2)\ (ra'[r]^2+za'[r]^2))}}$$

```
θ3=ArcTan[(ra[r])/(za[r]-zi-zo)];

Homotopy={
  {Piecewise[{
    {wz[r]tt+w[r],tt<ϑ[r]},
    {v Cos[θ3](tt-ϑ[r])+za[r],tt>ϑ[r]}}],
  Piecewise[{
   {wr[r]tt+r,tt<ϑ[r]},
   {v Sin[θ3](tt-ϑ[r])+ra[r],tt>ϑ[r]}}]
   }};

c=1;
v=c/ni;

no=1;
ni=1.5;

zo=40;
zi =50;
k=ni zi+no zo;

rmax=5;
density=1;

Surface:=ParametricPlot[{{w[2.5*r],2.5*r},{za[r],ra[r]},
{zo+1.1zi,ra[r/1.2]},{w[r/1.2],r/1.2},{za[r/1.2],ra[r/1.2]}},
{r,-1.2rmax,1.2rmax},AspectRatio→1/1, AxesOrigin→{0,0},
PlotStyle→{Directive[Black, Dashed,Thickness[0.0025]],
Directive[Red, Dashed,Thickness[0.0025]],Transparent, Black,Red}];
```

```
Lens:=Graphics[{EdgeForm[Directive[{Black,Thin}]],
        Directive[Gray,Opacity[0.3]],
        Polygon[Join[#[[5,1]],Reverse[#[[3,1]]]]&
    [Cases[Surface,_Line,∞]]]}]
```

```
Waves:=Table[ParametricPlot[Homotopy,{r,-rmax,rmax},
PlotRange→All,PlotStyle→Purple],{tt,-20,zo+ni zi,2*density}]
```

```
Rays:=Table[{ParametricPlot[Homotopy,{tt,-20,zo+ni zi},
PlotRange→All,PlotStyle→Directive[Blue, Thickness[0.0015]]]},
{r,-rmax,rmax}]
```

```
ShowOpt={PlotRange->All, AspectRatio→Automatic,
ImageSize→{640,Automatic}};
```

```
Show[Lens,Rays,Waves,Surface,ShowOpt]
```

3.5 Examples

Examples of the general catoptric surfaces and their eikonals given by equations (3.20) and (3.21) and (3.31) are presented in this section in figures 3.2, 3.3, 3.4, 3.5, 3.6, 3.7, 3.8, 3.9, 3.10, 3.11, 3.12, and 3.13. Please note, in these figures the rays are shown in blue and the wavefronts are shown in purple.

In general, the performance of these general catoptric surfaces is as expected. The OPL of all the rays is the same.

3.6 The general equation for stigmatic surfaces for images at infinity

In this section, we are interested in a collimated output wavefront, in other words, we are interested in setting the image at infinity. See figure 3.14.

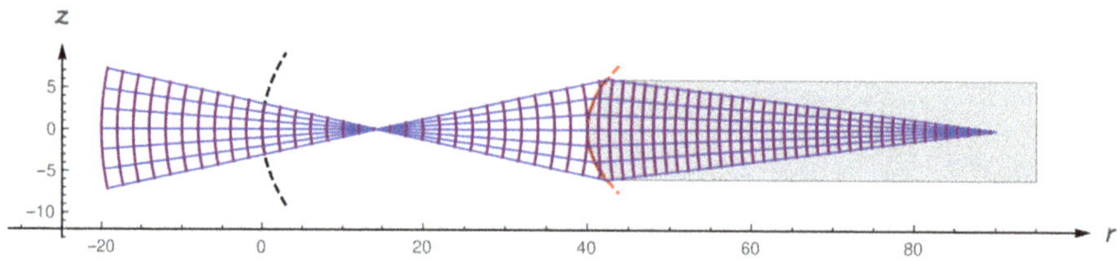

(a) A concave parabolic input wavefront

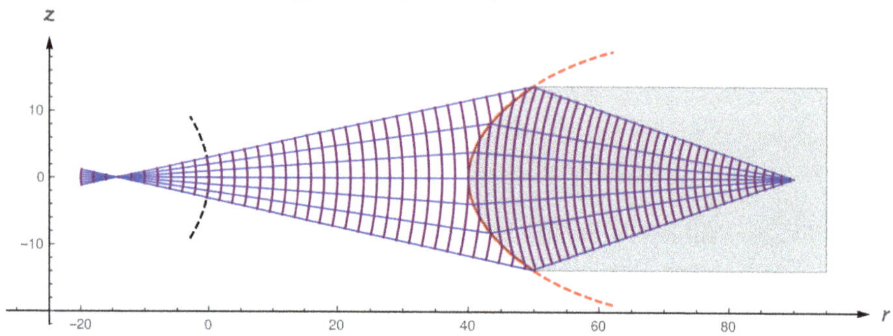

(b) A convex parabolic input wavefront

Figure 3.2. Surface design specifications for parabolic input wavefronts: (a) the input wavefront is $r^2/28$, refractive index $n = 1.5$, object distance $z_o = 40$ mm, image distance $z_i = 50$ mm, $\mathrm{r_{max}} = 3$ mm, $(r_a, z_a) =$ equations (3.20) and (3.21) and $H(t, r_a) =$ equation (3.31); (b) the input wavefront is $-r^2/28$, refractive index $n = 1.5$, object distance $z_o = 40$ mm, image distance $z_i = 50$ mm, $\mathrm{r_{max}} = 3$ mm, $(r_a, z_a) =$ equations (3.20) and (3.21), and $H(t, r_a) =$ equation (3.31).

The first step in achieving our goal is to obtain the optical path length for each ray, so we recall equation (3.3):

$$n_i\sqrt{(z_a - z_i - z_o)^2 + r_a^2} + n_o\sqrt{(r_a - r)^2 + (z_a - w)^2} = n_i z_i + n_o z_o, \tag{3.32}$$

where the left-hand side of equation (3.3) is the OPL of a non-axial ray, and the right-hand side is the OPL of an axial ray. As in the previous section, we use the Dupin–Malus theorem to express the refractive surface in terms of the input rays, so we recall equation (3.5):

$$r_a = r + \vartheta w_r, \qquad z_a = w + \vartheta w_z, \tag{3.33}$$

and equation (3.4):

$$w_r = -\frac{w'}{\sqrt{w'^2 + 1}}, \quad w_z = \frac{1}{\sqrt{w'^2 + 1}}, \tag{3.34}$$

as well as equation (3.6):

$$\vartheta \equiv \sqrt{(r_a - r)^2 + (z_a - w_a)^2}. \tag{3.35}$$

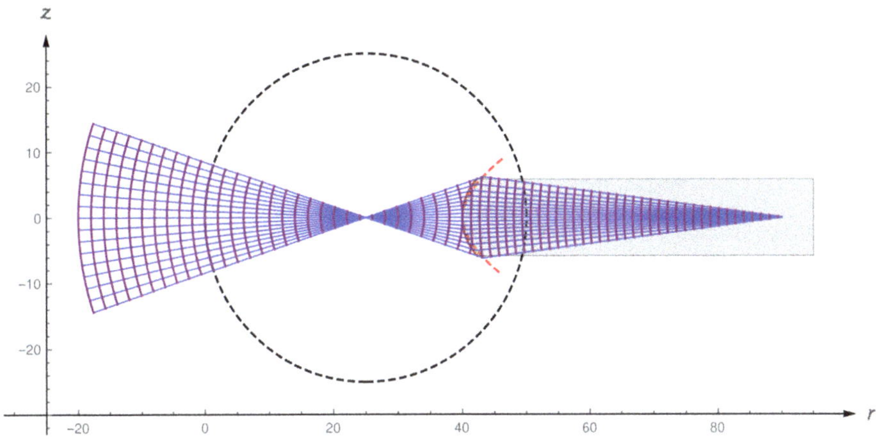

(a) A concave circular input wavefront

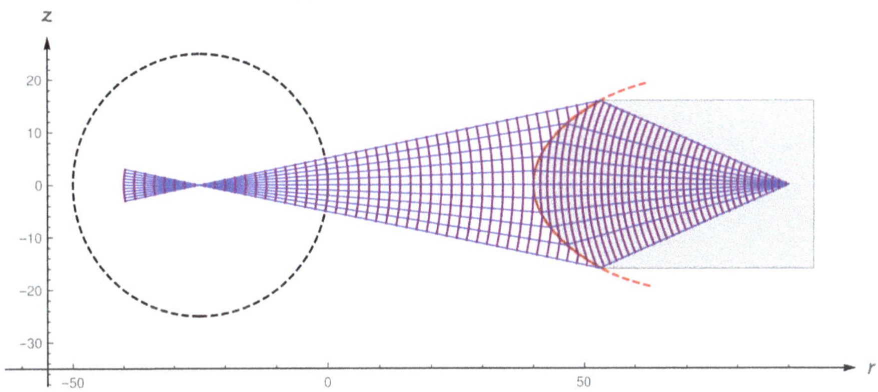

(b) A convex circular input wavefront

Figure 3.3. Surface design specifications for circular input wavefronts: (a) the input wavefront is $25 - \sqrt{25^2 - r^2}$, refractive index $n = 1.5$, object distance $z_o = 40$ mm, image distance $z_i = 50$ mm, $r_{\max} = 8$ mm, $(r_a, z_a) =$ equations (3.20) and (3.21) and $H(t, r_a) =$ equation (3.31); (b) the input wavefront is $-25 + \sqrt{25^2 - r^2}$, refractive index $n = 1.5$, object distance $z_o = 40$ mm, image distance $z_i = 50$ mm, $r_{\max} = 5$ mm, $(r_a, z_a) =$ equations (3.20) and (3.21), and $H(t, r_a) =$ equation (3.31).

We now substitute equations (3.4), (3.5), and (3.6) into equation (3.3), and as a result, we get

$$n_i\sqrt{(w + \vartheta w_z - z_i - z_o)^2 + (r + \vartheta w_r)^2} + n_o\vartheta - \left(n_i z_i + n_o z_o\right) = 0. \tag{3.36}$$

At this point, all we have done is follow a similar procedure to those of the previous sections, but now things may change, as we will compute the limit $z_i \to \infty$ over equation (3.36). This limit, which sets the image at infinity, will collimate the output rays:

$$\lim_{z_i \to \infty}\left[n_i\sqrt{(w + \vartheta w_z - z_i - z_o)^2 + (r + \vartheta w_r)^2} + n_o\vartheta - \left(n_i z_i + n_o z_o\right)\right]. \tag{3.37}$$

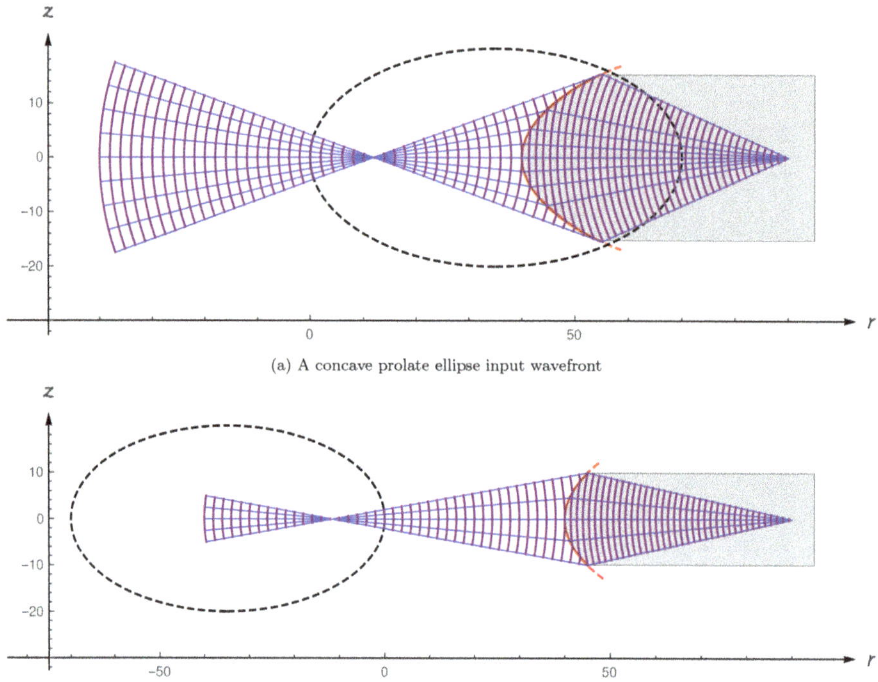

(a) A concave prolate ellipse input wavefront

(b) A convex prolate ellipse input wavefront

Figure 3.4. Surface design specifications for oblate elliptical input wavefronts: (a) the input wavefront is$b - \sqrt{(-a^2 * b^2 + b^2r^2)/ - a^2}$, refractive index $n = 1.5$, object distance $z_o = 40$ mm, image distance $z_i = 50$ mm, $r_{max} = 4$ mm, constant $a = 20$, constant $b = 35$, (r_a, z_a) equations (3.20) and (3.21) and $H(t, r_a) =$ equation (3.31); (b) the input wavefront is $-b + \sqrt{(-a^2 * b^2 + b^2r^2)/ - a^2}$, refractive index $n = 1.5$, object distance $z_o = 40$ mm, image distance $z_i = 50$ mm, $r_{max} = 2$ mm, constant $a = 20$, constant $b = 35$, $(r_a, z_a) =$ equations (3.20) and (3.21), and $H(t, r_a) =$ equation (3.31).

The computation of the limit gives as a result the following expression, equation (3.38):

$$-n_i w - n_i w_z \lim_{z_i \to \infty}(\vartheta) + n_i z_o - n_o z_o + n_o \lim_{z_i \to \infty}(\vartheta), \tag{3.38}$$

which is, in fact, zero, since it is the OPD, so

$$-n_i w - n_i w_z \lim_{z_i \to \infty}(\vartheta) + n_i z_o - n_o z_o + n_o \lim_{z_i \to \infty}(\vartheta) = 0. \tag{3.39}$$

Solving for ϑ in the last expression gives

$$\lim_{z_i \to \infty}(\vartheta) = \frac{-n_i w + z_o(n_i - n_o)}{n_i w_z - n_o}. \tag{3.40}$$

Once we obtain a proper expression for ϑ when $z_i \to \infty$, we can write the exact expressions for r_a and z_a in this scenario. We do this by substituting equation (3.40) into equation (3.5), so r_a changes to

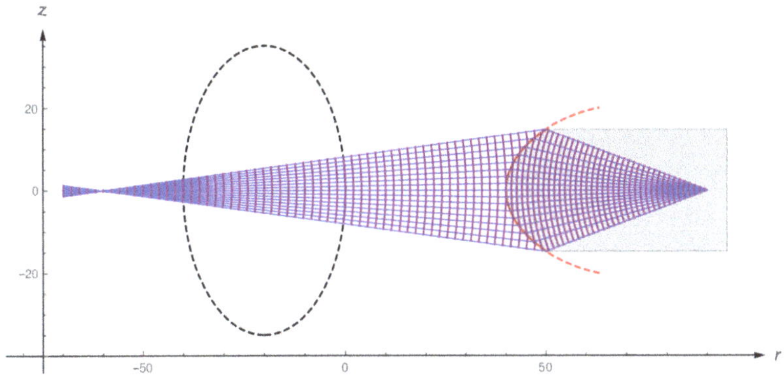

(a) A convex oblate ellipse input wavefront

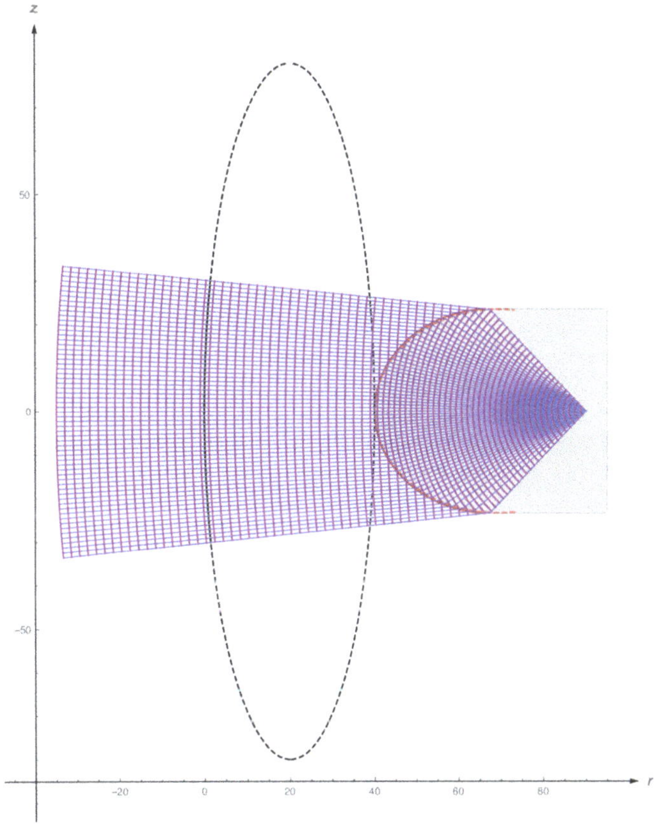

(b) A concave oblate ellipse input wavefront

Figure 3.5. Surface design specifications for prolate elliptical input wavefronts: (a) the input wavefront is $-b + \sqrt{(-a^2 * b^2 + b^2r^2)/ - a^2}$, refractive index $n = 1.5$, object distance $z_o = 40$ mm, image distance $z_i = 50$ mm, $\mathrm{r}_{\max} = 8$ mm, constant $a = 35$, constant $b = 20$, $(r_a, z_a) =$ equations (3.20) and (3.21) and $H(t, r_a) =$ equation (3.31); (b) the input wavefront is $b - \sqrt{(-a^2 * b^2 + b^2r^2)/ - a^2}$, refractive index $n - 1.5$, object distance $z_o = 40$ mm, image distance $z_i = 50$ mm, $\mathrm{r}_{\max} = 30$ mm, constant $a = 80$, constant $b = 20$, $(r_a, z_a) =$ equations (3.20) and (3.21), and $H(t, r_a) =$ equation (3.31).

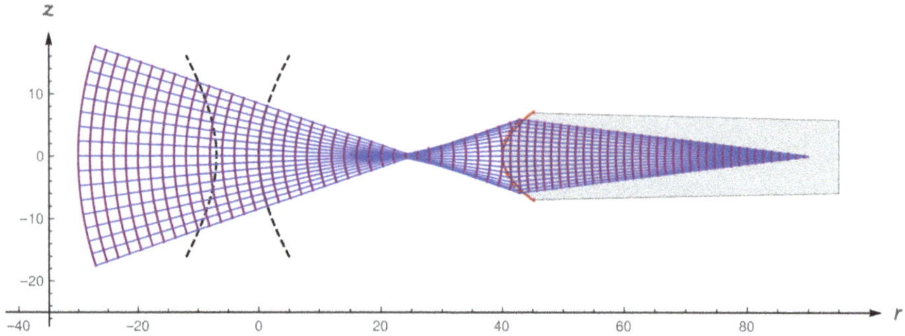

(a) A convex hyperbolic input wavefront

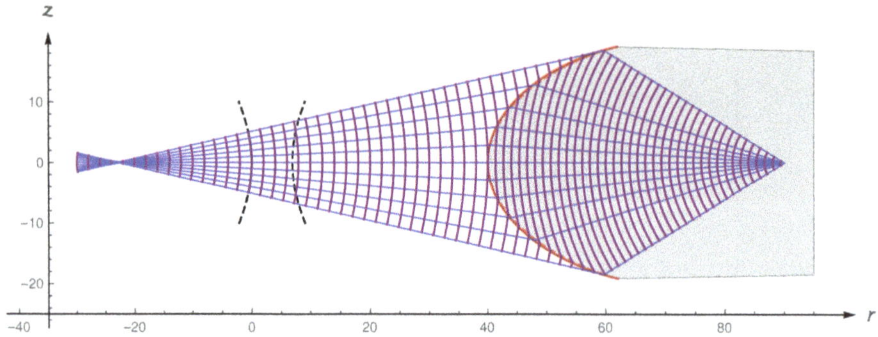

(b) A concave hyperbolic input wavefront

Figure 3.6. Surface design specifications for hyperbolic input wavefronts: (a) the input wavefront is $-b + \sqrt{(-a^2 * b^2 - b^2 r^2)/ - a^2}$, refractive index $n = 1.5$, object distance $z_o = 40$ mm, image distance $z_i = 50$ mm, $\mathrm{r}_{\max} = 8$ mm, constant $a = 16$, constant $b = 12$, $(r_a, z_a) =$ equations (3.20) and (3.21) and $H(t, r_a) =$ equation (3.31); (b) the input wavefront is $b - \sqrt{(-a^2 * b^2 - b^2 r^2)/ - a^2}$, refractive index $n = 1.5$, object distance $z_o = 40$ mm, image distance $z_i = 50$ mm, $\mathrm{r}_{\max} = 5$ mm, constant $a = 16$, constant $b = 12$, $(r_a, z_a) =$ equations (3.20) and (3.21), and $H(t, r_a) =$ equation (3.31).

$$\boxed{\lim_{z_i \to \infty}(r_a) = r + \frac{-n_i w + z_o(n_i - n_o)}{n_i\left(\frac{1}{\sqrt{w'^2 + 1}}\right) - n_o}\left(-\frac{w'}{\sqrt{w'^2 + 1}}\right),} \tag{3.41}$$

and z_a changes to

$$\boxed{\lim_{z_i \to \infty}(z_a) = w + \frac{-n_i w + z_o(n_i - n_o)}{n_i\left(\frac{1}{\sqrt{w'^2 + 1}}\right) - n_o}\left(\frac{1}{\sqrt{w'^2 + 1}}\right).} \tag{3.42}$$

Using (3.41) and (3.42), we can obtain a refractive surface with the desired property, i.e. one that collimates the wavefront of a given object. This deduction process was completely devoid of paraxial approximations or numerical methods. Equations (3.41) and (3.42) are rigorously analytical and close form. As with the

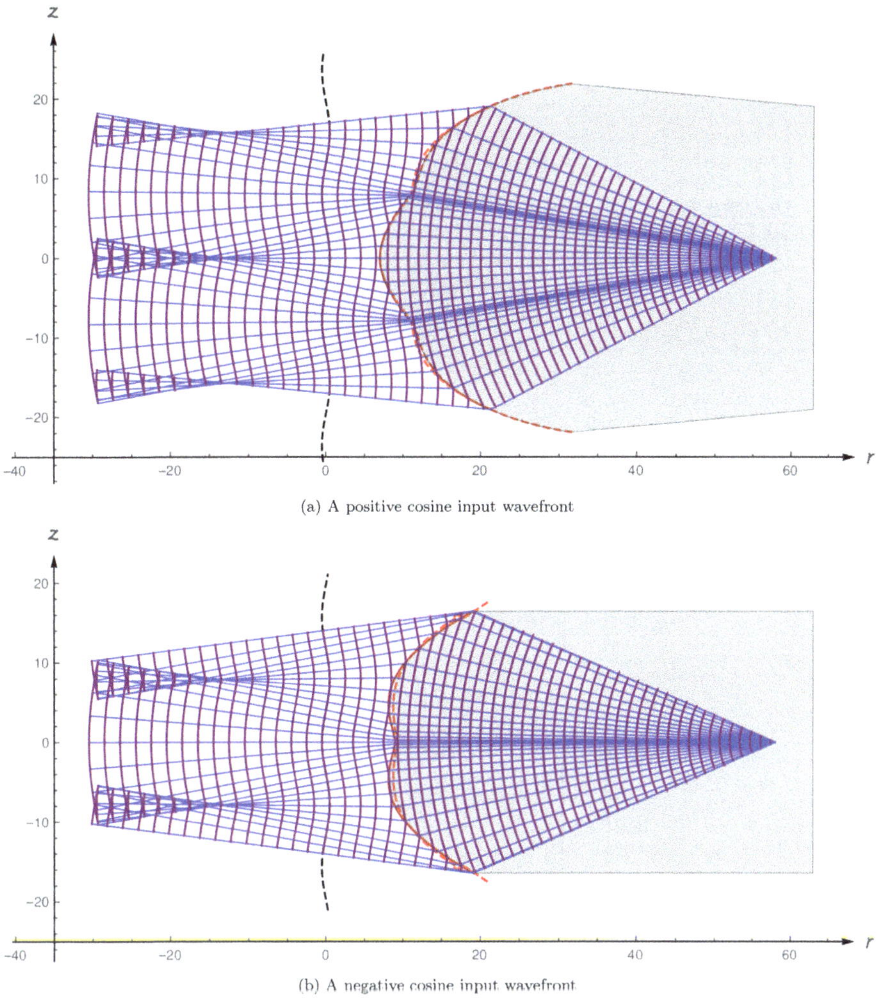

Figure 3.7. Surface design specifications for cosine input wavefronts: (a) the input wavefront is $\cos(2r/5)/2$, refractive index $n = 1.5$, object distance $z_o = 8$ mm, image distance $z_i = 50$ mm, $r_{max} = 17$ mm, $(r_a, z_a) =$ equations (3.20) and (3.21) and $H(t, r_a) =$ equation (3.31); (b) the input wavefront is $-\cos(2r/5)/2$, refractive index $n = 1.5$, object distance $z_o = 8$ mm, image distance $z_i = 50$ mm, $r_{max} = 14$ mm, $(r_a, z_a) =$ equations (3.20) and (3.21), and $H(t, r_a) =$ equation (3.31).

other equations obtained using Fermat's principle, there may be an overlap if and only if two or more rays strike a single point of the surface.

3.7 The wavefronts of images at infinity

We may compute the eikonal of the refractive surface when the image is far away by calculating the limit $z_i \to \infty$ over equations (3.29), and (3.30). Note that the input rays are not affected by this limit, so we focus on the cosine directors of the refracted ray. So, we begin with equation (3.29), $\lim_{z_i \to \infty}(\wp_r)$ which is given by

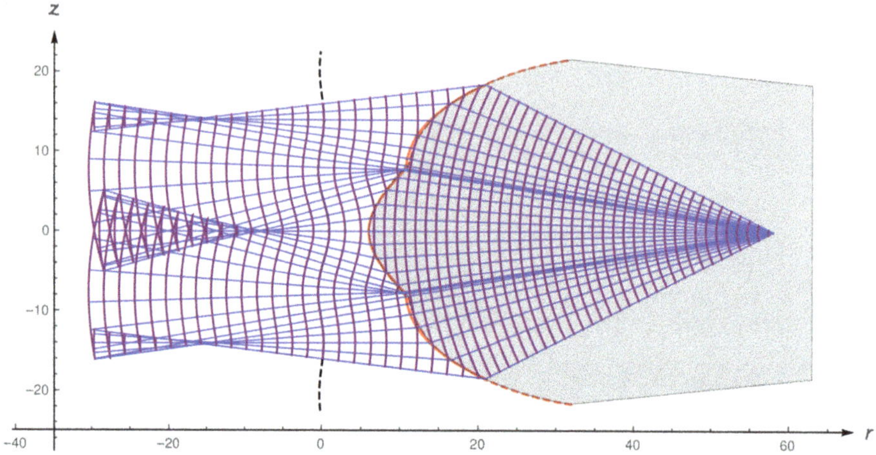

Figure 3.8. Surface design specifications for a Bessel zeroth-order input wavefront. The input wavefront is $J_o(r/2)$, refractive index $n = 1.5$, object distance $z_o = 8$ mm, image distance $z_i = 50$ mm, $r_{max} = 16$ mm, $(r_a, z_a) =$ equations (3.20) and (3.21), and $H(t, r_a) =$ equation (3.31).

$$\lim_{z_i\to\infty}(\wp_r) = \frac{n_o \lim\limits_{z_i\to\infty}\left(r_a'\right)\left[\lim\limits_{z_i\to\infty}\left(z_a'\right) - \lim\limits_{z_i\to\infty}\left(r_a'\right)w'\right]}{n_i\sqrt{w'^2+1}\left[\lim\limits_{z_i\to\infty}\left(r_a'^2\right) + \lim\limits_{z_i\to\infty}\left(z_a'^2\right)\right]} + \frac{\lim\limits_{z_i\to\infty}\left(z_a'\right)\sqrt{1 - \dfrac{n_o^2[\lim\limits_{z_i\to\infty}\left(z_a'\right) - \lim\limits_{z_i\to\infty}\left(r_a'\right)w']^2}{n_i^2(w'^2+1)\left[\lim\limits_{z_i\to\infty}\left(r_a'^2\right) + \lim\limits_{z_i\to\infty}\left(z_a'^2\right)\right]}}}{\sqrt{\lim\limits_{z_i\to\infty}\left(r_a'^2\right) + \lim\limits_{z_i\to\infty}\left(z_a'^2\right)}}, \tag{3.43}$$

and then with equation (3.30), $\lim\limits_{z_i\to\infty}(\wp_z)$ which turns into

$$\lim_{z_i\to\infty}(\wp_z) = \frac{n_o \lim_{z_i\to\infty}\left(z_a'\right)\left[\lim\limits_{z_i\to\infty}\left(z_a'\right) - \lim\limits_{z_i\to\infty}\left(r_a'\right)w'\right]}{n_i\sqrt{w'^2+1}\left[\lim\limits_{z_i\to\infty}\left(r_a'^2\right) + \lim\limits_{z_i\to\infty}\left(z_a'^2\right)\right]} - \frac{\lim\limits_{z_i\to\infty}\left(r_a'\right)\sqrt{1 - \dfrac{n_o^2\left[\lim\limits_{z_i\to\infty}\left(z_a'\right) - \lim\limits_{z_i\to\infty}\left(r_a'\right)w'\right]^2}{n_i^2(w'^2+1)\left[\lim\limits_{z_i\to\infty}\left(r_a'^2\right) + \lim\limits_{z_i\to\infty}\left(z_a'^2\right)\right]}}}{\sqrt{\lim\limits_{z_i\to\infty}\left(r_a'^2\right) + \lim\limits_{z_i\to\infty}\left(z_a'^2\right)}}, \tag{3.44}$$

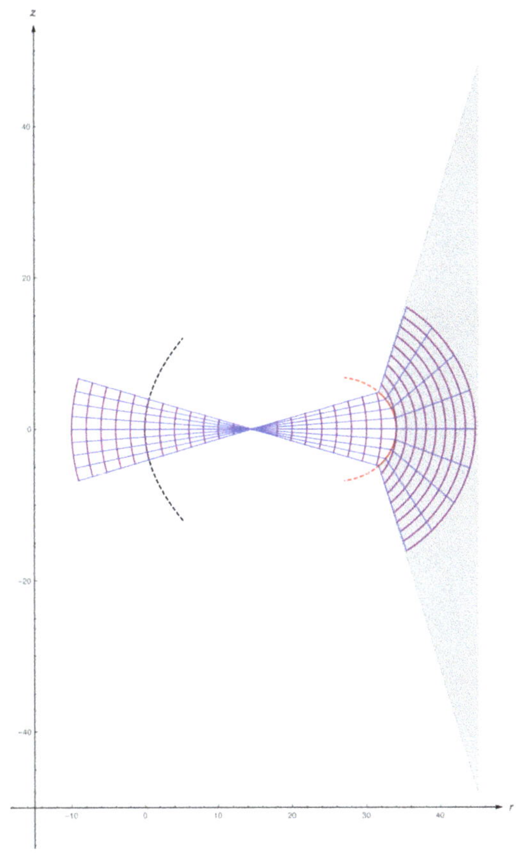

(a) A concave parabolic input wavefront -virtual image-

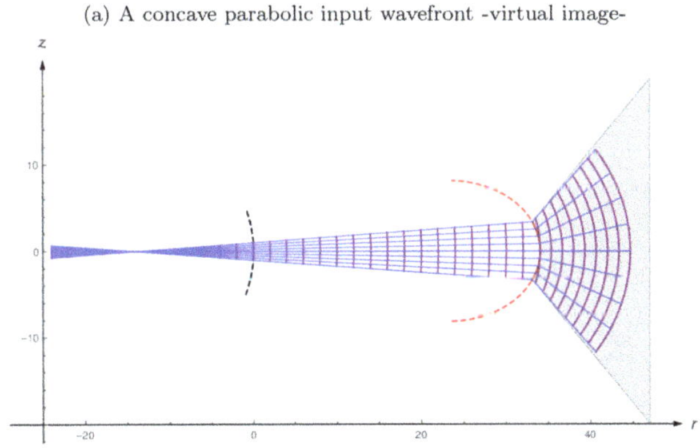

(b) A convex parabolic input wavefront -virtual image-

Figure 3.9. Surface design specifications for parabolic input wavefronts: (a) the input wavefront is $r^2/28$, refractive index $n = 1.5$, object distance $z_o = 10$ mm, image distance $z_i = 20$ mm, $\mathrm{r_{max}} = 4$ mm, (r_a, z_a) = equations (3.20) and (3.21) and $H(t, r_a)$= equation (3.31); (b) the input wavefront is $-r^2/28$, refractive index $n = 1.5$, object distance $z_o = 10$ mm, image distance $z_i = 20$ mm, $\mathrm{r_{max}} = 1$ mm, (r_a, z_a) = equations (3.20) and (3.21), and $H(t, r_a)$ = equation (3.31).

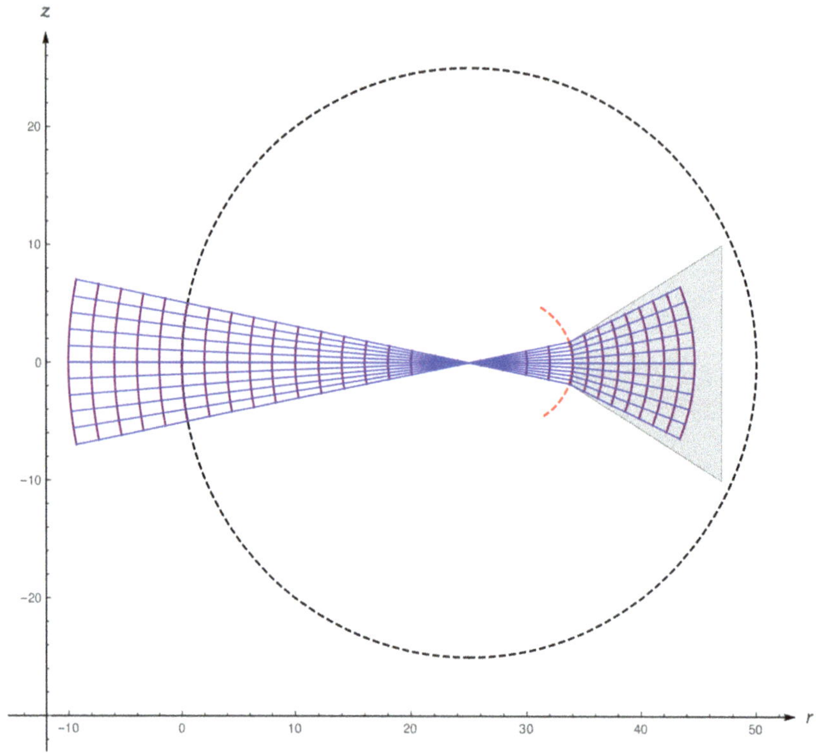

(a) A concave circular input wavefront -virtual image-

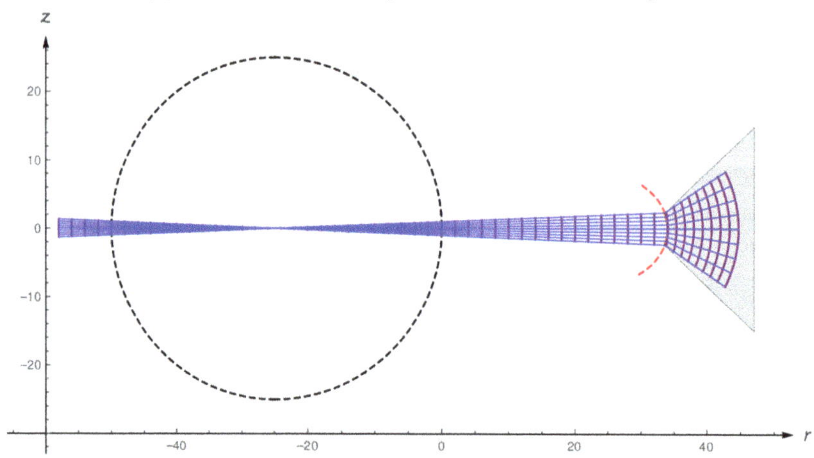

(b) A convex circular input wavefront -virtual image-

Figure 3.10. Surface design specifications for circular input wavefronts: (a) the input wavefront is $25 - \sqrt{25^2 - r^2}$, refractive index $n = 1.5$, object distance $z_o = 100$ mm, image distance $z_i = 20$ mm, $\mathrm{r}_{\max} = 5$ mm, (r_a, z_a) = equations (3.20) and (3.21) and $H(t, r_a)$ = equation (3.31); (b) the input wavefront is $-25 + \sqrt{25^2 - r^2}$, refractive index $n = 1.5$, object distance $z_o = 10$ mm, image distance $z_i = 20$ mm, $\mathrm{r}_{\max} = 1$ mm, (r_a, z_a) = equations (3.20) and (3.21), and $H(t, r_a)$ = equation (3.31).

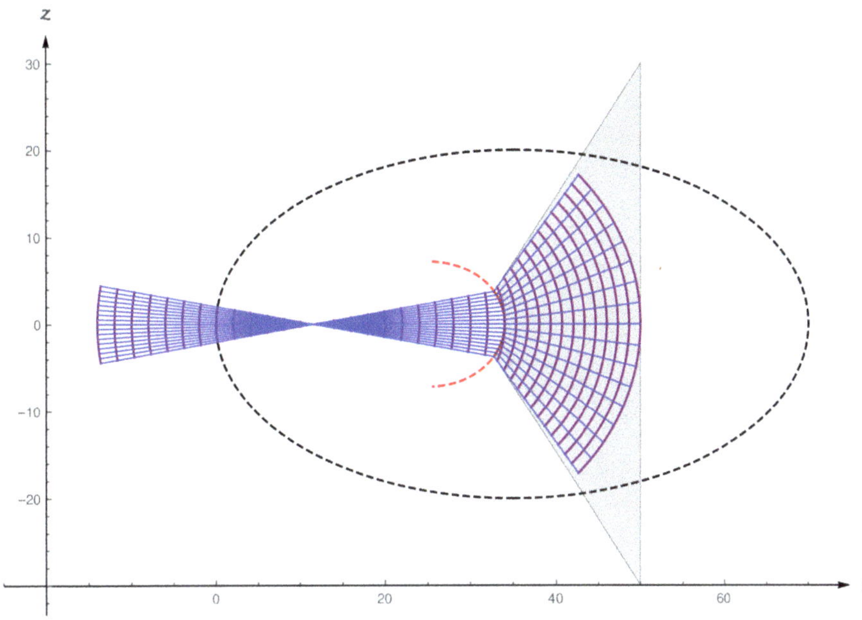

(a) A concave oblate ellipse input wavefront -virtual image-

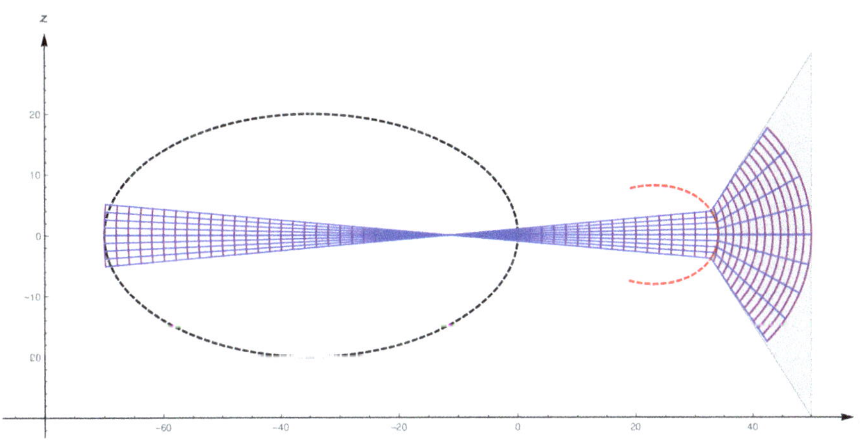

(b) A convex oblate ellipse input wavefront -virtual image-

Figure 3.11. Surface design specifications for oblate ellipse input wavefronts: (a) the input wavefront is $b - \sqrt{(-a^2 * b^2 + b^2 r^2)/ - a^2}$, refractive index $n = 1.5$, object distance $z_o = 10\,\text{mm}$, image distance $z_i = 20\,\text{mm}$, $\text{r}_{\text{max}} = 2\,\text{mm}$, constant $a = 20$, constant $b = 35$, (r_a, z_a) = equations (3.20) and (3.21), and $H(t, r_a)$ = equation (3.31); (b) the input wavefront is $-b + \sqrt{(-a^2 * b^2 + b^2 r^2)/ - a^2}$, refractive index $n = 1.5$, object distance $z_o = 10\,\text{mm}$, image distance $z_i = 20\,\text{mm}$, $\text{r}_{\text{max}} = 1\,\text{mm}$, constant $a = 20$, constant $b = 35$, (r_a, z_a) = equations (3.20) and (3.21), and $H(t, r_a)$ = equation (3.31).

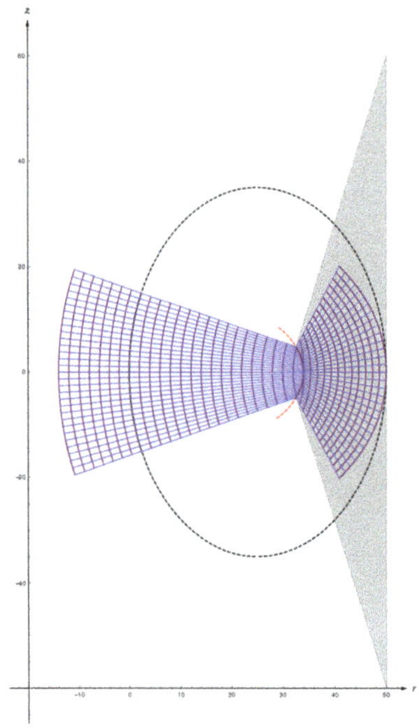

(a) A concave prolate ellipse -virtual image-

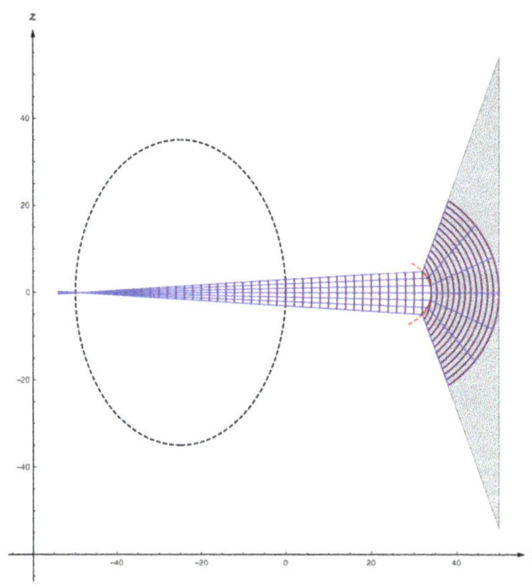

(b) A convex prolate ellipse input wavefront -virtual image-

Figure 3.12. Surface design specifications for prolate ellipse input wavefronts: (a) the input wavefront is $-b + \sqrt{(-a^2 * b^2 + b^2r^2)/ - a^2}$, refractive index $n = 1.5$, object distance $z_o = 10\,\text{mm}$, image distance $z_i = 20\,\text{mm}$, $\text{r}_{\text{max}} = 3\,\text{mm}$, constant $a = 35$, constant $b = 25$, $(r_a, z_a) =$ equations (3.20) and (3.21), and $H(t, r_a) =$ equation (3.31); (b) the input wavefront is $b - \sqrt{(-a^2 * b^2 + b^2r^2)/ - a^2}$, refractive index $n = 1.5$, object distance $z_o = 10\,\text{mm}$, image distance $z_i = 20\,\text{mm}$, $\text{r}_{\text{max}} = 15\,\text{mm}$, constant $a = 35$, constant $b = 25$, $(r_a, z_a) =$ equations (3.20) and (3.21), and $H(t, r_a) =$ equation (3.31).

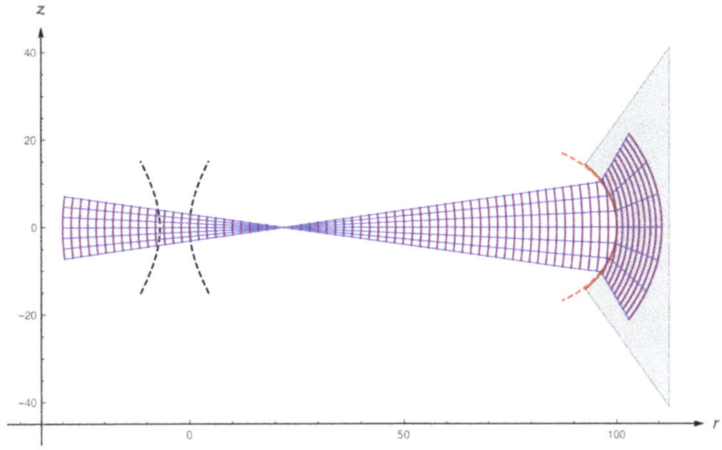

(a) A concave hyperbolic input wavefront -virtual image-

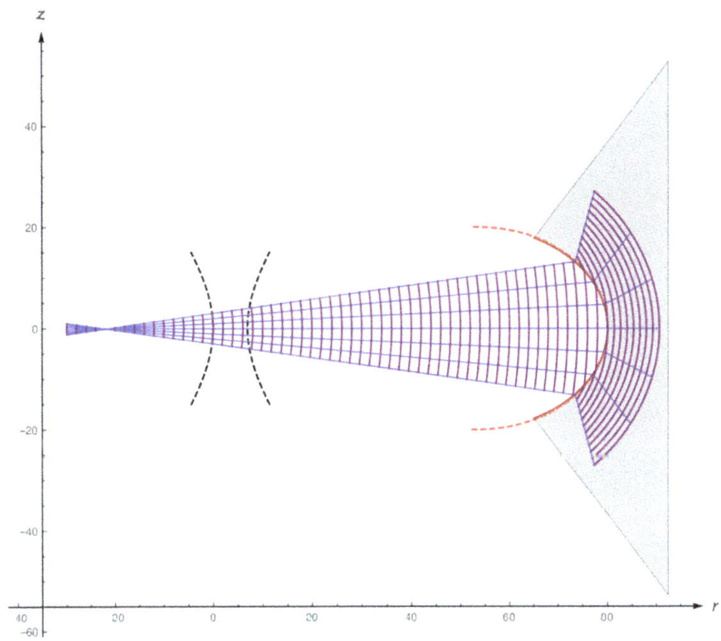

(b) A convex hyperbolic input wavefront -virtual image-

Figure 3.13. Surface design specifications for hyperbolic input wavefronts: (a) the input wavefront is $-b + \sqrt{(-a^2 * b^2 - b^2r^2)/ - a^2}$, refractive index $n = 1.5$, object distance $z_o = 40\,\text{mm}$, image distance $z_i = 50\,\text{mm}$, $\text{r}_{\text{max}} = 3\,\text{mm}$, constant $a = 16$, constant $b = 12$, $(r_a, z_a) =$ equations (3.20) and (3.21), and $H(t, r_a) =$ equation (3.31); (b) the input wavefront is $b - \sqrt{(-a^2 * b^2 - b^2r^2)/ - a^2}$, refractive index $n = 1.5$, object distance $z_o = 20\,\text{mm}$, image distance $z_i = 50\,\text{mm}$, $\text{r}_{\text{max}} = 3\,\text{mm}$, constant $a = 16$, constant $b = 12$, $(r_a, z_a) =$ equations (3.20) and (3.21), and $H(t, r_a) =$ equation (3.31).

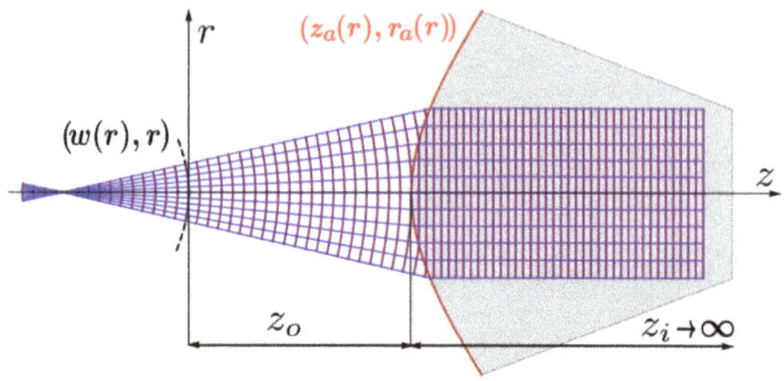

Figure 3.14. Basic configuration of an stigmatic refractive surface for images at infinity.

where $\lim_{z_i\to\infty}(r_a')$ is obtained by computing the derivative with respect to r over equation (3.41):

$$\begin{aligned}\lim_{z_i\to\infty}\left(r_a'\right) = &\frac{1}{(w'^2+1)\left(n_i - n_o\sqrt{w'^2+1}\right)^2}\\ &\times\left\{w'^4\left(n_i^2 - n_in_o\sqrt{w'^2+1} + n_o^2\right) + w'^2\right.\\ &\left[2n_i^2 + n_iw''\left(n_iw - n_iz_o + n_oz_o\right) - 3n_in_o\sqrt{w'^2+1} + 2n_o^2\right]\\ &+ n_i^2 - 2n_in_o\sqrt{w'^2+1} - w''\left(-n_iw + n_iz_o - n_oz_o\right)\\ &\left.\left(n_i - n_o\sqrt{w'^2+1}\right) + n_o^2\right\},\end{aligned} \tag{3.45}$$

and $\lim_{z_i\to\infty}(z_a')$ is obtained by computing the derivative with respect to r over equation (3.42):

$$\begin{aligned}&\lim_{z_i\to\infty}\left(z_a'\right)\\ &= \frac{n_ow'\left[w''\left(-n_iw + n_iz_o - n_oz_o\right) + w'^2\left(n_o\sqrt{w'^2+1} - n_i\right) - n_i + n_o\sqrt{w'^2+1}\right]}{\sqrt{w'^2+1}\left(n_i - n_o\sqrt{w'^2+1}\right)^2}.\end{aligned} \tag{3.46}$$

We now have the cosine directors at the input from equation (3.4) and the cosine directors at the output from equations (3.43) and (3.44). At this point, taking $n_o = 1$, the construction of the eikonal is straightforward:

$$\lim_{z_i \to \infty} H\left(t, r_a\right) = \begin{cases} [w + tw_z v_o, r + tw_r v_o], & t < \lim\limits_{z_i \to \infty}(\vartheta), \\ \begin{bmatrix} 1 \lim_{z_i \to \infty}(z_a) + v_i \lim_{z_i \to \infty}(\wp_z)\left\{t - \lim\limits_{z_i \to \infty}(\vartheta)\right\}, \\ \lim\limits_{z_i \to \infty}(r_a) + v_i \lim_{z_i \to \infty}(\wp_r)\left\{t - \lim\limits_{z_i \to \infty}(\vartheta)\right\} \end{bmatrix}, & t > \lim\limits_{z_i \to \infty}(\vartheta), \end{cases} \quad (3.47)$$

There seems to be no problem with the piecewise condition. It only separates the input rays from the output rays, since $\lim\limits_{z_i \to \infty}(\vartheta)$ is the distance that a ray needs to travel to reach the refractive surface. The eikonal for the cases $n_o \neq 1$ is left to the reader as an exercise.

3.8 Mathematica code

Figures 3.15, 3.16, 3.17, 3.18, 3.19, 3.20, and 3.21 show the outputs of the equations (3.41), (3.42), and (3.47) obtained using the algebraic algorithm which appears in this section.

```
In[4]:= ClearAll["Global`*"]

w[r_]:=-Cos[2r/5]/2

wr[r_]:=-w'[r]/Sqrt[w'[r]^2+1]

wz[r_]:=1/Sqrt[w'[r]^2+1]

zT= (zi+zo-w[r]);

ϑ[r_]:=(-ni w[r] +zo(ni-no))/(ni wz[r]-no)
```

$$\text{ra}[\text{r\_}]:=\text{r}+\frac{\text{-ni w}[\text{r}]\ +\text{zo}(\text{ni-no})}{\text{ni}\ \frac{1}{\sqrt{\text{w'}[\text{r}]^2+1}}\text{-no}}(-\frac{\text{w'}[\text{r}]}{\sqrt{\text{w'}[\text{r}]^2+1}})$$

$$\text{za}[\text{r\_}]:=\text{w}[\text{r}]+\frac{\text{-ni w}[\text{r}]\ +\text{zo}(\text{ni-no})}{\text{ni}\ \frac{1}{\sqrt{\text{w'}[\text{r}]^2+1}}\text{-no}}\frac{1}{\sqrt{\text{w'}[\text{r}]^2+1}}$$

$$\varrho\text{r}[\text{r\_}]:=\frac{\text{no ra'}[\text{r}]\ (\text{-ra'}[\text{r}]\ \text{w'}[\text{r}]+\text{za'}[\text{r}])}{\text{ni}\ \sqrt{1+\text{w'}[\text{r}]^2}\ (\text{ra'}[\text{r}]^2+\text{za'}[\text{r}]^2)}$$

$$+\frac{\text{za'}[\text{r}]}{\sqrt{\text{ra'}[\text{r}]^2+\text{za'}[\text{r}]^2}}\sqrt{\left(1-\frac{\text{no}^2\ (\text{-ra'}[\text{r}]\ \text{w'}[\text{r}]+\text{za'}[\text{r}])^2}{\text{ni}^2\ (1+\text{w'}[\text{r}]^2)\ (\text{ra'}[\text{r}]^2+\text{za'}[\text{r}]^2)}\right)}$$

$$\varrho\text{z}[\text{r\_}]:=\frac{\text{no za'}[\text{r}]\ (\text{-ra'}[\text{r}]\ \text{w'}[\text{r}]+\text{za'}[\text{r}])}{\text{ni}\ \sqrt{1+\text{w'}[\text{r}]^2}\ (\text{ra'}[\text{r}]^2+\text{za'}[\text{r}]^2)}$$

$$-\frac{\text{ra'}[\text{r}]}{\sqrt{\text{ra'}[\text{r}]^2+\text{za'}[\text{r}]^2}}\sqrt{\left(1-\frac{\text{no}^2\ (\text{-ra'}[\text{r}]\ \text{w'}[\text{r}]+\text{za'}[\text{r}])^2}{\text{ni}^2\ (1+\text{w'}[\text{r}]^2)\ (\text{ra'}[\text{r}]^2+\text{za'}[\text{r}]^2)}\right)}$$

```
θ3=Evaluate[Limit[ArcTan[(ra[r])/(za[r]-zi-zo)],zi→Infinity]];
Homotopy={
{Piecewise[{
{wz[r]tt+w[r],tt<ϑ[r]},{ v Cos[θ3](tt-ϑ[r])+za[r],tt>ϑ[r]}}],
Piecewise[{
{wr[r]tt+r,tt<ϑ[r]},{v  Sin[θ3](tt-ϑ[r])+ra[r],tt>ϑ[r]}}]
}};
c=1;
v=c/ni;
no=1;
```

```
ni=1.5;

zo=8

zi =52;

k=ni zi+no zo;

rmax=14;

density=1;


Surface:=ParametricPlot[{{w[1.5*r],1.5*r},{za[1.5r],ra[1.5r]},{zo+1.1zi,ra[r]},
{w[r],r},{za[r],ra[r]}},{r,-rmax,rmax},AspectRatio→1/1, Axes→{True,True},
AxesOrigin→{-35,-25}, PlotStyle→{Directive[Black, Dashed,Thickness[0.0025]],
Directive[Red, Dashed,Thickness[0.0025]],Transparent, Directive[Black,
Dashed,Thickness[0.0025]],Red},AxesLabel→{Style[r,FontSize→14,Bold],
Style[z,FontSize→14,Bold]},AxesStyle→Arrowheads[{0.015}]];


Lens:=Graphics[{EdgeForm[Directive[{Black,Thin}]],
          Directive[Gray,Opacity[0.3]],
          Polygon[Join[#[[5,1]],Reverse[#[[3,1]]]]&
         [Cases[Surface,_Line,∞]]]}]


Waves:=Table[ParametricPlot[Homotopy,{r,-rmax,rmax},
PlotRange→All,PlotStyle→Purple],{tt,-30,zo+ni zi,2*density}]


Rays:=Table[{ParametricPlot[Homotopy,{tt,-30,zo+ni zi},
PlotRange→All,PlotStyle→Directive[Blue, Thickness[0.0015]]]},
{r,-rmax,rmax}]


ShowOpt={PlotRange->All, AspectRatio→Automatic,
ImageSize→{640,Automatic}};


Show[Surface,Lens,Rays,Waves,ShowOpt]
```

3.9 Examples

Example of refractive surfaces and their eikonals given by equations (3.41), (3.42), and (3.47) are presented in this section. This gallery features figures 3.15, 3.16, 3.17, 3.18, 3.19, 3.20, and 3.21 along with the input parameters. These surfaces are generally as expected according to the theory.

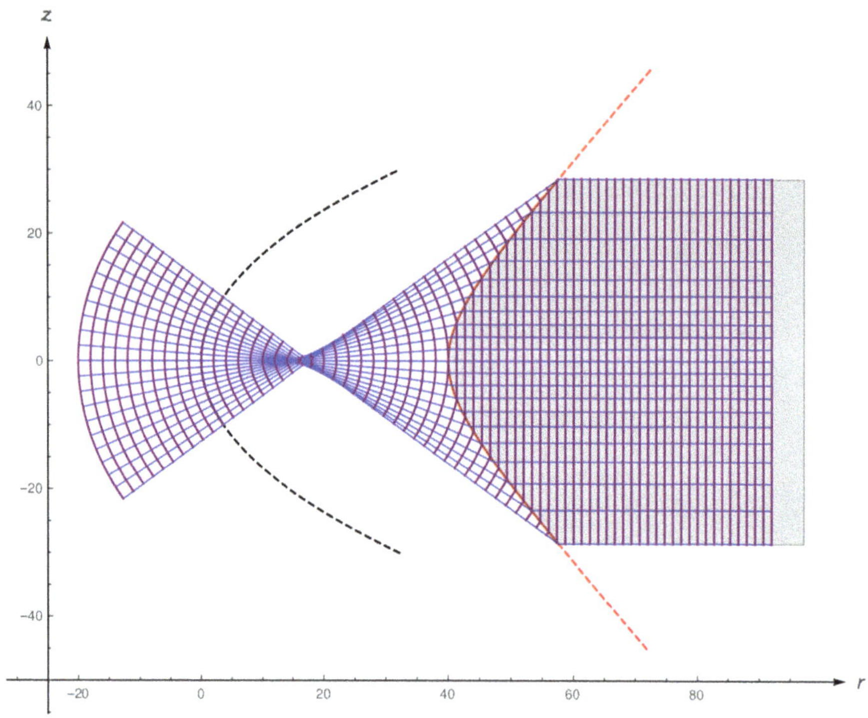

(a) A concave parabolic wavefront -image at infinity-

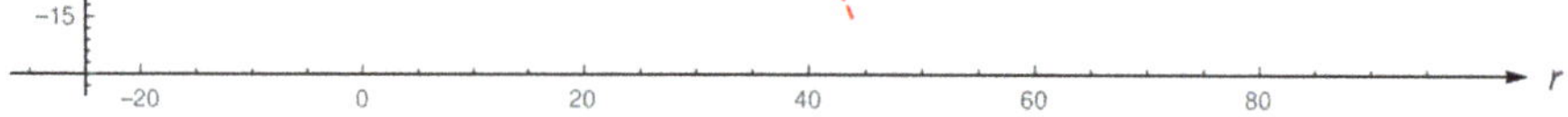

(b) A convex parabolic wavefront -image at infinity-

Figure 3.15. Surface design specifications for parabolic input wavefronts: (a) the input wavefront is $r^2/28$, refractive index $n = 1.5$, object distance $z_o = 40$ mm, image distance $z_i = 52$ mm, $\mathrm{r}_{\max} = 10$ mm, $(r_a, z_a) =$ equations (3.41) and (3.42) and $H(t, r_a) =$ equation (3.47); (b) the input wavefront is $-r^2/28$, refractive index $n = 1.5$, object distance $z_o = 40$ mm, image distance $z_i = 52$ mm, $\mathrm{r}_{\max} = 3$ mm, $(r_a, z_a) =$ equations (3.41) and (3.42), and $H(t, r_a) =$ equation (3.47).

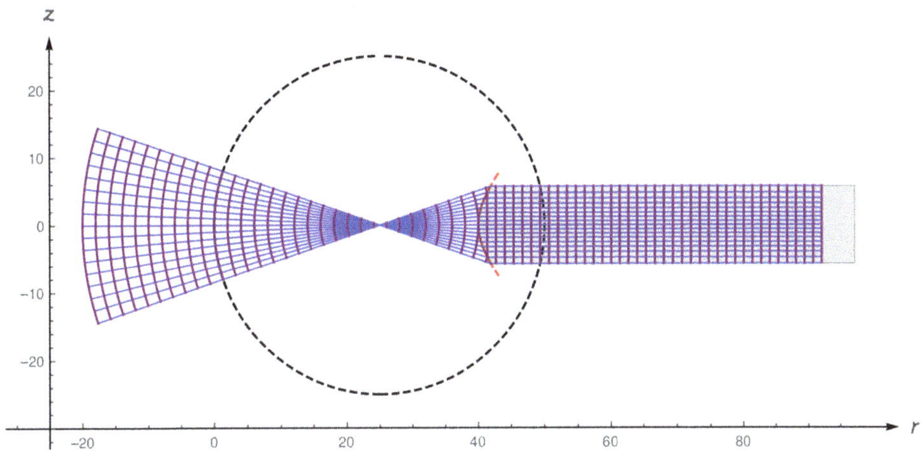

(a) A concave circular wavefront -image at infinity-

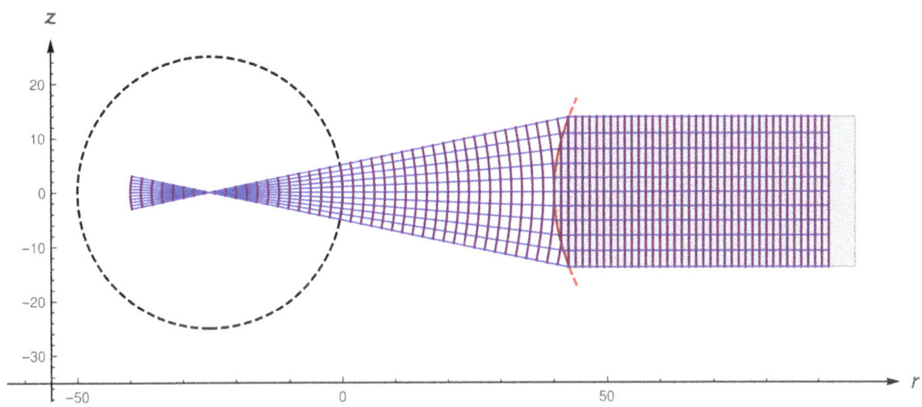

(b) A convex circular wavefront -image at infinity-

Figure 3.16. Surface design specifications for circular input wavefronts: (a) the input wavefront is $25 - \sqrt{25^2 - r^2}$, refractive index $n = 1.5$, object distance $z_o = 40$ mm, image distance $z_i = 52$ mm, $r_{max} = 8$ mm, (r_a, z_a) = equations (3.41) and (3.42), and $H(t, r_a)$ = equation (3.47); (b) the input wavefront is $-25 + \sqrt{25^2 - r^2}$, refractive index $n = 1.5$, object distance $z_o = 40$ mm, image distance $z_i = 52$ mm, $r_{max} = 5$ mm, (r_a, z_a) = equations (3.41) and (3.42), and $H(t, r_a)$ = equation (3.47).

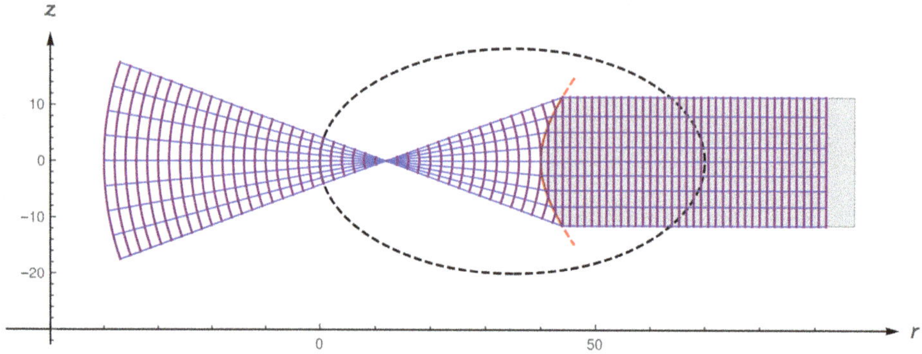

(a) A concave prolate ellipse wavefront -image at infinity-

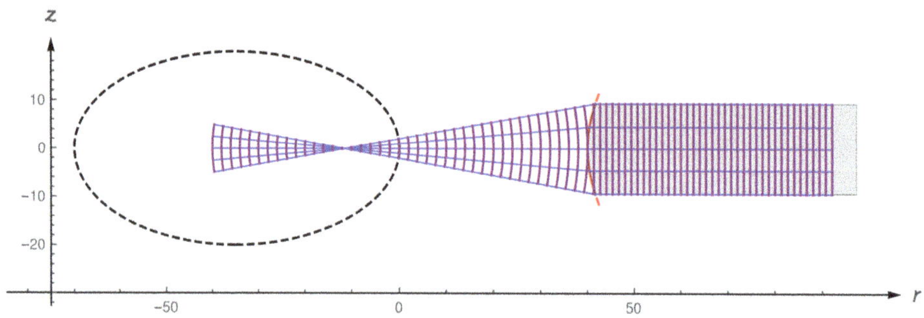

(b) A convex prolate ellipse wavefront -image at infinity-

Figure 3.17. Surface design specifications for prolate elliptical input wavefronts: (a) the input wavefront is $b - \sqrt{(-a^2 * b^2 + b^2r^2)/ - a^2}$, refractive index $n = 1.5$, object distance $z_o = 40\,\text{mm}$, image distance $z_i = 52\,\text{mm}$, $\text{r}_{\text{max}} = 4\,\text{mm}$, constant $a = 20$, constant $b = 35$, (r_a, z_a) = equations (3.41) and (3.42), and $H(t, r_a)$ = equation (3.47); (b) the input wavefront is $-b + \sqrt{(-a^2 * b^2 + b^2r^2)/ - a^2}$, refractive index $n = 1.5$, object distance $z_o = 40\,\text{mm}$, image distance $z_i = 50\,\text{mm}$, $\text{r}_{\text{max}} = 2\,\text{mm}$, constant $a = 20$, constant $b = 35$, (r_a, z_a) = equations (3.41) and (3.42), and $H(t, r_a)$ = equation (3.47).

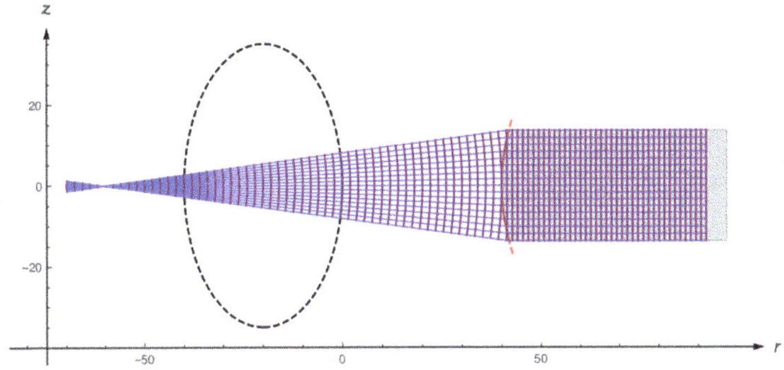

(a) A convex oblate ellipse wavefront -image at infinity-

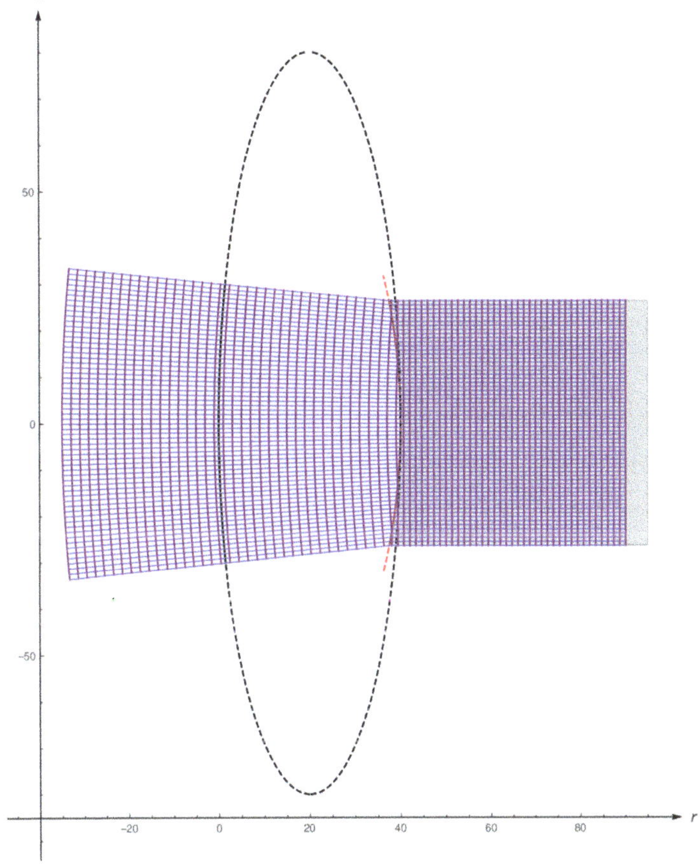

(b) A concave oblate ellipse wavefront -image at infinity-

Figure 3.18. Surface design specifications for oblate ellipse input wavefronts: (a) the input wavefront is $-b + \sqrt{(-a^2 * b^2 + b^2 r^2)/ - a^2}$, refractive index $n = 1.5$, object distance $z_o = 40$ mm, image distance $z_i = 52$ mm, $\mathrm{r}_{\max} = 8$ mm, constant $a = 35$, constant $b = 20$, (r_a, z_a) = equations (3.41) and (3.42), and $H(t, r_a)$ = equation (3.47); (b) the input wavefront is $b - \sqrt{(-a^2 * b^2 + b^2 r^2)/ - a^2}$, refractive index $n = 1.5$, object distance $z_o = 40$ mm, image distance $z_i = 50$ mm, $\mathrm{r}_{\max} = 30$ mm, constant $a = 80$, constant $b = 20$, (r_a, z_a) = equations (3.41) and (3.42), and $H(t, r_a)$ = equation (3.47).

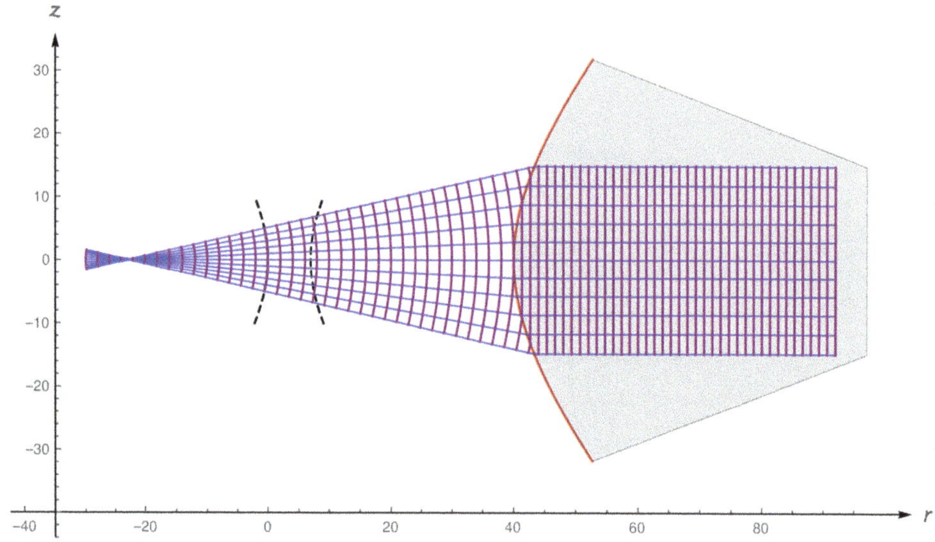

(a) A convex hyperbolic wavefront -image at infinity-

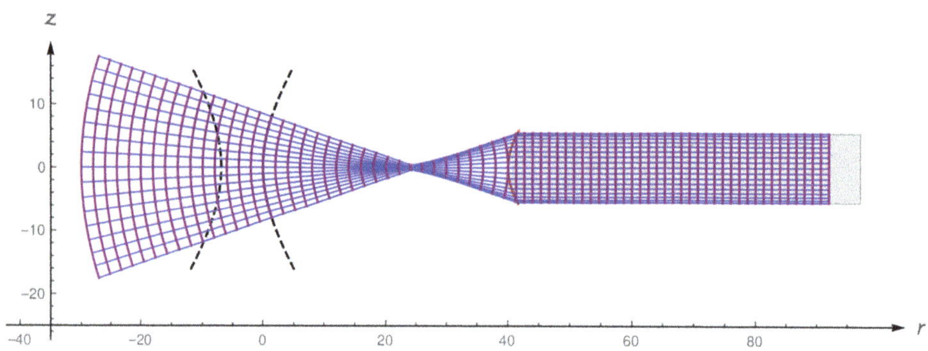

(b) A concave hyperbolic wavefront -image at infinity-

Figure 3.19. Surface design specifications for hyperbolic input wavefronts: (a) the input wavefront is $-b + \sqrt{(-a^2 * b^2 - b^2r^2)/ - a^2}$, refractive index $n = 1.5$, object distance $z_o = 40$ mm, image distance $z_i = 52$ mm, $\mathrm{r}_{\max} = 8$ mm, constant $a = 16$, constant $b = 12$, $(r_a, z_a) =$ equations (3.41) and (3.42), and $H(t, r_a) =$ equation (3.47); (b) the input wavefront is $b - \sqrt{(-a^2 * b^2 - b^2r^2)/ - a^2}$, refractive index $n = 1.5$, object distance $z_o = 40$ mm, image distance $z_i = 52$ mm, $\mathrm{r}_{\max} = 5$ mm, constant $a = 16$, constant $b = 12$, $(r_a, z_a) =$ equations (3.41) and (3.42), and $H(t, r_a) =$ equation (3.47).

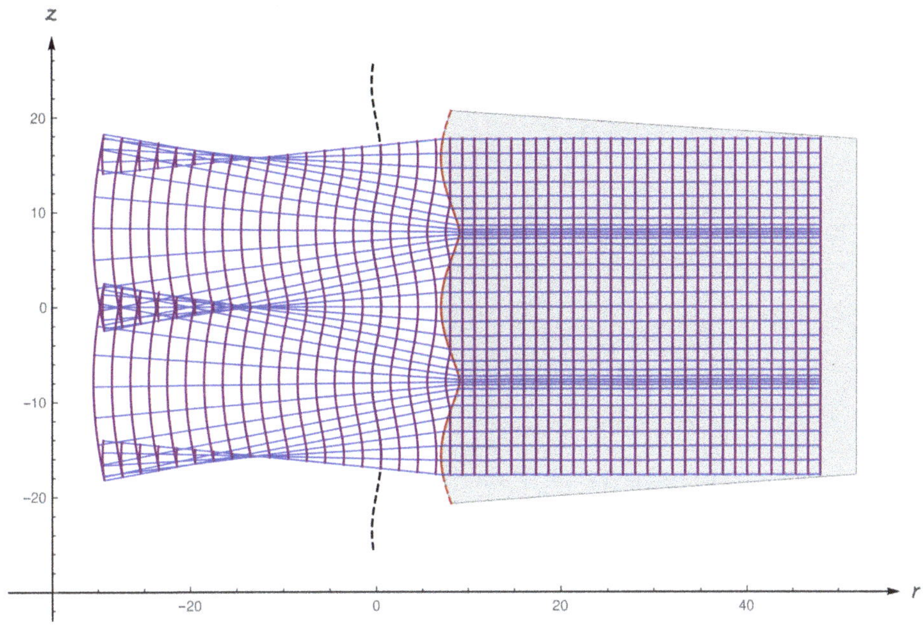

(a) A positive cosine wavefront -image at infinity-

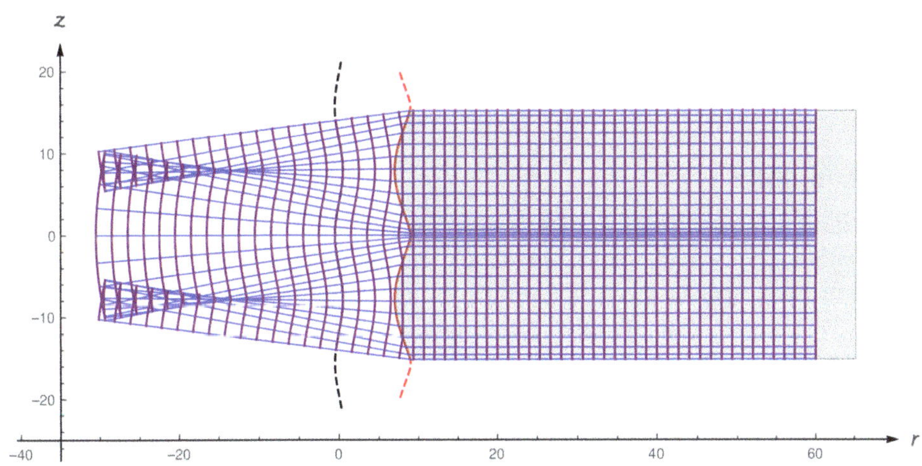

(b) A negative cosine wavefront -image at infinity-

Figure 3.20. Surface design specifications for cosine input wavefronts: (a) the input wavefront $\cos(2r/5)/2$, refractive index $n = 1.5$, object distance $z_o = 8$ mm, image distance $z_i = 52$ mm, $\mathrm{r_{max}} = 14$ mm, (r_a, z_a) = equations (3.41) and (3.42), and $H(t, r_a)$ = equation (3.47); (b) the input wavefront is $\cos(2r/5)/2$, refractive index $n = 1.5$, object distance $z_o = 8$ mm, image distance $z_i = 52$ mm, $\mathrm{r_{max}} = 17$ mm, (r_a, z_a) = equations (3.41) and (3.42), and $H(t, r_a)$ = equation (3.47).

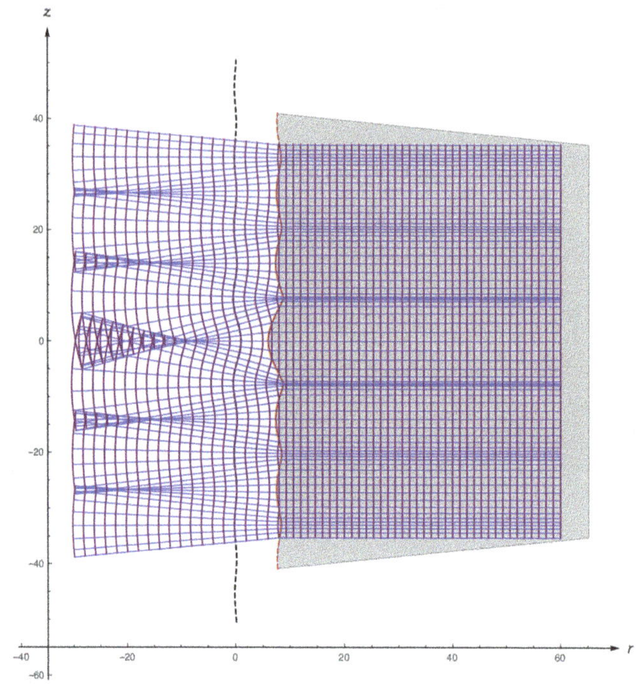

Figure 3.21. Surface design specifications for a Bessel zeroth-order input wavefront with the image at infinity. The input wavefront is $J_o(r/2)$, refractive index $n = 1.5$, object distance $z_o = 8$ mm, image distance $z_i = 52$ mm, $r_{max} = 36$ mm, (r_a, z_a) = equations (3.41) and (3.42), and $H(t, r_a)$ = equation (3.47).

3.10 End notes

In this chapter, we obtained a refractive surface such that it transforms a wavefront w into either a spherical wavefront or a plane wave. We did just not obtain the expression for a special case, but rather for the general case. Calculations in the software `Mathematica` validated the equations for the refractive surface described above, i.e. equations (3.20), (3.21), and (3.31) and equations (3.41), (3.42), and (3.47). An extension to a freeform wavefront is the natural continuation of this work and will be presented in chapter 7.

The proposed problem was solved analytically in a closed form and any numerical approaches and approximations were avoided. Therefore, it is obvious that everything we have presented in this chapter follows naturally from Fermat's principle. Thus, we can firmly argue that the mathematical equations presented here can be seen as theorems if Fermat's principle is seen as an axiom. And in fact the only axiom also called a postulate in geometrical optics is Fermat's principle. So, in conclusion to this chapter, the equations presented here are theorems for geometrical optics that will, in time, be presented like the Cartesian oval.

IOP Publishing

Optical Path Theory
Fundamentals to freeform adaptive optics
Rafael G González-Acuña and Héctor A Chaparro-Romo

Chapter 4

The aspheric transfer-function lens

In this chapter, we obtain the exact equations to design a lens that transforms an input wavefront into a given output wavefront. These equations are obtained using just Fermat's principle, and no numerical approximation is used in the procedure.

4.1 Transfer functions

An output signal is calculated for a given input signal by means of a mathematical representation of a transfer function; in other words, it is a product of the input signal. Control theory, for example, often uses transfer functions to describe the relationship between the inputs and outputs of systems described by time-invariant linear differential equations. From a more general perspective, lenses can be seen as transfer functions that take an input wavefront and produce an output wavefront. In generating analytical results for arbitrary inputs, standard approaches rely heavily on the assumption of paraxiality and iterative numerical optimizations.

In the previous chapters of this treatise, for a diversity of point sources and images, we developed a robust mathematical formulation for stigmatic surfaces using closed-form analytical expressions. In this chapter, we extend this formulation to answer the following question: how must a lens be shaped to transform an input aspherical wavefront into another aspherical wavefront, if both wavefronts are known?

Thus, our goal is the following: to obtain an analytical expression for the lens surfaces in terms of the wavefronts defined by the user. This problem is more challenging than the problems proposed in the past chapters, because the two wavefronts can now be aspheric. In the last chapters, one of the wavefronts was aspheric and the other was spheric or flat. This optical problem has been studied in various ways from a numerical point of view; please read the suggested literature presented in the bibliography of this chapter.

We are going to solve this problem using just Fermat's principle; remember that everything that is a natural implication of Fermat's principle is part of the optical path theory. We assume that the optical systems have axial symmetry and wavefronts described by smooth differentiable functions, just as in chapters 2 and 3. In chapter 8 we will return to solve this problem again, but from a freeform perspective.

Specifically, this lens can be used in optical shop testing to test complex optical system by generating the proper wavefront or to do null lens testing. Consequently, the problem of designing a lens that transforms a given wavefront into another wavefront is still an open problem that deserves to be studied in detail.

4.2 Mathematical model of the planar transfer-function lens

The geometry of the problem is shown in figure 4.1. The input and output wavefronts are given by the functions w_a and w_b, respectively, where r is the radial variable. The input wavefront parameters are all represented by a, whereas the data representing the output wavefronts have the subscript b. The coordinate system's origin is at the vertex of the first wavefront. In this discussion, we assume that both surfaces depend on the independent variable r.

Two surfaces are unknown, and they are represented by the curves (z_a, r_a) and (z_b, r_b), respectively. In this case, z_a and z_b are the sagittas of the first and second surfaces, respectively; and r_a and r_b are the radial coordinates of each surface. Inside the lens, the refractive index is n, while within the input space it is n_a, and within the output space it is n_b; z_o is the distance between the first wavefront and the first surface. As shown in figure 4.1, the lens's axial thickness is τ, and the distance between the second surface and the output wavefront is z_i. See figure 4.1.

Initially, we look at the wavefront propagation incident on the lens, and then at the refracted wavefront propagation to the final wavefront. The strategy is as follows: the first refractive surface (z_a, r_a) is designed to collimate the rays coming from the incident wavefront w_a. The flat wavefront propagates along the optical axis inside the lens, and the second surface (z_b, r_b) is now designed to transform the collimated rays into the desired wavefront w_b. Our device is reversible, so we call it a bidirectional transfer lens.

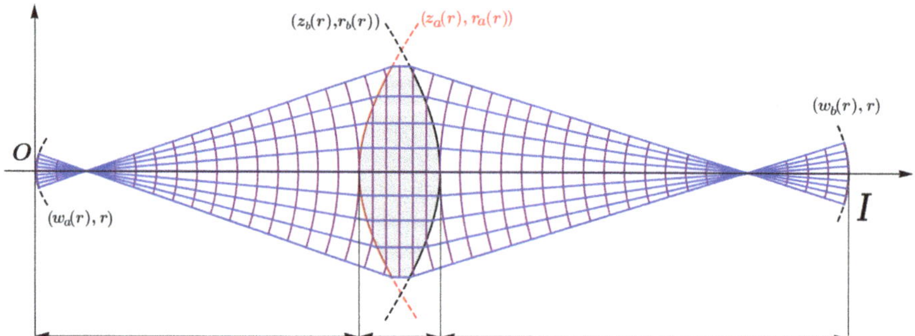

Figure 4.1. Basic configuration of a lens whose optical properties allow the transfer of a given wavefront to another desired wavefront, both wavefronts being prescribed.

To start the design process for the first surface, we consider a stigmatic refractive surface, such as a Cartesian oval, as the basis of our design.

Stigmatically, the Cartesian oval is formed by all of the rays originating at the object point that converge at the image point and which have the same optical path length. A refractive surface is required to collimate rays from the incident wavefront. To do this, we define an auxiliary, distant image, determine the equation for the surface, and finally set a limit $z_{in} \to \infty$. Hence, we begin by determining the finite distance z_{in}.

In comparison with an arbitrary non-axial ray, the OPL of the axial ray is

$$\boxed{n\sqrt{(z_a - z_{in} - z_o)^2 + r_a^2} + n_a\sqrt{(r_a - r)^2 + (z_a - w_a)^2} = nz_{in} + n_a z_o,} \tag{4.1}$$

where the left-hand side of equation (4.1) is the OPL of a non-axial ray, and the right-hand side is the OPL of an axial ray. Since the rays have the same direction as the wavefront, according to the Malus–Dupin theorem, the cosine directors of the input rays are given by

$$w_{a_r} = -\frac{w_a'}{\sqrt{w_a'^2 + 1}}, \qquad w_{a_z} = \frac{1}{\sqrt{w_a'^2 + 1}}, \tag{4.2}$$

where $w_a' \equiv \mathrm{d}w_a/\mathrm{d}r$ is the derivative of w_a with respect to r. By analysing the input rays, we can write the the equations for the first surface, as follows:

$$r_a = r + \vartheta_a w_{a_r}, \qquad z_a = w_a + \vartheta_a w_{a_z}, \tag{4.3}$$

where ϑ_a is the distance that each ray of the input wavefront w_a travels to reach the first surface, namely

$$\vartheta_a \equiv \sqrt{(r_a - r)^2 + (z_a - w_a)^2}. \tag{4.4}$$

Note that ϑ_a is expressed in terms of the unknown parameters r_a and z_a. We need to determine an expression for ϑ_a in terms of the known parameters. In order to restore parallelism, we should compute the limit when $z_i \to \infty$ over equation (4.1).

$$\lim_{z_{in}\to\infty}\left[n\sqrt{(w_a + \vartheta_a w_{a_z} - z_{in} - z_o)^2 + (r + \vartheta_a w_{a_r})^2} + n_a\vartheta_a - \left(nz_{in} + n_a z_o\right) = 0\right] \tag{4.5}$$

leads to the following relationship:

$$-nw_a - n\vartheta_a w_{a_z} + nz_o - n_a z_o + n_a\vartheta_a = 0. \tag{4.6}$$

Solving for ϑ_a, we get

$$\vartheta_a = \frac{-nw_a + z_o(n - n_a)}{nw_{a_z} - n_a}. \tag{4.7}$$

Now that we have ϑ_a, we can express the first surface properly; r_a is given by

$$\boxed{r_a = r - \frac{w_a'(-nw_a + nz_o - n_o z_o)}{\sqrt{w_a'^2 + 1}\left(\frac{n}{\sqrt{w_a'^2 + 1}} - n_o\right)}}, \tag{4.8}$$

and z_a is given by

$$\boxed{z_a = w_a + \frac{-nw_a + nz_o - n_o z_o}{\sqrt{w_a'^2 + 1}\left(\frac{n}{\sqrt{w_a'^2 + 1}} - n_o\right)}}. \tag{4.9}$$

Using equations (4.8) and (4.9) we can obtain the refractive surface with the desired property, i.e. a refractive surface that collimates a given wavefront. We emphasise that no paraxial approximation or numerical method has been used in this deduction process.

4.2.1 Construction of the second surface

We can now focus on the second surface, whose analysis is very similar to that of the first, but with some minor modifications. As shown in figure 4.1, we can calculate the cosine directors of the wavefront w_b as

$$w_{b_r} = -\frac{w_b'}{\sqrt{w_b'^2 + 1}}, \qquad w_{b_z} = \frac{1}{\sqrt{w_b'^2 + 1}}, \tag{4.10}$$

where $w_b' \equiv \mathrm{d}w_b/\mathrm{d}r$ is the derivative of w_b with respect to r.

The distance that each ray of the output wavefront w_b must travel to reach the second surface is ϑ_b. The parameter ϑ_b is very similar to ϑ_a, but it works with the refraction indices of the lens n and the output space n_b and is given by

$$\vartheta_b = \frac{-nw_b + z_i(n - n_b)}{nw_{b_z} - n_b}. \tag{4.11}$$

The second surface is just like the first, but it is inverted and displaced by an axial distance $z_o + \tau$.

$$r_b = r + \vartheta_b w_{b_r}, \qquad z_b = z_o + \tau + z_i - w_b - w_{b_z}\vartheta_b. \tag{4.12}$$

so, using ϑ_b we can express r_b as

$$\boxed{r_b = r - \frac{w_b'(-nw_b + nz_i - n_i z_i)}{\sqrt{w_b'^2 + 1}\left(\frac{n}{\sqrt{w_b'^2 + 1}} - n_i\right)}}, \tag{4.13}$$

and, we can express z_b as,

$$z_b = -w_b + \tau + z_i + z_o - \frac{-nw_b + nz_i - n_iz_i}{\sqrt{w_b'^2 + 1}\left(\frac{n}{\sqrt{w_b'^2 + 1}} - n_i\right)}. \tag{4.14}$$

Note that the negative signs of $-w_{b_z}\vartheta_b$ and $-w_b$ in equation (4.14) are responsible for the inversion of the second surface. Thus, equation (4.14) expresses the second surface as required, namely, a surface that receives collimated rays and generates the wavefront w_b at the output plane. τ appears in equation (4.14) because the coordinate system is placed at the vertex of the first wavefront; τ is a necessary offset for the output wavefront and the second surface.

Similarly to the behavior of the Cartesian oval with spherical waterfronts, these surfaces collimate a given wavefront inside a medium with a refractive index of n. As with Cartesian ovals, theoretically, neither of these surfaces can be physically limited. However, in practice, the size of the lens is determined by the points at which both surfaces touch, and is proportional to τ.

Finally, we assert that the solution to equations (4.13) and (4.14) is unique, given the boundary conditions. It can be deduced that. The key to this is the limit $z_i \to \infty$, which results in an algebraic equation by removing the square roots. So equations (4.13) and (4.14) can be seen as natural implications of the optical path theory and Fermat's principle.

4.3 Ray tracing light passing through the transfer-function lens

It is now time to calculate the paths of the rays in such a way that the system is stigmatic. We start by calculating the first surface, which guarantees parallelism in the interior of the lens, and then we calculate the second surface, which guarantees the desired stigmatism.

4.3.1 Equations for first-surface ray tracing

In order to get the eikonal of the transfer-function lens on the plane, we need to focus first on the refraction at the first surface, so we recall Snell's law at the first surface,

$$\vec{\mathbf{v}}_2 = \frac{n_o}{n}[\vec{\mathbf{v}}_1 - (\vec{\mathbf{n}}_a \cdot \vec{\mathbf{v}}_1)\vec{\mathbf{n}}_a] - \vec{\mathbf{n}}_a\sqrt{1 - \frac{n_o^2}{n^2}(\vec{\mathbf{n}}_a \times \vec{\mathbf{v}}_1)^2} \quad \text{for} \quad \vec{\mathbf{v}}_2, \vec{\mathbf{v}}_1, \vec{\mathbf{n}}_a \in \mathbb{R}^2, \tag{4.15}$$

where $\vec{\mathbf{v}}_1$ is the unit vector of the incident ray, $\vec{\mathbf{v}}_2$ is the unit vector of the refracted ray, and $\vec{\mathbf{n}}_a$ is the normal vector of the first surface.

$$\vec{\mathbf{v}}_1 = \frac{-w_a'\vec{\mathbf{e}}_1 + \vec{\mathbf{e}}_2}{\sqrt{w_a'^2 + 1}}, \qquad \vec{\mathbf{v}}_2 = \wp_r\vec{\mathbf{e}}_1 + \wp_r\vec{\mathbf{e}}_2, \qquad \vec{\mathbf{n}}_a = \frac{z_a'\vec{\mathbf{e}}_1 - r_a'\vec{\mathbf{e}}_2}{\sqrt{r_a' + z_a'^2}}, \tag{4.16}$$

where $\vec{\mathbf{e}}_1$ denotes the r direction and $\vec{\mathbf{e}}_2$ denotes the z direction. To obtain the normal vector at the first surface, we need to compute the derivatives of the first surface, so, for r_a' we take the derivative of equation (4.8),

$$r_a' = \frac{1}{(w_a'^2 + 1)\left(n - n_o\sqrt{w_a'^2 + 1}\right)^2} \times \left\{ w_a'^4\left(n^2 - nn_o\sqrt{w_a'^2 + 1} + n_o^2\right) \right.$$
$$+ w_a'^2\left(2n^2 + nw_a''\left(nw_a - nz_o + n_oz_o\right) - 3nn_o\sqrt{w_a'^2 + 1} + 2n_o^2\right)$$
$$+ n^2 - 2nn_o\sqrt{w_a'^2 + 1} - w_a''\left(-nw_a + nz_o - n_oz_o\right) \tag{4.17}$$
$$\left. \left(n - n_o\sqrt{w_a'^2 + 1}\right) + n_o^2 \right\},$$

and for z_a', we take the derivative of equation (4.9),

$$z_a' = \frac{1}{\sqrt{w_a'^2 + 1}\left(n - n_o\sqrt{w_a'^2 + 1}\right)^2}$$
$$\times \left\{ n_ow_a'(w_a''(-nw_a + nz_o - n_oz_o) \right. \tag{4.18}$$
$$\left. + w_a'^2\left(n_o\sqrt{w_a'^2 + 1} - n\right) - n + n_o\sqrt{w_a'^2 + 1}\right) \right\}.$$

Once we have computed r_a' and z_a', we focus on Snell's law, equation (4.15). First, we compute the term $(\vec{\mathbf{n}}_a \times \vec{\mathbf{v}}_1)^2$:

$$(\vec{\mathbf{n}}_a \times \vec{\mathbf{v}}_1) = \frac{z_a' - r_a'w_a'}{\sqrt{w_a'^2 + 1}\sqrt{r_a'^2 + z_a'^2}}(\vec{\mathbf{e}}_1 \times \vec{\mathbf{e}}_2); \tag{4.19}$$

squaring it yields

$$(\vec{\mathbf{n}}_a \times \vec{\mathbf{v}}_1)^2 = \frac{(z_a' + r_a'w_a')^2}{(w_a'^2 + 1)(r_a'^2 + z_a'^2)}. \tag{4.20}$$

We are now in a position to directly compute the term of the square root in equation (4.15):

$$-\vec{\mathbf{n}}_a\sqrt{1 + \frac{n_o^2}{n^2}(\vec{\mathbf{n}}_a \times \vec{\mathbf{v}}_1)^2} = -\frac{(z_a'\vec{\mathbf{e}}_1 - \vec{\mathbf{e}}_2)}{\sqrt{1 + z_a'^2}}\sqrt{1 - \frac{n_o^2}{n^2}\frac{(z_a' - r_a'w_a')^2}{(w_a'^2 + 1)(r_a'^2 + z_a'^2)}}, \tag{4.21}$$

and then examine the term $\vec{\mathbf{v}}_1 - (\vec{\mathbf{n}}_a \cdot \vec{\mathbf{v}}_1)\vec{\mathbf{n}}_a$ of equation (4.15):

$$\vec{\mathbf{v}}_1 - (\vec{\mathbf{n}}_a \cdot \vec{\mathbf{v}}_1)\vec{\mathbf{n}}_a = \frac{-w_a'\vec{\mathbf{e}}_1 + \vec{\mathbf{e}}_2}{\sqrt{w_a'^2 + 1}} - \left[-\frac{r_a' + w_a'z_a'}{\sqrt{w_a'^2 + 1}\sqrt{r_a'^2 + z_a'^2}}\right]$$
$$\frac{(z_a'\vec{\mathbf{e}}_1 - r_a'\vec{\mathbf{e}}_2)}{\sqrt{r_a'^2 + z_a'^2}}. \tag{4.22}$$

By simplifying and multiplying both sides by $\frac{n_o}{n}$, we obtain

$$\frac{n_o}{n}[\vec{\mathbf{v}}_1 - (\vec{\mathbf{n}}_a \cdot \vec{\mathbf{v}}_1)\vec{\mathbf{n}}_a] = \frac{n_o}{n}\frac{r_a'(z_a' - r_a'w_a')}{\sqrt{w_a'^2 + 1}\,(r_a'^2 + z_a'^2)}\vec{\mathbf{e}}_1 + \frac{n_o}{n}\frac{z_a'(z_a' - r_a'w_a')}{\sqrt{w_a'^2 + 1}\,(r_a'^2 + z_a'^2)}\vec{\mathbf{e}}_2. \tag{4.23}$$

Substituting equations (4.21) and (4.23) into equation (4.15) and separating components yields

$$\wp_r = \frac{n_o r_a'(z_a' - r_a'w_a')}{n\sqrt{w_a'^2 + 1}\,(r_a'^2 + z_a'^2)} + \frac{z_a'\sqrt{1 - \dfrac{n_o^2(z_a' - r_a'w_a')^2}{n^2(w_a'^2 + 1)(r_a'^2 + z_a'^2)}}}{\sqrt{r_a'^2 + z_a'^2}} \tag{4.24}$$

and

$$\wp_z = \frac{n_o z_a'(z_a' - r_a'w_a')}{n\sqrt{w_a'^2 + 1}\,(r_a'^2 + z_a'^2)} - \frac{r_a'\sqrt{1 - \dfrac{n_o^2(z_a' - r_a'w_a')^2}{n^2(w_a'^2 + 1)(r_a'^2 + z_a'^2)}}}{\sqrt{r_a'^2 + z_a'^2}}. \tag{4.25}$$

At this point we have the cosine directors of the input rays w_{a_r} and w_{a_z} from equation (4.2). We also have the directors of the refracted rays inside the lens from equations (4.24) and (4.25), but note that we set the OPL such that $z_{in} \to \infty$, so it is clear that for the first surface,

$$\wp_r = 0, \quad \wp_z = 1. \tag{4.26}$$

The demonstration is left to the reader as an exercise.

4.3.2 Equations for second-surface ray tracing

For the second surface, we use almost the same procedure as for the first surface. So, we use Snell's law at the second surface, which is given by

$$\boxed{-\vec{\mathbf{v}}_2 = \frac{n_i}{n}[\vec{\mathbf{v}}_3 - (\vec{\mathbf{n}}_b \cdot \vec{\mathbf{v}}_3)\vec{\mathbf{n}}_a] - \vec{\mathbf{n}}_b\sqrt{1 - \frac{n_i^2}{n^2}(\vec{\mathbf{n}}_b \times \vec{\mathbf{v}}_1)^2} \quad \text{for} \quad \vec{\mathbf{v}}_2, \vec{\mathbf{v}}_3, \vec{\mathbf{n}}_b \in \mathbb{R}^2,} \tag{4.27}$$

where $\vec{\mathbf{v}}_3$ is the unit vector of the incident ray, $\vec{\mathbf{v}}_2$ is the unit vector of the refracted ray, and $\vec{\mathbf{n}}_b$ is the normal vector of the first surface, as can be seen in figure 4.1:

$$\vec{\mathbf{v}}_3 = \frac{-w_b'\vec{\mathbf{e}}_1 + \vec{\mathbf{e}}_2}{\sqrt{w_b'^2 + 1}}, \qquad \vec{\mathbf{v}}_2 = \wp_r\vec{\mathbf{e}}_1 + \wp_r\vec{\mathbf{e}}_2, \qquad \vec{\mathbf{n}}_b = \frac{z_b'\vec{\mathbf{e}}_1 - r_b'\vec{\mathbf{e}}_2}{\sqrt{r_b' + z_b'^2}}, \tag{4.28}$$

where $\vec{\mathbf{e}}_1$ denotes the r direction and $\vec{\mathbf{e}}_2$ denotes the z direction. To compute the normal vector at the second surface, we compute the derivatives of the second surface; by taking the derivative of equation (4.13), we get r_b':

$$r_b' = \frac{1}{(w_b'^2+1)\left(n - n_i\sqrt{w_b'^2+1}\right)^2}\left\{w_b'^4\left(n^2 - nn_i\sqrt{w_b'^2+1} + n_i^2\right)\right.$$
$$+ w_b'^2\left(2n^2 + nw_b''\left(nw_b - nz_i + n_iz_i\right) - 3nn_i\sqrt{w_b'^2+1} + 2n_i^2\right) \tag{4.29}$$
$$\left. + n^2 - 2nn_i\sqrt{w_b'^2+1} - w_b''\left(-nw_b + nz_i - n_iz_i\right)\left(n - n_i\sqrt{w_b'^2+1}\right) + n_i^2\right\},$$

and by taking the derivative of equation (4.14), we get z_b':

$$z_b' = \frac{1}{\sqrt{w_b'^2+1}\left(n - n_i\sqrt{w_b'^2+1}\right)^2}\left\{n_iw_b'\right.$$
$$\times\left(w_b''\left(nw_b - nz_i + n_iz_i\right) + w_b'^2\left(n - n_i\sqrt{w_b'^2+1}\right) + n\right. \tag{4.30}$$
$$\left.\left. - n_i\sqrt{w_b'^2+1}\right)\right\}.$$

We now focus on the cross-product of Snell's law at the second surface, equation (4.27):

$$(\vec{\mathbf{n}}_b \times \vec{\mathbf{v}}_3) = \frac{z_b' - r_b'w_b'}{\sqrt{w_b'^2+1}\sqrt{r_b'^2+z_b'^2}}(\vec{\mathbf{e}}_1 \times \vec{\mathbf{e}}_2); \tag{4.31}$$

we square the last expression and obtain

$$(\vec{\mathbf{n}}_b \times \vec{\mathbf{v}}_3)^2 = \frac{(z_b' + r_b'w_b')^2}{(w_b'^2+1)(r_b'^2+z_b'^2)}, \tag{4.32}$$

so the squared root of equation (4.27) is

$$-\vec{\mathbf{n}}_b\sqrt{1 + \frac{n_o^2}{n_i^2}(\vec{\mathbf{n}}_b \times \vec{\mathbf{v}}_3)^2} = -\frac{(z_b'\vec{\mathbf{e}}_1 - \vec{\mathbf{e}}_2)}{\sqrt{1+z_b'^2}}\sqrt{1 - \frac{n_o^2}{n_i^2}\frac{(z_b' - r_b'w_b')^2}{(w_b'^2+1)(r_b'^2+z_b'^2)}}. \tag{4.33}$$

The term $\vec{\mathbf{v}}_3 - (\vec{\mathbf{n}}_b \cdot \vec{\mathbf{v}}_3)\vec{\mathbf{n}}_b$ of equation (4.27) is given by

$$\vec{\mathbf{v}}_3 - (\vec{\mathbf{n}}_b \cdot \vec{\mathbf{v}}_3)\vec{\mathbf{n}}_b = \frac{-w_b'\vec{\mathbf{e}}_1 + \vec{\mathbf{e}}_2}{\sqrt{w_b'^2+1}} - \left[-\frac{r_b' + w_b'z_b'}{\sqrt{w_b'^2+1}\sqrt{r_b'^2+z_b'^2}}\right]\frac{(z_b'\vec{\mathbf{e}}_1 - r_b'\vec{\mathbf{e}}_2)}{\sqrt{r_b'^2+z_b'^2}}. \tag{4.34}$$

We multiply the last expression by $\frac{n_i}{n}$, and reorder its terms to obtain

$$\frac{n_i}{n}[\vec{\mathbf{v}}_3 - (\vec{\mathbf{n}}_b \cdot \vec{\mathbf{v}}_3)\vec{\mathbf{n}}_b] = \frac{n_i}{n}\frac{r_b'(z_b' - r_b'w_b')}{\sqrt{w_b'^2+1}\,(r_b'^2+z_b'^2)}$$
$$\vec{\mathbf{e}}_1 + \frac{n_i}{n}\frac{z_b'(z_b' - r_b'w_b')}{\sqrt{w_b'^2+1}\,(r_b'^2+z_b'^2)}\vec{\mathbf{e}}_2. \tag{4.35}$$

Finally, summing equation (4.33) and (4.35), we get the cosine directors of the rays refracted by the second surface, from the perspective of the output wavefront as the input wavefront:

$$-\wp_r = \frac{n_o r_b'(z_b' - r_b' w_b')}{n_i\sqrt{w_b'^2 + 1}\,(r_b'^2 + z_b'^2)} + \frac{z_b'\sqrt{1 - \dfrac{n_o^2(z_b' - r_b' w_b')^2}{n_i^2(w_b'^2 + 1)(r_b'^2 + z_b'^2)}}}{\sqrt{r_b'^2 + z_b'^2}} \tag{4.36}$$

and

$$-\wp_r = \frac{n_o z_b'(z_b' - r_b' w_b')}{n_i\sqrt{w_b'^2 + 1}\,(r_b'^2 + z_b'^2)} - \frac{r_b'\sqrt{1 - \dfrac{n_o^2(z_b' - r_b' w_b')^2}{n_i^2(w_b'^2 + 1)(r_b'^2 + z_b'^2)}}}{\sqrt{r_b'^2 + z_b'^2}}. \tag{4.37}$$

Note that we use the same letters to describe these cosine directors, but with a different sign from those seen in equations (4.36) and (4.37). Therefore, it is clear that

$$\wp_r = 0, \quad \wp_z = 1. \tag{4.38}$$

Thus, at this point, we have the input cosine directors, which are w_{a_r} and w_{a_z}, from equation (4.2). The cosine directors for the refracted ray inside the lens are $\wp_\mathrm{r}$ and $\wp_\mathrm{z}$ from equations (4.24) and (4.25), and the output cosine directors are w_{b_r} and w_{b_z} from equation (4.10). But observe that since the rays are collimated inside the lens, $r_a = r_b$, thus from equations (4.8) and (4.13)

$$r - \frac{w_a'(-n w_a + n z_o - n_o z_o)}{\sqrt{w_a'^2 + 1}\left(\dfrac{n}{\sqrt{w_a'^2 + 1}} - n_o\right)} = r - \frac{w_b'(-n w_b + n z_i - n_i z_i)}{\sqrt{w_b'^2 + 1}\left(\dfrac{n}{\sqrt{w_b'^2 + 1}} - n_i\right)}, \tag{4.39}$$

and eliminating r from both sides gives

$$\frac{w_a'(-n w_a + n z_o - n_o z_o)}{\sqrt{w_a'^2 + 1}\left(\dfrac{n}{\sqrt{w_a'^2 + 1}} - n_o\right)} = \frac{w_b'(-n w_b + n z_i - n_i z_i)}{\sqrt{w_b'^2 + 1}\left(\dfrac{n}{\sqrt{w_b'^2 + 1}} - n_i\right)}. \tag{4.40}$$

It is very important to be aware of equation (4.40), $r_a = r_b$, in order to understand the eikonal presented in `Mathematica` code in the next subsection.

4.4 Mathematica code

This section presents the Mathematica code for equations (4.8), (4.9), (4.13), and (4.14); the outputs of the algorithm presented in this section are shown in figures 4.2, 4.3, 4.4, 4.5, 4.6, and 4.7.

```
In[5]:= ClearAll["Global`*"]

w[r_]:=r^2/12

wr[r_]:=-w'[r]/Sqrt[w'[r]^2+1]

wz[r_]:=1/Sqrt[w'[r]^2+1]

ϑ[r_]:=(n zo-no zo-n w[r])/(-no+n wz[r])

ra[r_]:=r+ϑ[r]wr[r]

za[r_]:=w[r]+ϑ[r]wz[r]

ϱr[r_]:=(no ra'[r] (-ra'[r] w'[r]+za'[r]))/(n Sqrt[1+w'[r]^2] (ra'[r]^2 +za'[r]^2))
    +za'[r]/Sqrt[ra'[r]^2+za'[r]^2] Sqrt[(1-(no^2 (-ra'[r] w'[r]+za'[r])^2)/(n^2 (1+w'[r]^2) (ra'[r]^2+za'[r]^2)))]

ϱz[r_]:=(no za'[r] (-ra'[r] w'[r]+za'[r]))/(n Sqrt[1+w'[r]^2] (ra'[r]^2+za'[r]^2))
    -ra'[r]/Sqrt[ra'[r]^2+za'[r]^2] Sqrt[(1-(no^2 (-ra'[r] w'[r]+za'[r])^2)/(n^2 (1+w'[r]^2) (ra'[r]^2+za'[r]^2)))]
```

```
wb[r_]:=r^2/24

wbb[r_]:=zo +τ+zi-wb[r]

wrb[r_]:=-wb'[r]/Sqrt[wb'[r]^2+1]

wzb[r_]:=1/Sqrt[wb'[r]^2+1]

ϑb[r_]:=(n zi-ni zi-n wb[r])/(-ni+n wzb[r])

rb[r_]:=r+ϑb[r]wrb[r]

zb[r_]:=wbb[r]-ϑb[r]wzb[r]

θ3=0;

no=1;
n=1.5;
ni=1;
zo=40;
zi=50;
mmaUndτmmaUnd=10;
c=1;
v=c/n;
rmax=2;
density=0.5;

Surface:=ParametricPlot[{{w[r],r},{za[r],ra[r]},
{zb[NSolve[ra[r]==rb[rr],rr][[3]][[1,2]]],
rb[NSolve[ra[r]==rb[rr],rr][[3]][[1,2]]]},
{wbb[NSolve[ra[r]==rb[rr],rr][[3]][[1,2]]],
NSolve[ra[r]==rb[rr],rr][[3]][[1,2]]},
{w[2r],2r},{za[1.3r],ra[1.3r]},
{zb[NSolve[ra[1.3r]==rb[rr],rr][[3]][[1,2]]],
rb[NSolve[ra[1.3r]==rb[rr],rr][[3]][[1,2]]]},
```

```
{wbb[NSolve[ra[1.5r]==rb[rr],rr][[3]][[1,2]]],
NSolve[ra[1.5r]==rb[rr],rr][[3]][[1,2]]}},
{r,-rmax,rmax},AspectRatio→1/1, AxesOrigin→{-5,-20},
PlotStyle→{{Black,Thick},Red,Black,
{Black,Thick},{Black,Dashed},{Red,Dashed},
{Black,Dashed},{Black,Dashed}},Axes→True,
AxesLabel→{Style[z,Larger,Bold],Style[r,Larger,Bold]},
AxesStyle→Arrowheads[{0.015}]]


Lens:=Graphics[{EdgeForm[Directive[{Black,Thin}]],
        Directive[Gray,Opacity[0.3]],
        Polygon[Join[#[[2,1]],Reverse[#[[3,1]]]]&
    [Cases[Surface,_Line,∞]]]}]


Waves:=Table[ParametricPlot[{
{Piecewise[{
{wz[r]tt+w[r],tt<ϑ[r]},{ v Cos[θ3](tt-ϑ[r])+za[r],ϑ[r]
+n (zb[NSolve[ra[r]==rb[rr],rr][[3]][[1,2]]]-za[r])>tt>ϑ[r]},
{zb[NSolve[ra[r]==rb[rr],rr][[3]][[1,2]]]+
wzb[NSolve[ra[r]==rb[rr],rr][[3]][[1,2]]](tt-ϑ[r]
-n (zb[NSolve[ra[r]==rb[rr],rr][[3]][[1,2]]]-za[r])),tt>ϑ[r]
+n (zb[NSolve[ra[r]==rb[rr],rr][[3]][[1,2]]]-za[r])}}],
Piecewise[{
{wr[r]tt+r,tt<ϑ[r]},{v Sin[θ3](tt-ϑ[r])+ra[r],ϑ[r]
+n (zb[NSolve[ra[r]==rb[rr],rr][[3]][[1,2]]]-za[r])>tt>ϑ[r]},
{rb[NSolve[ra[r]==rb[rr],rr][[3]][[1,2]]]
-wrb[NSolve[ra[r]==rb[rr],rr][[3]][[1,2]]](tt-ϑ[r]
-n (zb[NSolve[ra[r]==rb[rr],rr][[3]][[1,2]]]-za[r])),tt>ϑ[r]
+n (zb[NSolve[ra[r]==rb[rr],rr][[3]][[1,2]]]-za[r])}}]},
{Piecewise[{
{wz[r]tt+w[r],tt<ϑ[r]},{ v Cos[θ3](tt-ϑ[r])+za[r],ϑ[r]
+n (zb[NSolve[ra[r]==rb[rr],rr][[3]][[1,2]]]-za[r])>tt>ϑ[r]},
{zb[NSolve[ra[r]==rb[rr],rr][[3]][[1,2]]]
+wzb[NSolve[ra[r]==rb[rr],rr][[3]][[1,2]]](tt-ϑ[r]
-n (zb[NSolve[ra[r]==rb[rr],rr][[3]][[1,2]]]-za[r])),tt>ϑ[r]
+n (zb[NSolve[ra[r]==rb[rr],rr][[3]][[1,2]]]-za[r])}}],
Piecewise[{
{-(wr[r]tt+r),tt<ϑ[r]},{-(v Sin[θ3](tt-ϑ[r])+ra[r]),ϑ[r]
+n (zb[NSolve[ra[r]==rb[rr],rr][[3]][[1,2]]]-za[r])>tt>ϑ[r]},
{-(rb[NSolve[ra[r]==rb[rr],rr][[3]][[1,2]]]
-wrb[NSolve[ra[r]==rb[rr],rr][[3]][[1,2]]](tt-ϑ[r]
-n (zb[NSolve[ra[r]==rb[rr],rr][[3]][[1,2]]]-za[r]))),tt>ϑ[r]
+n (zb[NSolve[ra[r]==rb[rr],rr][[3]][[1,2]]]-za[r])}}]
```

```
}},{r,0,rmax},
PlotRange→All,PlotStyle→Purple],{tt,0,zo+n τ+zi,1.25/density}]



Rays:=Table[ParametricPlot[{
{Piecewise[{
{wz[r]tt+w[r],tt<ϑ[r]},{ v Cos[θ3](tt-ϑ[r])+za[r],ϑ[r]
+n (zb[NSolve[ra[r]==rb[rr],rr][[3]][[1,2]]]-za[r])>tt>ϑ[r]},
{zb[NSolve[ra[r]==rb[rr],rr][[3]][[1,2]]]
+wzb[NSolve[ra[r]==rb[rr],rr][[3]][[1,2]]](tt-ϑ[r]
-n (zb[NSolve[ra[r]==rb[rr],rr][[3]][[1,2]]]-za[r])),tt>ϑ[r]
+n (zb[NSolve[ra[r]==rb[rr],rr][[3]][[1,2]]]-za[r])}}],
Piecewise[{
{wr[r]tt+r,tt<ϑ[r]},{v Sin[θ3](tt-ϑ[r])+ra[r],ϑ[r]
+n (zb[NSolve[ra[r]==rb[rr],rr][[3]][[1,2]]]-za[r])>tt>ϑ[r]},
{rb[NSolve[ra[r]==rb[rr],rr][[3]][[1,2]]]
-wrb[NSolve[ra[r]==rb[rr],rr][[3]][[1,2]]](tt-ϑ[r]
-n (zb[NSolve[ra[r]==rb[rr],rr][[3]][[1,2]]]-za[r])),tt>ϑ[r]
+n (zb[NSolve[ra[r]==rb[rr],rr][[3]][[1,2]]]-za[r])}}]},
{Piecewise[{
{wz[r]tt+w[r],tt<ϑ[r]},{ v Cos[θ3](tt-ϑ[r])+za[r],ϑ[r]
+n (zb[NSolve[ra[r]==rb[rr],rr][[3]][[1,2]]]-za[r])>tt>ϑ[r]},
{zb[NSolve[ra[r]==rb[rr],rr][[3]][[1,2]]]
+wzb[NSolve[ra[r]==rb[rr],rr][[3]][[1,2]]](tt-ϑ[r]
-n (zb[NSolve[ra[r]==rb[rr],rr][[3]][[1,2]]]-za[r])),tt>ϑ[r]
+n (zb[NSolve[ra[r]==rb[rr],rr][[3]][[1,2]]]-za[r])}}],
Piecewise[{
{-(wr[r]tt+r),tt<ϑ[r]},{-(v Sin[θ3](tt-ϑ[r])+ra[r]),ϑ[r]
+n (zb[NSolve[ra[r]==rb[rr],rr][[3]][[1,2]]]-za[r])>tt>ϑ[r]},
{-(rb[NSolve[ra[r]==rb[rr],rr][[3]][[1,2]]]
-wrb[NSolve[ra[r]==rb[rr],rr][[3]][[1,2]]](tt-ϑ[r]
-n (zb[NSolve[ra[r]==rb[rr],rr][[3]][[1,2]]]-za[r]))),tt>ϑ[r]
+n (zb[NSolve[ra[r]==rb[rr],rr][[3]][[1,2]]]-za[r])}}]
}},{tt,0,zo+n τ +zi},
PlotRange→All,PlotStyle→Blue],{r,0,rmax,density}]



Show[Surface,Lens,Waves, Rays,PlotRange->All,
AspectRatio→Automatic, ImageSize→{960,Automatic}]
```

4.5 Examples

The examples computed in this section are presented in figures 4.2, 4.3, 4.4, 4.5, 4.6, and 4.7. From all the examples computed, we can confirm that equations (4.8), (4.9), (4.13), and (4.14) behave as expected. In all cases, the output wavefront is a plane, as predicted.

It is important to note that the analytical formalism described above has a restriction: for the rays to not self-intersect as they travel toward the lens, the wavefront must be sufficiently smooth and differentiable. If this restriction is not met, overlapping wavefronts will occur at some points. This mathematical restriction can be intuited from equations (4.8), (4.9), (4.13), and (4.14), where the on-axis

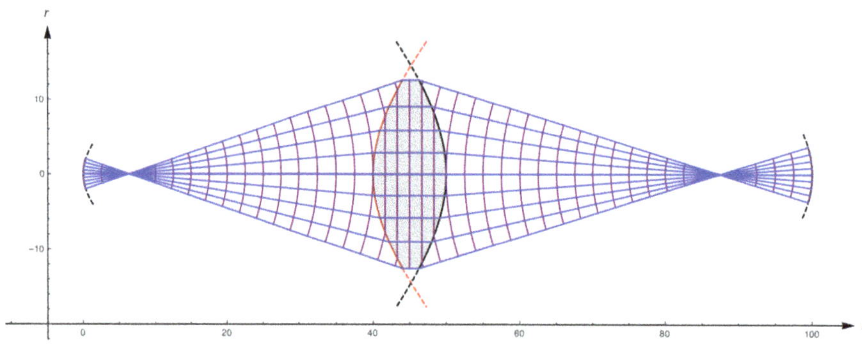

(a) A concave parabolic input wavefront

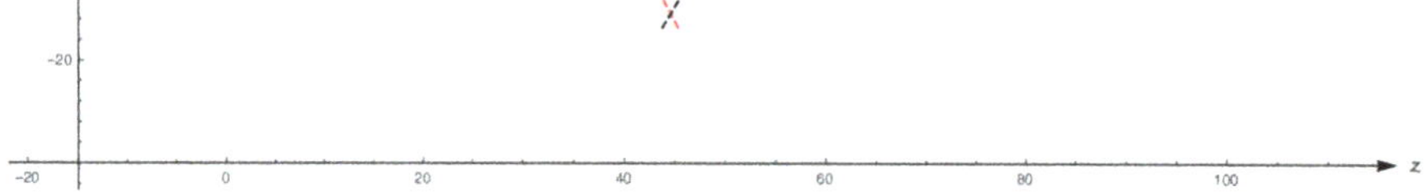

(b) A convex parabolic input wavefront

Figure 4.2. Lens design specifications for arbitrary parabolic input wavefronts; the system produces concave parabolic output wavefronts: (a) the input wavefront is $r^2/12$, the output wavefront is $r^2/24$, refractive index $n = 1.5$, central thickness $\tau = 10$ mm, object distance $z_o = 40$ mm, image distance $z_i = 50$ mm, $r_{max} = 2$ mm, (r_a, z_a) = equations (4.8) and (4.9), (r_b, z_b) = equations (4.13) and (4.14); (b) the input wavefront is $-r^2/12$, the output wavefront is $r^2/24$, refractive index $n = 1.5$, central thickness $\tau = 10$ mm, object distance $z_o = 40$ mm, image distance $z_i = 50$ mm, $r_{max} = 1.5$ mm, (r_a, z_a) = equations (4.8) and (4.9), (r_b, z_b) = equations (4.13) and (4.14).

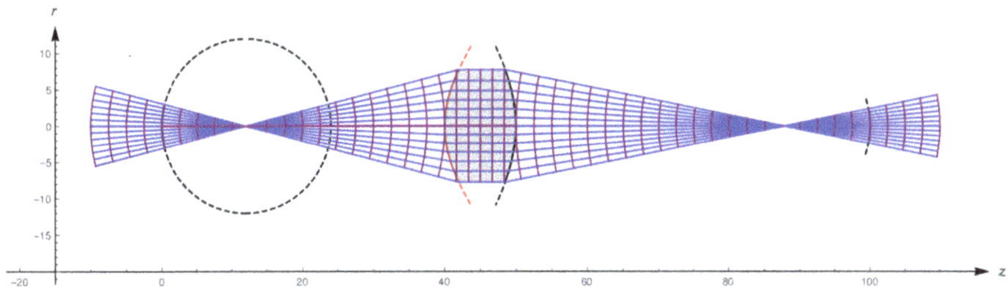

(a) A concave circular input wavefront

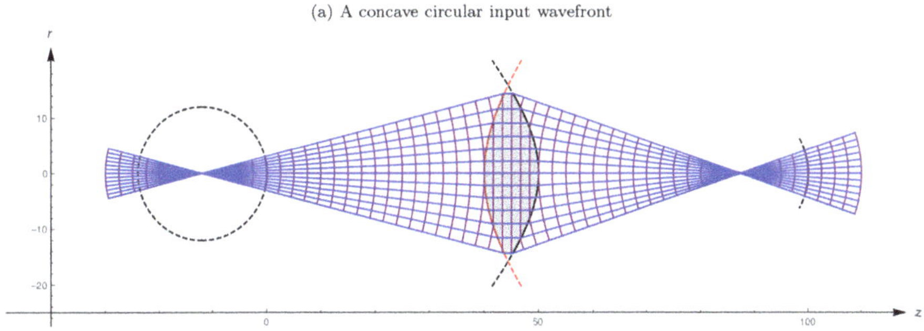

(b) A convex circular input wavefront

Figure 4.3. Lens design specifications for arbitrary circular input wavefronts; the system produces concave parabolic output wavefronts: (a) the input wavefront is $12 - \sqrt{12^2 - r^2}$, the output wavefront is $r^2/24$, refractive index $n = 1.5$, central thickness $\tau = 10$ mm, object distance $z_o = 40$ mm, image distance $z_i = 50$ mm, $\mathrm{r_{max}} = 3$ mm, (r_a, z_a) = equations (4.8) and (4.9), (r_b, z_b) = equations (4.13) and (4.14); (b) the input wavefront is $-12 + \sqrt{12^2 - r^2}$, the output wavefront is $r^2/24$, refractive index $n = 1.5$, central thickness $\tau = 10$ mm, object distance $z_o = 40$ mm, image distance $z_i = 50$ mm, $\mathrm{r_{max}} = 3$ mm, (r_a, z_a) = equations (4.8) and (4.9), (r_b, z_b) = equations (4.13) and (4.14).

ray and any other off-axis ray are compared. Neither equations (4.8), (4.9), (4.13), nor (4.14) account for the off-axis rays used in the calculation. It is impossible to fulfil this condition if two rays intersect, as all of their optical paths must be the same. For example, consider an input wavefront that is continuous and has no overlap, and that two different rays (perpendicular to the wavefront) touch each other at a single point on the refractive surface. Then, according to (4.8), (4.9), (4.13), and (4.14), the refractive surface will overlap itself, since, at a single point, it needs to collimate two different rays with two different input angles. Thus, this refractive surface does not share the same topological properties as the input wavefront. In other words, they are not homeomorphic to each other. Clearly, there are cases in which overlaps does not occur, in which each ray touches a different point on the refractive surface, and in such cases, the input wavefront is homeomorphic to the refractive surface. It is only possible to have a homeomorphism between the first and second surfaces if each ray touches a different point on each refractive surface. The first and second surfaces may overlap if this is not the

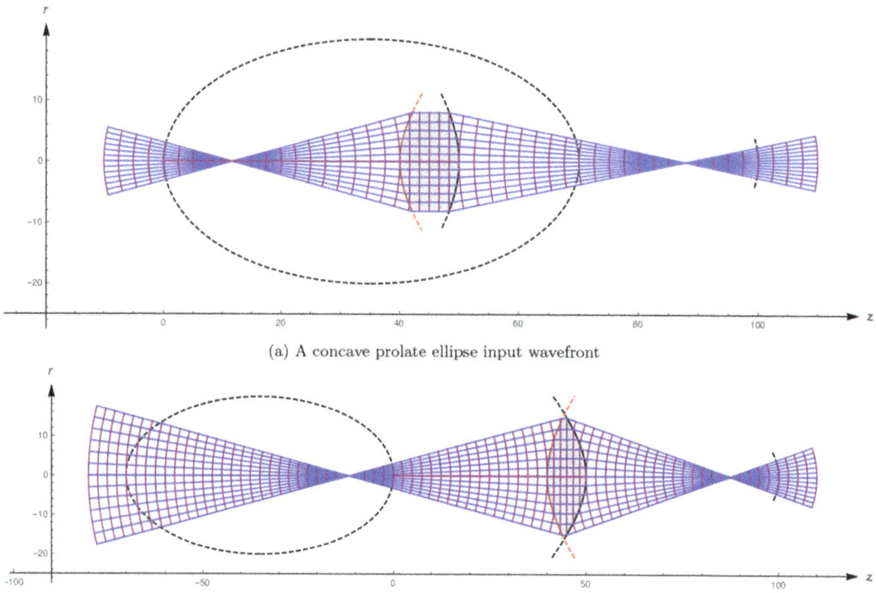

(a) A concave prolate ellipse input wavefront

(b) A convex prolate ellipse input wavefront

Figure 4.4. Lens design specifications for arbitrary prolate elliptical input wavefronts; the system produces concave parabolic output wavefronts: (a) the input wavefront is $b - \sqrt{(-a^2*b^2 + b^2r^2)/ - a^2}$, the output wavefront is $r^2/24$, refractive index $n = 1.5$, central thickness $\tau = 10$ mm, object distance $z_o = 40$ mm, image distance $z_i = 50$ mm, $r_{max} = 3$ mm, constant $a = 20$; constant $b = 35$; (r_a, z_a) = equations (4.8) and (4.9), (r_b, z_b) = equations (4.13) and (4.14); (b) the input wavefront is $-b + \sqrt{(-a^2*b^2 + b^2r^2)/ - a^2}$, the output wavefront is $r^2/24$, refractive index $n = 1.5$, central thickness $\tau = 10$ mm, object distance $z_o = 40$ mm, image distance $z_i = 50$ mm, $r_{max} = 3$ mm, constant $a = 20$; constant $b = 35$; (r_a, z_a) = equations (4.8) and (4.9), (r_b, z_b) = equations (4.13) and (4.14).

case, and then homeomorphism is lost. It follows that if this is the case, then all wavefronts in the real domain are continuous functions without overlaps, so there is also a homeomorphism between them.

4.6 End notes

According to this study, it is possible to analytically determine the shape of a lens's refractive surfaces in order to couple two arbitrary wavefronts. Observations suggest that every ray has the same optical path length from beginning to end and that Snell's law applies to each of the two surfaces. In addition, the formal approach does not rely on numerical methods that assume the paraxial regime. It is permissible to use arbitrary input/output waves, with the proviso that they need to be smooth enough for the rays not to cross each other on propagation. We tested equations (4.8), (4.9), (4.13), and (4.14) for a large number of wavefronts and conditions, and the results were always as expected.

In addition to providing a new paradigm for optical systems analysis from an analytical perspective, the results are highly significant. The work presented here

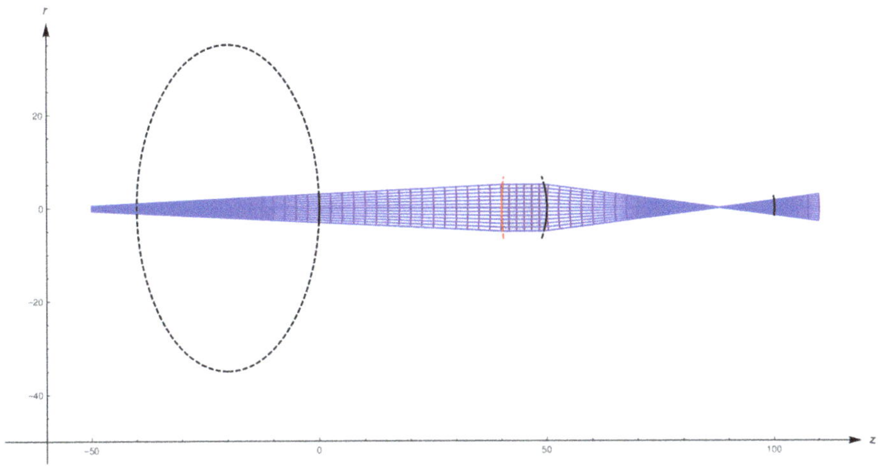

(a) A concave oblate ellipse input wavefront

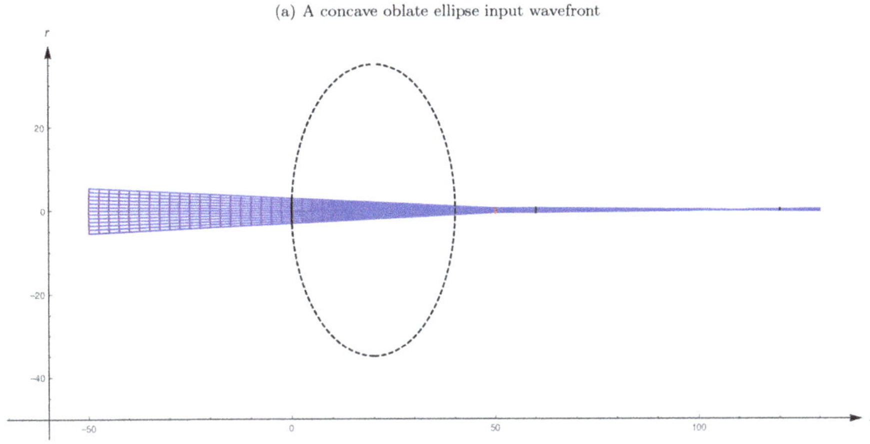

(b) A convex oblate ellipse input wavefront

Figure 4.5. Lens design specifications for arbitrary oblate elliptical input wavefronts: the system produces concave parabolic output wavefronts: (a) the input wavefront is $-b + \sqrt{(-a^2*b^2 + b^2r^2)/ - a^2}$, the output wavefront is $r^2/24$, refractive index $n = 1.5$, central thickness $\tau = 10$ mm, object distance $z_o = 40$ mm, image distance $z_i = 50$ mm, $\mathrm{r_{max}} = 3$ mm, constant $a = 35$; constant $b = 20$; (r_a, z_a) = equations (4.8) and (4.9), (r_b, z_b) = equations (4.13) and (4.14); (b) the input wavefront is $b - \sqrt{(-a^2*b^2 + b^2r^2)/ - a^2}$, the output wavefront is $r^2/24$, refractive index $n = 1.5$, central thickness $\tau = 10$ mm, object distance $z_o = 40$ mm, image distance $z_i = 50$ mm, $\mathrm{r_{max}} = 15$ mm, constant $a = 20$; constant $b = 35$; (r_a, z_a) = equations (4.8) and (4.9), (r_b, z_b) = equations (4.13) and (4.14).

relies on axially symmetric systems to simplify the investigation, but in chapter 8, we address freeform wavefronts.

A unique aspect of our analytical closed-form solution is that we can produce it efficiently. Accordingly, there is only one possible solution for the given set of input values, just like the results presented in the previous chapters. This fact can be quickly confirmed using the steps presented in the deduction of (4.8), (4.9), (4.13), and (4.14).

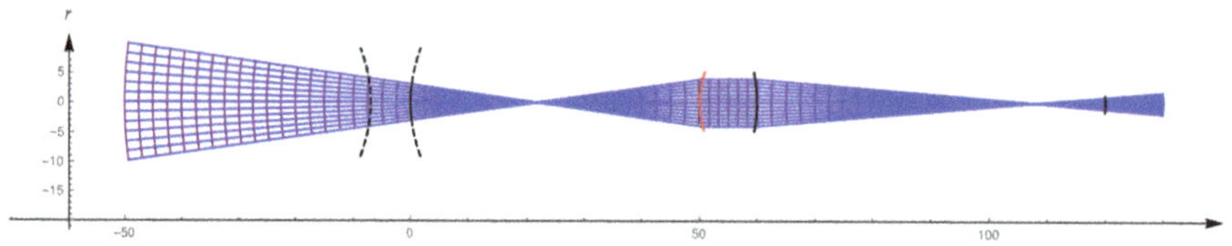

(a) A concave hyperbolic input wavefront, a concave parabolic output wavefront

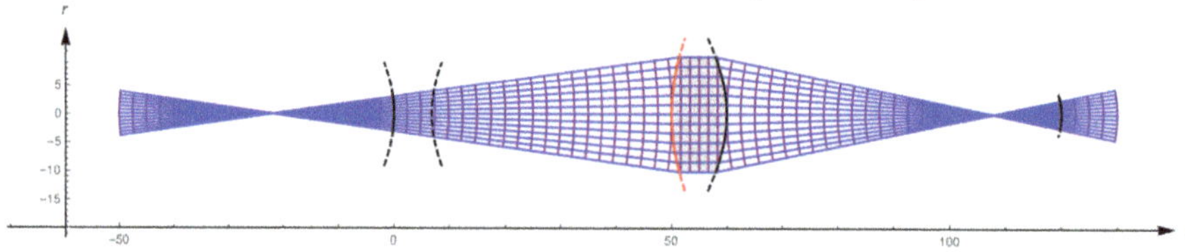

(b) A convex hyperbolic input wavefront, a concave parabolic output wavefront

Figure 4.6. Lens design specifications for arbitrary hyperbolic input wavefronts; the system produces concave parabolic output wavefronts: (a) the input wavefront is $-b + \sqrt{(-a^2*b^2 - b^2r^2)/ - a^2}$, the output wavefront is $r^2/24$, refractive index $n = 1.5$, central thickness $\tau = 10$ mm, object distance $z_o = 50$ mm, image distance $z_i = 60$ mm, $r_{max} = 3$ mm, constant $a = 16$; constant $b = 12$; $(r_a, z_a) =$ equations (4.8) and (4.9), $(r_b, z_b) =$ equations (4.13) and (4.14); (b) the input wavefront is $b + \sqrt{(-a^2*b^2 - b^2r^2)/ - a^2}$, the output wavefront is $r^2/24$, refractive index $n = 1.5$, central thickness $\tau = 10$ mm, object distance $z_o = 50$ mm, image distance $z_i = 60$ mm, $r_{max} = 3$ mm, constant $a = 16$; constant $b = 12$; $(r_a, z_a) =$ equations (4.8) and (4.9), $(r_b, z_b) =$ equations (4.13) and (4.14).

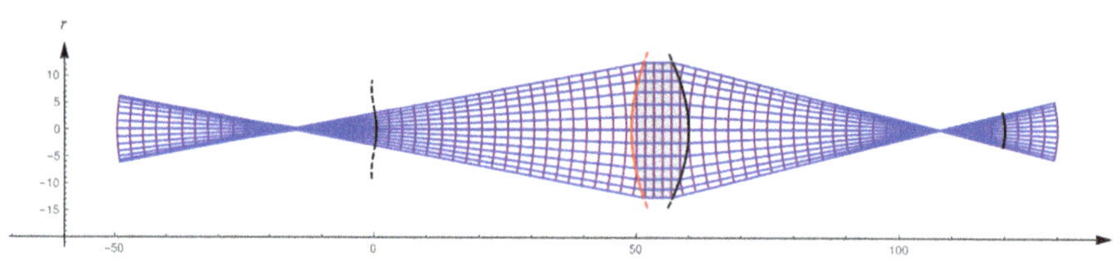

(a) A positive cosine input wavefront, a concave parabolic output wavefront

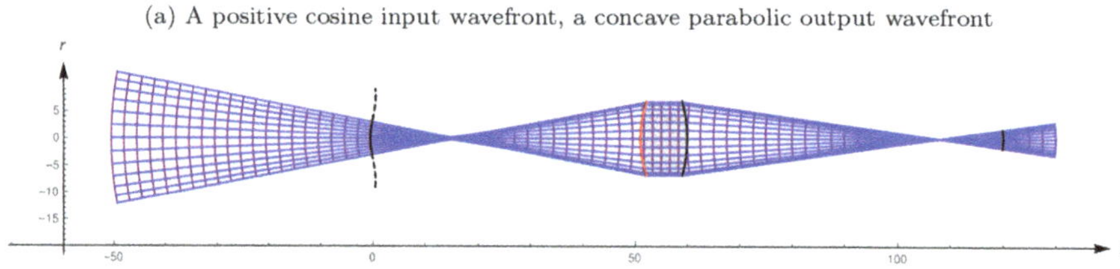

(b) A negative cosine input wavefront, a concave parabolic output wavefront

Figure 4.7. Lens design specifications for arbitrary cosine input wavefronts; the system produces concave parabolic output wavefronts: (a) the input wavefront is $\cos(2r/5)/2$, the output wavefront is $r^2/24$, refractive index $n = 1.5$, central thickness $\tau = 10$ mm, object distance $z_o = 50$ mm, image distance $z_i = 60$ mm, $r_{max} = 3$ mm, $(r_a, z_a) =$ equations (4.8) and (4.9), $(r_b, z_b) =$ equations (4.13) and (4.14); (b) the input wavefront is $-\cos(2r/5)/2$, the output wavefront is $r^2/24$, refractive index $n = 1.5$, central thickness $\tau = 10$ mm, object distance $z_o = 50$ mm, image distance $z_i = 60$ mm, $r_{max} = 3$ mm, $(r_a, z_a) =$ equations (4.8) and (4.9), $(r_b, z_b) =$ equations (4.13) and (4.14).

Chapter 5

General equation for the aspheric wavefront generator lens

We present a lens that refracts a given arbitrary input wavefront to a single image point. The aforementioned lens is designed using a fully analytical closed-form equation. This equation has been tested for several scenarios, and its performance is as expected. These lenses can be seen as null lenses.

5.1 Introduction

In chapter 1 we described the Fizeau interferometer. A Fizeau interferometer is an interferometric arrangement whereby two reflecting surfaces are placed facing each other (see figure 1.17). Recall that interferometry is a family of techniques that consist of combining light from different receivers, telescopes, or radio antennas to obtain a higher-resolution image by applying the superposition principle. It is used in multiple sciences and technologies, such as astronomy, optical metrology, etc. but it can also be used to test optical elements such as mirrors or lenses. Fizeau interferometers can be used to test such lenses; the interferometer's light source is typically a laser beam, which passes through the interferometer and is refracted by the lens (see figure 1.17) and then reflected by the mirror. Finally, the light travels in the opposite direction and the interferometric pattern is displayed at the output of the Fizeau interferometer.

If the lens and the mirror are such that the optical path lengths (OPLs) of all the rays are the same then no fringe pattern forms at the output of the Fizeau interferometer. On the other hand, if the OPLs of the rays are not all the same, then the optical path difference (OPD) is non-zero, and nonuniform fringes appear at the output of the Fizeau interferometer. These fringes are directly related to the OPD, which in turn is related to the phase difference. When lenses/mirrors are tested, a Fizeau interferometer can be very useful to see whether they have the shape that they supposed to have, since a mirror/lens pair that produces a uniform fringe at

doi:10.1088/978-0-7503-4705-1ch5

the output means that the rays striking the mirror have an OPD of zero. Such an array is commonly known as a null lens. If null lenses or null mirror have surface defects then we get nonuniform fringe patterns at the output (see figure 1.18). Note that in figure 1.18 the null lens has a spherical input wavefront, and the output wavefront is such that when the light strikes the mirror, the normal vectors of the mirror surface are parallel to the striking rays.

In this chapter, we are going to obtain the general equation for the aspheric wavefront generator lens using just Fermat's principle; these lenses can potentially be used as null lenses. Here, 'the general equation for the aspheric wavefront generator lens' means the equation that describes the shape of a wavefront generator lens that, given a wavefront, transforms it into a spherical wavefront using refraction alone. The wavefront can be produced after reflection in a test mirror, just as in the Fizeau interferometer. We will obtain the general equation for the aspheric wavefront generator lens without using paraxial approximations or iterations, by following the same approach that we have been using since chapter 2.

5.2 Mathematical model for adaptive optics for finite images

Figure 5.1 displays the diagram of the problem presented in the introduction. The goal is to design a lens such that, given an input wavefront, the light coming from the wavefront converges into a single image point. This wavefront can be seen as the wavefront of the light returning from the mirror under test, i.e. the wavefront as reflected by the mirror. Note that we can do this because the equations that we are going to propose are only based on Fermat's principle and are stigmatic: all the rays have the same OPL.

The wavefront is expressed by the function w, where r is the only independent variable in the model. The first refractive surface of the lens is given by (z_a, r_a), where z_a is the sagitta of the surface, and r_a is the radial component. The distance from the vertex of the wavefront w to the first refractive surface's vertex is z_o. The vertex of this first surface is located at the origin of the coordinate system. The central thickness of the lens is τ. The second surface is denoted by (z_b, r_b), where z_b is the sagitta of the surface, and r_b is the radial component. We consider that the first surface is also given. Thus, the only unknowns of the model are z_b and r_b. The distance from the second surface's vertex to the image point is z_i. We consider that

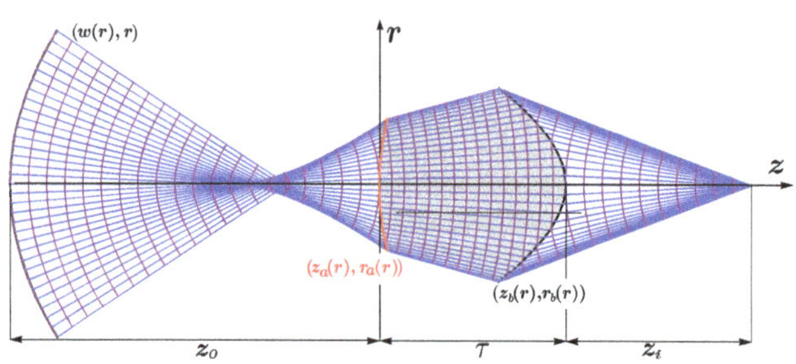

Figure 5.1. Basic configuration of a lens whose optical properties allow stigmatic images to be obtained given a prescribed input wavefront.

the refractive index is constant in the input wavefront space and is given by n_o. The refractive index inside the lens is n. Finally, in the image space, the refractive index (which is also constant) is n_i.

5.2.1 The first surface in terms of the input rays

Our strategy is to express all the wavefronts and surfaces in terms of r, which is the only independent variable of the model. The input wavefront (which, without loss of generality, can be seen as the wavefront reflected by the mirror) is a function of r, so the cosine directors are given by

$$w_r = -\frac{w'}{\sqrt{w'^2 + 1}}, \qquad w_z = \frac{1}{\sqrt{w'^2 + 1}}, \tag{5.1}$$

where $w' \equiv dw/dr$ is the derivative of w with respect to r, and w_r and w_z are the directions of the cosines in the r and z directions, respectively. Note that these cosine directors are along the normal; the result is predicted by Fermat's principle and, as we have previously mentioned, by the Dupin–Malus theorem. So, we can express the first surface in terms of the cosine directors of the input rays as follows:

$$r_a = r + w_r\vartheta_a, \qquad z_a = w + w_z\vartheta_a, \tag{5.2}$$

where ϑ_a is the distance that each ray travels from the input wavefront to the first surface of the lens. Also, ϑ_a depends on the shape of the first surface. Since the first surface is given by the user, there is no general expression for ϑ_a.

5.2.2 About the parameter ϑ_a

In this section, we focus on ϑ_a. We recall that ϑ_a is the distance each input ray needs to travel from the wavefront to the first surface. We can always solve for ϑ_a, given w and the sagitta of the first surface. The sagitta of the first surface can be expressed by

$$z_a(r_a) = w + w_z\vartheta_a. \tag{5.3}$$

By substituting equation (5.2) into the last expression, we obtain

$$z_a(r + w_r\vartheta_a) = w + w_z\vartheta_a, \tag{5.4}$$

and now we can finally solve for ϑ_a. So, for example, if we set the first surface to be a parabola with focus $-f_a$, we can use equation (5.4) to obtain ϑ_a:

$$-\frac{(w_r\vartheta_a + r)^2}{4f_a} = w + w_z\vartheta_a. \tag{5.5}$$

Finally, we obtain ϑ_a by solving equation (5.5):

$$\vartheta_a = -\frac{-2\sqrt{f_a}\sqrt{f_a w_z^2 - ww_r^2 + rw_rw_z} + 2f_a w_z + rw_r}{w_r^2}. \tag{5.6}$$

We can apply a similar approach for other curves that model the first surface.

5.2.3 The second surface in terms of the input rays

For the next step in the design of the desired lens, we apply Snell's law at the first surface:

$$\boxed{\vec{\mathbf{v}}_2 = \frac{n_o}{n_i}[\vec{\mathbf{v}}_1 - (\vec{\mathbf{n}}_o \cdot \vec{\mathbf{v}}_1)\vec{\mathbf{n}}_o] - \vec{\mathbf{n}}_o\sqrt{1 - \frac{n_o^2}{n_i^2}(\vec{\mathbf{n}}_o \times \vec{\mathbf{v}}_1)^2} \quad \text{for } \vec{\mathbf{v}}_2, \vec{\mathbf{v}}_1, \vec{\mathbf{n}}_o \in \mathbb{R}^2,} \tag{5.7}$$

where $\vec{\mathbf{v}}_1$ is the unit vector of the incident ray, $\vec{\mathbf{v}}_2$ is unit vector of the refracted ray, and $\vec{\mathbf{n}}_o$ is the normal vector of the first surface (see figure 5.1).

$$\vec{\mathbf{v}}_1 = \frac{-w'\vec{\mathbf{e}}_1 + \vec{\mathbf{e}}_2}{\sqrt{w'^2 + 1}}, \qquad \vec{\mathbf{v}}_2 = \wp_r\vec{\mathbf{e}}_1 + \wp_z\vec{\mathbf{e}}_2, \qquad \vec{\mathbf{n}}_o = \frac{z_a'\vec{\mathbf{e}}_1 - r_a'\vec{\mathbf{e}}_2}{\sqrt{r_a' + z_a'^2}}, \tag{5.8}$$

where $\vec{\mathbf{e}}_1$ denotes the r direction and $\vec{\mathbf{e}}_2$ denotes the z direction. We need to replace the unit vectors in equation (5.7). We begin with $(\vec{\mathbf{n}}_o \times \vec{\mathbf{v}}_1)^2$:

$$(\vec{\mathbf{n}}_o \times \vec{\mathbf{v}}_1) = \frac{z_a' - r_a'w'}{\sqrt{w'^2 + 1}\sqrt{r_a'^2 + z_a'^2}}(\vec{\mathbf{e}}_1 \times \vec{\mathbf{e}}_2), \tag{5.9}$$

so

$$(\vec{\mathbf{n}}_o \times \vec{\mathbf{v}}_1)^2 = \frac{(z_a' + r_a'w')^2}{(w'^2 + 1)(r_a'^2 + z_a'^2)}. \tag{5.10}$$

The square root in equation (5.7) can now be expressed as

$$-\vec{\mathbf{n}}_o\sqrt{1 + \frac{n_o^2}{n_i^2}(\vec{\mathbf{n}}_o \times \vec{\mathbf{v}}_1)^2} = -\frac{(z_a'\vec{\mathbf{e}}_1 - \vec{\mathbf{e}}_2)}{\sqrt{1 + z_a'^2}}\sqrt{1 - \frac{n_o^2}{n_i^2}\frac{(z_a' - r_a'w')^2}{(w'^2 + 1)(r_a'^2 + z_a'^2)}}. \tag{5.11}$$

We now turn to $\vec{\mathbf{v}}_1 - (\vec{\mathbf{n}}_o \cdot \vec{\mathbf{v}}_1)\vec{\mathbf{n}}_o$ in equation (5.7):

$$\vec{\mathbf{v}}_1 - (\vec{\mathbf{n}}_o \cdot \vec{\mathbf{v}}_1)\vec{\mathbf{n}}_o = \frac{-w'\vec{\mathbf{e}}_1 + \vec{\mathbf{e}}_2}{\sqrt{w'^2 + 1}} - \left[-\frac{r_a' + w'z_a'}{\sqrt{w'^2 + 1}\sqrt{r_a'^2 + z_a'^2}}\right]\frac{(z_a'\vec{\mathbf{e}}_1 - r_a'\vec{\mathbf{e}}_2)}{\sqrt{r_a'^2 + z_a'^2}}. \tag{5.12}$$

By simplifying and multiplying by $\frac{n_o}{n_i}$, we obtain

$$\frac{n_o}{n_i}[\vec{\mathbf{v}}_1 - (\vec{\mathbf{n}}_o \cdot \vec{\mathbf{v}}_1)\vec{\mathbf{n}}_o] = \frac{n_o}{n_i}\frac{r_a'(z_a' - r_a'w')}{\sqrt{w'^2 + 1}\,(r_a'^2 + z_a'^2)}\vec{\mathbf{e}}_1 + \frac{n_o}{n_i}\frac{z_a'(z_a' - r_a'w')}{\sqrt{w'^2 + 1}\,(r_a'^2 + z_a'^2)}\vec{\mathbf{e}}_2. \tag{5.13}$$

By summing equations (5.11) and (5.13), and separating the Cartesian components, we get the cosine directors of the rays inside the lens:

$$\wp_r = \frac{n_o r_a'(z_a' - r_a'w')}{n_i\sqrt{w'^2 + 1}\,(r_a'^2 + z_a'^2)} + \frac{z_a'\sqrt{1 - \dfrac{n_o^2(z_a' - r_a'w')^2}{n_i^2(w'^2 + 1)(r_a'^2 + z_a'^2)}}}{\sqrt{r_a'^2 + z_a'^2}} \tag{5.14}$$

and

$$\wp_z = \frac{n_o z_a'(z_a' - r_a'w')}{n_i\sqrt{w'^2 + 1}\,(r_a'^2 + z_a'^2)} - \frac{r_a'\sqrt{1 - \dfrac{n_o^2(z_a' - r_a'w')^2}{n_i^2(w'^2 + 1)(r_a'^2 + z_a'^2)}}}{\sqrt{r_a'^2 + z_a'^2}}. \tag{5.15}$$

We can now express the second surface in terms of these cosine directors, without loss of generality, the following relationship holds:

$$\frac{r_b - r_a}{\sqrt{(z_b - z_a)^2 + (r_b - r_a)^2}} = \frac{n_o r_a'(z_a' - r_a'w')}{n_i\sqrt{w'^2 + 1}\,(r_a'^2 + z_a'^2)} + \frac{z_a'\sqrt{1 - \dfrac{n_o^2(z_a' - r_a'w')^2}{n_i^2(w'^2 + 1)(r_a'^2 + z_a'^2)}}}{\sqrt{r_a'^2 + z_a'^2}} \tag{5.16}$$

and

$$\frac{z_b - z_a}{\sqrt{(z_b - z_a)^2 + (r_b - r_a)^2}} = \frac{n_o z_a'(z_a' - r_a'w')}{n_i\sqrt{w'^2 + 1}\,(r_a'^2 + z_a'^2)} - \frac{r_a'\sqrt{1 - \dfrac{n_o^2(z_a' - r_a'w')^2}{n_i^2(w'^2 + 1)(r_a'^2 + z_a'^2)}}}{\sqrt{r_a'^2 + z_a'^2}}. \tag{5.17}$$

Since, in fact, the right-hand sides of equations (5.16) and (5.17) are the cosine directors, therefore

$$\frac{r_b - r_a}{\sqrt{(z_b - z_a)^2 + (r_b - r_a)^2}} = \wp_r \tag{5.18}$$

and

$$\frac{z_b - z_a}{\sqrt{(z_b - z_a)^2 + (r_b - r_a)^2}} = \wp_z. \tag{5.19}$$

We can now simplify the left-hand sides of equations (5.18) and (5.19) using the parameter ϑ:

$$\vartheta \equiv \sqrt{(z_b - z_a)^2 + (r_b - r_a)^2}\,. \tag{5.20}$$

Let ϑ be the distance travelled by the ray inside the lens. Substituting ϑ into (5.18) and (5.19) yields

$$\frac{z_b - z_a}{\vartheta} = \wp_z \tag{5.21}$$

and

$$\frac{r_b - r_a}{\vartheta} = \wp_r. \tag{5.22}$$

Using the previous equations, we can simply solve for z_b and r_b,

$$r_b = r_a + \vartheta \wp_r \tag{5.23}$$

and

$$z_b = z_a + \vartheta \wp_z. \tag{5.24}$$

The above expression is the solution we were aiming for. It describes, point by point, the second surface of the lens that is stigmatic given an arbitrary input wavefront w. The only problem in this equation is that we do not know the length of ϑ. However, we can find the solution by combining this result with Fermat's principle. We will do this in the next section!

A final remark about $\wp_r$ and $\wp_z$: r is the only independent variable used. This may not be clear in equations (5.14) and (5.15), but if we substitute them into equations (5.1) and (5.2), the relation to r is confirmed. Thus, for the vertical direction we find that equation (5.14) is equal to

$$\begin{aligned}
\wp_r = &\left\{ n\sqrt{w'^2+1}\left[\left(-\frac{w'\vartheta a'}{\sqrt{w'^2+1}} + \frac{\vartheta a w'^2 w''}{\left(w'^2+1\right)^{3/2}} - \frac{\vartheta a w''}{\sqrt{w'^2+1}} + 1\right)^2\right.\right.\\
&\left.\left.+\left(\frac{\vartheta a'}{\sqrt{w'^2+1}} + w' - \frac{\vartheta a w' w''}{\left(w'^2+1\right)^{3/2}}\right)^2\right]^{-1}\right\} \times \left\{ n_0 \left(-\frac{w'\vartheta a'}{\sqrt{w'^2+1}} + \frac{\vartheta a w'^2 w''}{\left(w'^2+1\right)^{3/2}} - \frac{\vartheta a w''}{\sqrt{w'^2+1}} + 1\right)\right.\\
&\left.\left(\frac{\vartheta a'}{\sqrt{w'^2+1}} + w' + w'\left(\frac{w'\vartheta a'}{\sqrt{w'^2+1}} - \frac{\vartheta a w'^2 w''}{\left(w'^2+1\right)^{3/2}} + \frac{\vartheta a w''}{\sqrt{w'^2+1}} - 1\right) - \frac{\vartheta a w' w''}{\left(w'^2+1\right)^{3/2}}\right)\right\}\\
&+\left\{\sqrt{\left(-\frac{w'\vartheta a'}{\sqrt{w'^2+1}} + \frac{\vartheta a w'^2 w''}{\left(w'^2+1\right)^{3/2}} - \frac{\vartheta a w''}{\sqrt{w'^2+1}} + 1\right)^2 + \left(\frac{\vartheta a'}{\sqrt{w'^2+1}} + w' - \frac{\vartheta a w' w''}{\left(w'^2+1\right)^{3/2}}\right)^2}\right\}^{-1}\\
&\times\left[\frac{\vartheta a'}{\sqrt{w'^2+1}} + w' - \frac{\vartheta a w' w''}{\left(w'^2+1\right)^{3/2}}\right]\\
&\times\sqrt{1 - \frac{n_0^2\left(\frac{\vartheta a'}{\sqrt{w'^2+1}} + w' + w'\left(\frac{w'\vartheta a'}{\sqrt{w'^2+1}} - \frac{\vartheta a w'^2 w''}{\left(w'^2+1\right)^{3/2}} + \frac{\vartheta a w''}{\sqrt{w'^2+1}} - 1\right) - \frac{\vartheta a w' w''}{\left(w'^2+1\right)^{3/2}}\right)^2}{n^2\left(w'^2+1\right)\left(\left(-\frac{w'\vartheta a'}{\sqrt{w'^2+1}} + \frac{\vartheta a w'^2 w''}{\left(w'^2+1\right)^{3/2}} - \frac{\vartheta a w''}{\sqrt{w'^2+1}} + 1\right)^2 + \left(\frac{\vartheta a'}{\sqrt{w'^2+1}} + w' - \frac{\vartheta a w' w''}{\left(w'^2+1\right)^{3/2}}\right)^2\right)}},
\end{aligned} \tag{5.25}$$

and for the horizontal direction, we find that equation (5.15) is equal to

$$
\begin{aligned}
\wp_z = {} & \left\{ n\sqrt{w'^2+1}\left[\left(-\frac{w'\vartheta_a'}{\sqrt{w'^2+1}} + \frac{\vartheta_a w'^2 w''}{\left(w'^2+1\right)^{3/2}} - \frac{\vartheta_a w''}{\sqrt{w'^2+1}} + 1\right)^2 \right.\right. \\
& \left.\left. + \left(\frac{\vartheta_a'}{\sqrt{w'^2+1}} + w' - \frac{\vartheta_a w' w''}{\left(w'^2+1\right)^{3/2}}\right)^2\right]^{-1}\right\} \times \left\{ n_o\left(\frac{\vartheta_a'}{\sqrt{w'^2+1}} + w' - \frac{\vartheta_a w' w''}{\left(w'^2+1\right)^{3/2}}\right)\right. \\
& \left.\left(\frac{\vartheta_a'}{\sqrt{w'^2+1}} + w' + w'\left(\frac{w'\vartheta_a'}{\sqrt{w'^2+1}} - \frac{\vartheta_a w'^2 w''}{\left(w'^2+1\right)^{3/2}} + \frac{\vartheta_a w''}{\sqrt{w'^2+1}} - 1\right) - \frac{\vartheta_a w' w''}{\left(w'^2+1\right)^{3/2}}\right)\right\} \\
& + \left\{\sqrt{\left(-\frac{w'\vartheta_a'}{\sqrt{w'^2+1}} + \frac{\vartheta_a w'^2 w''}{\left(w'^2+1\right)^{3/2}} - \frac{\vartheta_a w''}{\sqrt{w'^2+1}} + 1\right)^2 + \left(\frac{\vartheta_a'}{\sqrt{w'^2+1}} + w' - \frac{\vartheta_a w' w''}{\left(w'^2+1\right)^{3/2}}\right)^2}\right\}^{-1} \\
& \times \left[-\frac{w'\vartheta_a'}{\sqrt{w'^2+1}} + \frac{\vartheta_a w'^2 w''}{\left(w'^2+1\right)^{3/2}} - \frac{\vartheta_a w''}{\sqrt{w'^2+1}} + 1\right] \\
& \times \sqrt{1 - \frac{n_o^2\left(\frac{\vartheta_a'}{\sqrt{w'^2+1}} + w' + w'\left(\frac{w'\vartheta_a'}{\sqrt{w'^2+1}} - \frac{\vartheta_a w'^2 w''}{\left(w'^2+1\right)^{3/2}} + \frac{\vartheta_a w''}{\sqrt{w'^2+1}} - 1\right) - \frac{\vartheta_a w' w''}{\left(w'^2+1\right)^{3/2}}\right)^2}{n^2\left(w'^2+1\right)\left(\left(-\frac{w'\vartheta_a'}{\sqrt{w'^2+1}} + \frac{\vartheta_a w'^2 w''}{\left(w'^2+1\right)^{3/2}} - \frac{\vartheta_a w''}{\sqrt{w'^2+1}} + 1\right)^2 + \left(\frac{\vartheta_a'}{\sqrt{w'^2+1}} + w' - \frac{\vartheta_a w' w''}{\left(w'^2+1\right)^{3/2}}\right)^2\right)}}.
\end{aligned}
\tag{5.26}
$$

Remember that $\wp_r$ and $\wp_z$ are cosine directors that always hold for $\wp_r^2 + \wp_z^2 = 1$. Therefore, equations (5.25) and (5.26) hold for $\wp_r^2 + \wp_z^2 = 1$ as well as equations (5.14) and (5.15).

5.2.4 The optical path and the solution

In the last section, we learned that ϑ needs to be expressed in terms of the unknowns z_b and r_b. We can express it in terms of the known parameters by comparing the axial and non-axial rays' OPLs using Fermat's principle. Of course, we can do this because we have expressed the first and second surfaces in terms of the input rays and the refracted rays, respectively. So, we set the OPL of the non-axial ray as follows:

$$
\begin{aligned}
\mathrm{OPL} = {} & n\sqrt{(z_a - z_b)^2 + (r_a - r_b)^2} + n_o\sqrt{(z_a - w)^2 + (r_a - r)^2} \\
& + n_i\sqrt{(z_b - z_i - t)^2 + r_b{}^2},
\end{aligned}
\tag{5.27}
$$

and for the axial ray, we define

$$\mathrm{OPL} = -n_o z_o + n\tau + n_i z_i. \tag{5.28}$$

Equating both OPLs, we obtain

$$\begin{aligned} -n_o z_o + n\tau + n_i z_i &= n\sqrt{(z_a - z_b)^2 + (r_a - r_b)^2} \\ &\quad + n_o\sqrt{(z_a - w)^2 + (r_a - r)^2} + n_i\sqrt{(z_b - z_i - \tau)^2 + r_b{}^2}. \end{aligned} \tag{5.29}$$

The last expression is the equation we need in order to obtain ϑ, so we substitute equations (5.20), (5.23), and (5.24) into it:

$$\begin{aligned} -n_o z_o + n\tau + n_i z_i = n\vartheta + {} & n_o\sqrt{(z_a - w)^2 + (r_a - r)^2} \\ & + n_i\sqrt{(z_a + \vartheta\wp_z - z_i - \tau)^2 + (r_a + \vartheta\wp_r)^2}. \end{aligned} \tag{5.30}$$

To reduce the size of equation (5.30), we define the following parameter:

$$f \equiv -n_o z_o + n\tau + n_i z_i - n_o\sqrt{(z_a - w)^2 + (r_a - r)^2}; \tag{5.31}$$

we then replace f in equation (5.30):

$$f = n\vartheta + n_i\sqrt{(z_a + \vartheta\wp_z - z_i - \tau)^2 + (r_a + \vartheta\wp_r)^2}, \tag{5.32}$$

and by defining ζ, we reduce the size of the previous expression:

$$\zeta = z_a - z_i - \tau. \tag{5.33}$$

Substituting equation (5.33) into equation (5.32) yields

$$f = n\vartheta + n_i\sqrt{(\vartheta\wp_z + \zeta)^2 + (r_a + \vartheta\wp_r)^2}. \tag{5.34}$$

Squaring both sides gives

$$(f - n\vartheta)^2 = n_i^2[(\vartheta\wp_z + \zeta)^2 + (r_a + \vartheta\wp_r)^2]. \tag{5.35}$$

Expanding both sides results in

$$f^2 - 2nf\vartheta + n^2\vartheta^2 = n_i^2[\vartheta^2\wp_z^2 + 2\vartheta\wp_z\zeta + \zeta^2 + r_a^2 + 2r_a\vartheta\wp_r + \vartheta^2\wp_r^2]. \tag{5.36}$$

By reordering the terms, we obtain

$$(n^2 - n_i^2)\vartheta^2 - 2\vartheta[\wp_z\zeta n_i^2 + \wp_r r_a n_i^2 + nf] + f^2 - n_i^2\zeta^2 - n_i^2 r_a^2 = 0, \tag{5.37}$$

and finally, solving for ϑ gives

$$\begin{aligned} \vartheta = {} & \frac{nf + n_i^2\zeta\wp_z + n_i^2 r_a\wp_r}{(n^2 + n_i^2)} \\ & \frac{-\sqrt{\left(-nf - n_i^2\zeta\wp_z - n_i^2 r_a\wp_r\right)^2 - \left(n^2 + n_i^2\right)\left(f^2 - n_i^2\zeta^2 - n_i^2 r_a^2\right)}}{\left(n^2 + n_i^2\right)}. \end{aligned} \tag{5.38}$$

If we use equations (5.24), (5.23), and (5.38) for ϑ, we can obtain the second surface in terms of the given parameters,

$$r_b = r_a + \left\{ \frac{nf + n_i^2 \zeta \wp_z + n_i^2 r_a \wp_r}{\left(n^2 + n_i^2\right)} - \frac{\sqrt{\left(-nf - n_i^2 \zeta \wp_z - n_i^2 r_a \wp_r\right)^2 - \left(n^2 + n_i^2\right)\left(f^2 - n_i^2 \zeta^2 - n_i^2 r_a^2\right)}}{\left(n^2 + n_i^2\right)} \right\} \times \left\{ \frac{n_o r_a' (z_a' - r_a' w')}{n_i \sqrt{w'^2 + 1}\,(r_a'^2 + z_a'^2)} + \frac{z_a' \sqrt{1 - \frac{n_o^2 (z_a' - r_a' w')^2}{n_i^2 (w'^2 + 1)(r_a'^2 + z_a'^2)}}}{\sqrt{r_a'^2 + z_a'^2}} \right\} \tag{5.39}$$

and

$$z_b = z_a + \left\{ \frac{nf + n_i^2 \zeta \wp_z + n_i^2 r_a \wp_r}{\left(n^2 + n_i^2\right)} - \frac{\sqrt{\left(-nf - n_i^2 \zeta \wp_z - n_i^2 r_a \wp_r\right)^2 - \left(n^2 + n_i^2\right)\left(f^2 - n_i^2 \zeta^2 - n_i^2 r_a^2\right)}}{\left(n^2 + n_i^2\right)} \right\} \times \left\{ \frac{n_o z_a' (z_a' - r_a' w')}{n_i \sqrt{w'^2 + 1}\,(r_a'^2 + z_a'^2)} - \frac{r_a' \sqrt{1 - \frac{n_o^2 (z_a' - r_a' w')^2}{n_i^2 (w'^2 + 1)(r_a'^2 + z_a'^2)}}}{\sqrt{r_a'^2 + z_a'^2}} \right\}. \tag{5.40}$$

Once again, note that we have not used any approximation to obtain equations (5.39) and (5.40). In fact equations (5.39) and (5.40) are analytic closed-form solutions to the proposed problem. Equations (5.39) and (5.40) are just further natural extensions of Fermat's principle. We used neither paraxial approximation, nor iteration, nor optimization, and no iteration was allowed. From beginning to end, in the derivation of equations (5.39) and (5.40) we just used Fermat's principle; equations (5.39) and (5.40) were predicted by the optical path theory. Remember that Fermat's principle is the axiom and equations (5.39) and (5.40) are theorems dictated by the fundamental axiom of the optical path theory. It is also important to remark that for a given set of input values, there is only one solution, which is given by equations (5.39) and (5.40). So again, we have presented a uniqueness theorem. The optical path theory follows a single deterministic path of light.

5.3 The wavefront generator lens for images at finite distances

To get the eikonal of the aspheric wavefront generator lens, we need the cosine directors at the input, at the lens, and at the output. The first two sets of cosine directors are given by equations (5.1), (5.14), and (5.15). The cosine directors at the output can be computed using Snell's law. But we can also define them, without loss of generality, using the angle θ:

$$\theta = \arctan\left[\frac{r_b}{z_b - \tau - z_o}\right], \tag{5.41}$$

where θ is the angle subtended by the output ray and the optical axis (see figure 5.1). With this angle in mind, and using equations (5.1), (5.14), and (5.15), it is easy to construct the eikonal $H(t, r_a)$ that follows Fermat's principle:

$$H(t, r_a) = (k_r(t, r_a), k_z(t, r_a)). \tag{5.42}$$

Taking $n_o = n_i = 1$, we obtain

$$\boxed{H(t, r_a) = \begin{cases} (z_o + tw_z, tw_r), & t < \vartheta_a, \\ (v(t - \vartheta_a)\wp_z, v(t - \vartheta_a)\wp_r), & \vartheta_a < t < \vartheta_a + n\vartheta, \\ (z_b + (t - \vartheta_a + n\vartheta)\cos\theta, r_b + (t - \vartheta_a + n\vartheta)\sin\theta), & t > \vartheta_a + n\vartheta. \end{cases}} \tag{5.43}$$

where $v = c/n$ is the speed of light inside the lens, c is the speed of light in vacuum (outside the lens $n_o = n_i = 1$), ϑ_a is the condition that separates the eikonals of the first and second regions, and $\vartheta_a + n\vartheta$ is the condition that separates the eikonals of the second and third regions.

5.4 Mathematica code

This section presents the `Mathematica` code for the aspheric wavefront generator lens based on equations (5.39), (5.40), and (5.43).

```
In[6]:=  w[r_]:=R-√R^2-r^2+zo

         ϑa[r_]:= (-r wr[r]+R wz[r]-w[r] wz[r])
                  / (-√(R² wz[r]²+2 R wr[r] (w[r] wr[r]-r wz[r])-(w[r] wr[r]-r wz[r])²))

         wr[r_]:=- w'[r] / √(w'[r]²+1)

         wz[r_]:= 1 / √(w'[r]²+1)

         ra[r_]:=r+ϑa[r]wr[r]

         za[r_]:=w[r]+ϑa[r]wz[r]

         ℘r[r_]:= (no ra'[r] (-ra'[r] w'[r]+za'[r])) / (n √(1+w'[r]²) (ra'[r]²+za'[r]²))
```

$$+\frac{za'[r]}{\sqrt{ra'[r]^2+za'[r]^2}}\sqrt{(1-\frac{no^2\ (-ra'[r]\ w'[r]+za'[r])^2}{n^2\ (1+w'[r]^2)\ (ra'[r]^2+za'[r]^2)})}$$

$$\varrho z[r\_]:=\frac{no\ za'[r]\ (-ra'[r]\ w'[r]+za'[r])}{n\ \sqrt{1+w'[r]^2}\ (ra'[r]^2+za'[r]^2)} -\frac{ra'[r]}{\sqrt{ra'[r]^2+za'[r]^2}}\sqrt{(1-\frac{no^2\ (-ra'[r]\ w'[r]+za'[r])^2}{n^2\ (1+w'[r]^2)\ (ra'[r]^2+za'[r]^2)})}$$

$$f[r\_]:=-zo\ no+\tau\ n+zi\ ni-no\sqrt{(ra[r]-r)^2+(za[r]-w[r])^2}$$

$$zt[r\_]:=za[r]-\tau-zi$$

$$\vartheta[r\_]:=\frac{1}{n^2-ni^2}(n\ f[r]+ni^2\ (ra[r]\ \varrho r[r]+zt[r]\ \varrho z[r]) -\sqrt{(ni^2\ (f[r]^2\ \varrho r[r]^2+(zt[r]\ (n-ni\ \varrho r[r])+f[r]\ \varrho z[r])\ (zt[r]\ (n+ni\ \varrho r[r])+f[r]\ \varrho z[r]) +2\ ra[r]\ \varrho r[r]\ (n\ f[r]+ni^2\ zt[r]\ \varrho z[r])+ra[r]^2\ (n^2-ni^2\ \varrho z[r]^2))))}$$

$$rb[r\_]:=ra[r]+\vartheta[r]\varrho r[r]$$

$$zb[r\_]:=za[r]+\vartheta[r]\varrho z[r]$$

$\theta 1$=ArcCos[wz[r]];

c1=ϑa[r];

$\theta 3$=ArcTan[(rb[r])/(zb[r]-τ-zi)];

c2=ϑa[r]+n ϑ[r];

tt=68;

c=1;

```
v=c/n;

no=1;

ni=1;

n=1.5;

τ=10;

zi =20;

zo=-20;

rmax=7;

R=12;

Surface:=ParametricPlot[{
{w[r],r},
{za[r],ra[r]},
{zb[r],rb[r]},
{w[2r],2r},
{za[2.3r],ra[2.3r]},
{zb[2.3r],rb[2.3r]},
{-16-w[2r],2r}},{r,-rmax,rmax},
 PlotStyle→{
{Black,Thick},
Red,
Black,
{Black,Thin,Dashed},
{Red,Dashed},
{Black,Thick,Dashed},
```

```
{Black,Thin,Dashed}},AspectRatio→1/1, AxesOrigin→{-45,-25},
Axes→True,AxesLabel→{Style[z,Larger,Bold],Style[r,Larger,Bold]},
AxesStyle→Arrowheads[{0.015}]]
```

```
Lens:=Graphics[{EdgeForm[Directive[{Black,Thin}]],
        Directive[Gray,Opacity[0.3]],
        Polygon[Join[#[[2,1]],Reverse[#[[3,1]]]]&
    [Cases[Surface,_Line,∞]]]}]
```

```
Waves:=Table[{ParametricPlot[{
Piecewise[{{wz[r]tt+w[r],tt<ϑa[r]},
                    { v ϱz[r](tt-c1)+za[r],ϑa[r]+n ϑ[r]>tt>ϑa[r]},
                    {zb[r]+(tt-c2)Cos[θ3],tt>ϑa[r]+n ϑ[r]}}],
Piecewise[{{wr[r]tt+r,tt<ϑa[r]},
                    {v ϱr[r](tt-ϑa[r])+ra[r],ϑa[r]+n ϑ[r]>tt>ϑa[r]},
                    {rb[r]+(tt-c2)Sin[θ3],tt>ϑa[r]+n ϑ[r]}}]},
{r,-rmax,rmax},
PlotStyle→{Purple,Thickness[0.0015]}]},{tt,-20,70,2}]
```

```
Rays:=Table[{ParametricPlot[{
Piecewise[{{wz[r]tt+w[r],tt<ϑa[r]},
                    { v ϱz[r](tt-c1)+za[r],ϑa[r]+n ϑ[r]>tt>ϑa[r]},
                    {zb[r]+(tt-c2)Cos[θ3],tt>ϑa[r]+n ϑ[r]}}],
Piecewise[{{wr[r]tt+r,tt<ϑa[r]},
                    {v ϱr[r](tt-ϑa[r])+ra[r],ϑa[r]+n ϑ[r]>tt>ϑa[r]},
                    {rb[r]+(tt-c2)Sin[θ3],tt>ϑa[r]+n ϑ[r]}}]},
{tt,-20,70},
PlotStyle→{Blue,Thickness[0.0015]}]},{r, rmax,rmax}]
```

```
Show[Surface,Lens,Waves,Rays,Surface,PlotRange->All,
  AspectRatio→Automatic]
```

5.5 Examples

This section presents examples of the application of equations (5.39), (5.40), and (5.43) in `Mathematica` in figures 5.2, 5.3, 5.4, and 5.5.

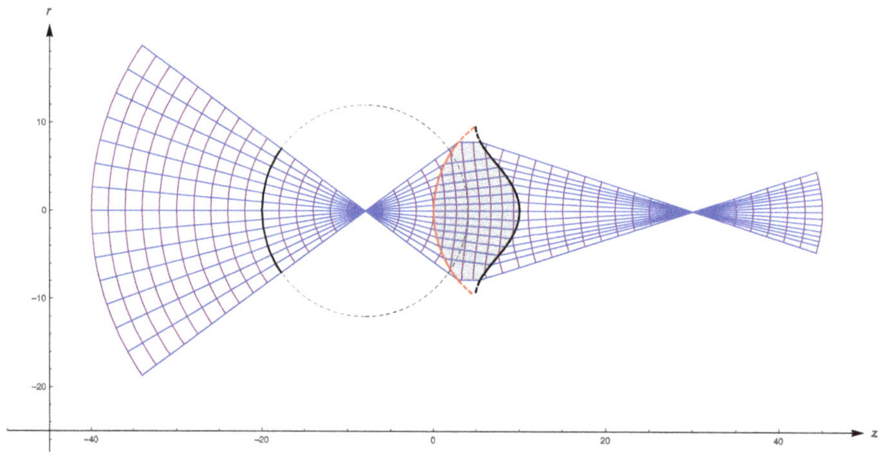

Figure 5.2. Ray-traced graph of a stigmatic lens; the incoming wavefront is a circle and the image is at a finite distance. Observe the natural spherical aberration that the incoming wave presents, while the output delivered by the system concentrates all the rays at a single point, thus guaranteeing the desired stigmatic behaviour. The calculations were made using equations (5.39), (5.40), and (5.43) and by substituting the corresponding ϑ_a equation for the incoming wavefront. The parameters of the lens design's input wavefront are: $R - \sqrt{R^2 - r^2}$,

$$\vartheta_a = -r\ wr(r) + Rwz(r) - w(r)\ wz(r) - \sqrt{R^2\ wz(r)^2 + 2R\ wr(r)\ (w(r)\ wr(r) - r\ wz(r)) - (w(r)\ wr(r) - r\ wz(r))^2},$$

refractive index $n = 1.5$, object distance: $zo = -20$ mm, image distance: $zi = 20$ mm, thickness: $\tau = 10$ mm, radius: $R = 12$ mm, and $r_{max} = 12$ mm.

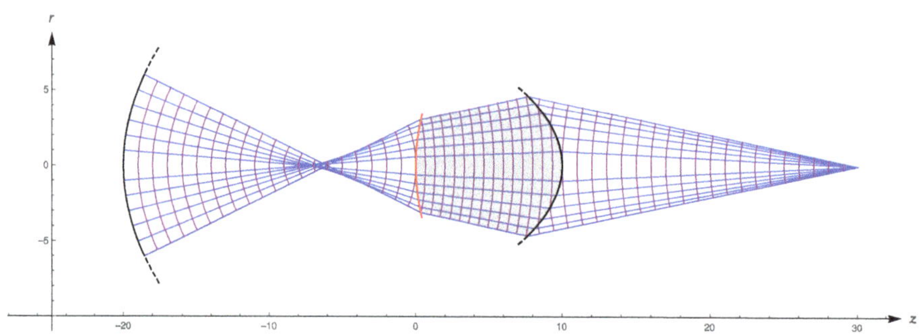

Figure 5.3. Ray-traced graph of a stigmatic lens; the incoming wavefront is a parabola and the image is at a finite distance. Observe the natural spherical aberration that the incoming wave presents, while the output delivered by the system concentrates all the rays at a single point, thus guaranteeing the desired stigmatic behaviour. The calculations were made using equations (5.39), (5.40), and (5.43) and by substituting the corresponding ϑ_a equation for the incoming wavefront. The parameters of the lens design's input wavefront are: $(r^2/25) + zo$, $\vartheta_a = \frac{\left(-r\ wr(r) - 2f_a wz(r) + \sqrt{f_a}\sqrt{-w(r)wr(r)^2 + rwr(r)wz(r) + f_a wz(r)^2}\right)}{wr(r)^2}$, refractive index : $n = 1.5$, object distance: $zo = -20$ mm, image distance: $zi = 20$ mm, thickness: $\tau = 10$ mm; the parameters of the parabola are: $f_a = -7$ mm, $r_{max} = 6$ mm.

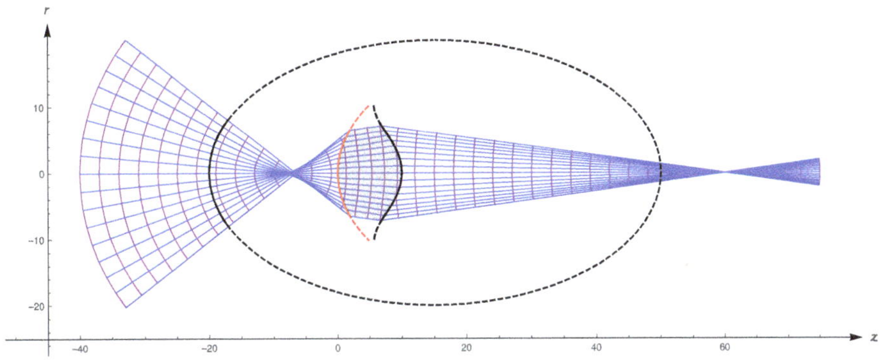

Figure 5.4. Ray-traced graph of a stigmatic lens; the incoming wavefront is an ellipse and the image is at a finite distance. Observe the natural spherical aberration that the incoming wave presents, while the output delivered by the system concentrates all the rays at a single point, thus guaranteeing the desired stigmatic behaviour. The calculations were made using equations (5.39), (5.40), and (5.43) and by substituting the corresponding ϑ_a equation for the incoming wavefront. The parameters of the lens input wavefront are: $b - \sqrt{(-a^2 * b^2 + b^2 r^2)/ - a^2} + zo$,

$$\vartheta_a = -\frac{1}{\left(b^2 wr(r)^2 + a^2 wz(r)^2\right)} \times (b^2 r wr(r) - a^2 b wz(r) + a^2 w(r) wz(r) + \sqrt{a^2 b^2 (-w(r)^2 wr(r)^2 + 2rw(r)wr(r)wz(r) + (a^2 - r^2)wz(r)^2 + 2bwr(r)(w(r)wr(r) - rwz(r)))}\Big),$$

refractive index : $n = 1.5$, object distance: $zo = -20$ mm, image distance: $zi = 50$ mm, thickness: $\tau = 10$ mm, constant: $a = 20$ mm, constant: $b = 35$ mm, $r_{max} = 8$ mm.

5.6 Mathematical model for wavefront generator lenses for images at infinity

We now turn to the computation of a wavefront generator lens for images at infinity. This task shares similarities with the lens presented in section 5.2. The idea is similar: given an input wavefront w, the lens must be such that (this time) the light is collimated at the output (see figure 5.6).

The input wavefront w is given, and r is the only independent variable of the model. The first refractive surface of the lens is given by (z_a, r_a) and it is also given. The distance from the centre of the wavefront w to the first refractive surface's centre is denoted by z_o. The centre of this first surface is located at the origin of the coordinate system. The lens has a central thickness of τ. The second surface is denoted by (z_b, r_b), where z_b and r_b are unknowns. The gap from the second surface's centre to the image point is z_i. Let n_o, n, and n_i be the refractive indexes of the object, lens, and image spaces, respectively.

Due to the similarity between the task of this section and the task of section 5.2, it is useful to recall the cosine directors of the input wavefront w,

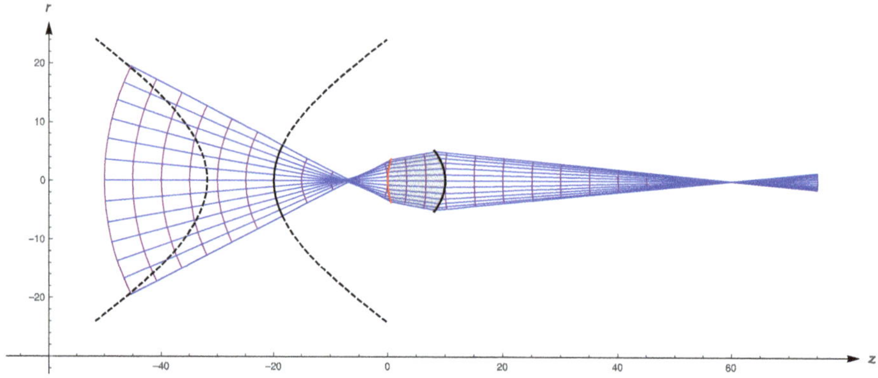

Figure 5.5. Ray-traced graph of a stigmatic lens; the incoming wavefront is a hyperbola and the image is at a finite distance. Observe the natural spherical aberration that the incoming wave presents, while the output delivered by the system concentrates all the rays at a single point, thus guaranteeing the desired stigmatic behaviour. The calculations were made using equations (5.39), (5.40), and (5.43) and by substituting the corresponding ϑ_a equation for the incoming wavefront. The parameters of the lens design's input wavefront are: $b - \sqrt{(-a^2 * b^2 - b^2 r^2)/ - a^2} + zo$,

$$\vartheta_a = -\frac{1}{b^2 wr(r)^2 + a^2 wz(r)^2} \times (b^2 r wr(r) - a^2 b wz(r) + a^2 w(r) wz(r) + \sqrt{(a^2 b^2(-w(r)^2 wr(r)^2 + 2r w(r) wr(r) wz(r) + (a^2 - r^2) wz(r)^2 + 2b wr(r)(w(r) wr(r) - r wz(r))))}),$$

refractive index : $n = 1.5$, object distance: $zo = -20$ mm, image distance: $zi = 50$ mm, thickness: $\tau = 10$ mm, constant: $a = 20$ mm, constant: $b = 35$ mm, $r_{max} = 12$ mm.

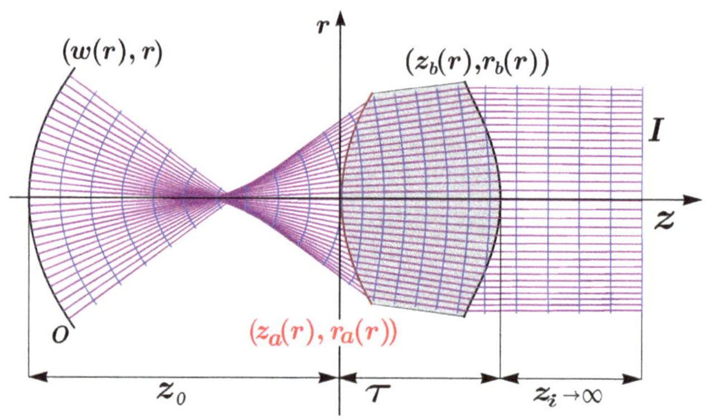

Figure 5.6. Basic configuration of a lens whose optical properties allow stigmatic images to be obtained given a prescribed input wavefront; the image is at infinity.

$$w_r = -\frac{w'}{\sqrt{w'^2+1}}, \qquad w_z = \frac{1}{\sqrt{w'^2+1}}, \tag{5.44}$$

where $w' \equiv dw/dr$ is the derivative of w with respect to r, and w_r and w_z are the cosine directors in the r and z directions, respectively. Remember that we can express the first surface in terms of the cosine directors, so recalling equation (5.2),

$$r_a = r + w_r\vartheta_a, \qquad z_a = w + w_z\vartheta_a. \tag{5.45}$$

To achieve our goal we need to set the image to infinity; for this, we need to work with the OPL. Recalling equation (5.29),

$$\begin{aligned} -n_o z_o + n\tau + n_i z_i = n&\sqrt{(z_a - z_b)^2 + (r_a - r_b)^2} \\ &+ n_o\sqrt{(z_a - w)^2 + (r_a - r)^2} + n_i\sqrt{(z_b - z_i - \tau)^2 + r_b{}^2}. \end{aligned} \tag{5.46}$$

We obtain the distance between both surfaces by recalling equation (5.20),

$$\vartheta \equiv \sqrt{(z_b - z_a)^2 + (r_b - r_a)^2}. \tag{5.47}$$

By substituting equations (5.2) and (5.20) into equation (5.29), we get

$$\begin{aligned} -n_o z_o + n\tau + n_i z_i &= n\vartheta \\ &+ n_o\sqrt{(z_a - w)^2 + (r_a - r)^2} + n_i\sqrt{(z_a + \vartheta\wp_z - z_i - \tau)^2 + (r_a + \vartheta\wp_r)^2}. \end{aligned} \tag{5.48}$$

We are now in a position to compute $\lim\limits_{z_i\to\infty}$ over equation (5.48):

$$\begin{aligned} &\lim_{z_i\to\infty}\Big\{ -n_o z_o + n\tau + n_i z_i = n\vartheta \\ &+ n_o\sqrt{(z_a - w)^2 + (r_a - r)^2} + n_i\sqrt{(z_a + \vartheta\wp_z - z_i - \tau)^2 + (r_a + \vartheta\wp_r)^2}\Big\} \\ &-n\lim_{z_i\to\infty}(\vartheta) + n\tau + n_i\left[z_a + \lim_{z_i\to\infty}(\vartheta)\wp_z - \tau\right] - n_o\sqrt{(r - r_a)^2 + (w - z_a)^2} \\ &- n_o z_o. \end{aligned} \tag{5.49}$$

Recalling that the result of equation (5.49) is the OPD as $z_i \to \infty$, then

$$\begin{aligned} -n\lim_{z_i\to\infty}(\vartheta) + n\tau + n_i\left[z_a + \lim_{z_i\to\infty}(\vartheta)\wp_z - \tau\right] - n_o\sqrt{(r - r_a)^2 + (w - z_a)^2} \\ - n_o z_o = 0; \end{aligned} \tag{5.50}$$

using the last expression, we can solve for $\lim_{z_i\to\infty}(\vartheta)$:

$$\lim_{z_i\to\infty}(\vartheta) = \frac{-n\tau - n_i z_a + n_i\tau + n_o\sqrt{(r - r_a)^2 + (w - z_a)^2} + n_o z_o}{n_i\wp_z - n}. \tag{5.51}$$

Therefore, we can express the parameters of the second surface as

$$\lim_{z_i\to\infty}(r_b) = r_a + \lim_{z_i\to\infty}(\vartheta)\wp_r \tag{5.52}$$

and

$$\lim_{z_i\to\infty}(z_b) = z_a + \lim_{z_i\to\infty}(\vartheta)\wp_z, \tag{5.53}$$

or, using equations (5.51), (5.52), and (5.53), as

$$\boxed{\begin{aligned}\lim_{z_i\to\infty}(r_b) = r_a + &\left\{\frac{-n\tau - n_i z_a + n_i\tau + n_o\sqrt{(r-r_a)^2+(w-z_a)^2} + n_o z_o}{n_i\wp_z - n}\right\}\\ &\times\left\{\frac{n_o r_a'(z_a'-r_a'w')}{n_i\sqrt{w'^2+1}}(r_a'^2+z_a'^2) + \frac{z_a'\sqrt{1-\dfrac{n_o^2(z_a'-r_a'w')^2}{n_i^2(w'^2+1)(r_a'^2+z_a'^2)}}}{\sqrt{r_a'^2+z_a'^2}}\right\}\end{aligned}} \tag{5.54}$$

and

$$\boxed{\begin{aligned}\lim_{z_i\to\infty}(z_b) = z_a + &\left\{\frac{-n\tau - n_i z_a + n_i\tau + n_o\sqrt{(r-r_a)^2+(w-z_a)^2} + n_o z_o}{n_i\wp_z - n}\right\}\\ &\times\left\{\frac{n_o z_a'(z_a'-r_a'w')}{n_i\sqrt{w'^2+1}}(r_a'^2+z_a'^2) - \frac{r_a'\sqrt{1-\dfrac{n_o^2(z_a'-r_a'w')^2}{n_i^2(w'^2+1)(r_a'^2+z_a'^2)}}}{\sqrt{r_a'^2+z_a'^2}}\right\}.\end{aligned}} \tag{5.55}$$

At this point, it should not be a surprise that we have not used any paraxial or iteration to obtain equations (5.54) and (5.55). The paradigm presented in this book only applies the optical path theory, which in turn is only based on Fermat's principle. Therefore, we do not use any paraxial, and iteration is not allowed. Equations (5.54) and (5.55) are the unique solution of the presented problem given a set of input values, and this is revealed by the procedure used to obtain equations (5.54) and (5.55). This does not mean that equations (5.54) and (5.55) are not general; in fact, equations (5.54) and (5.55) are the general closed-form solution of the problem, because the set of input values is given by the user.

5.7 Wavefront of the wavefront generator lens for images at infinity

The procedure used to obtain the eikonal of the wavefront generator lens for images at infinity is similar to that of the wavefront generator lens for finite images described in section 4.3. We only need to compute the limit $z_i \to \infty$ over the angle θ using equation (5.41):

$$\lim_{z_i\to\infty}(\theta) = \arctan\left[\frac{\lim_{z_i\to\infty}(r_b)}{\lim_{z_i\to\infty}(z_b) - \tau - (z_i \to \infty)}\right] = 0. \tag{5.56}$$

Thus, the cosine directors at the output are zero in the radial direction and one on the optical axis, therefore equation (5.43) tends toward the limit $z_i \to \infty$ when it is computed:

$$\boxed{\lim_{z_i\to\infty} H(t, r_a) = \begin{cases} [w + tw_z, r + tw_r], & t < \vartheta_a, \\ (v(t - \vartheta_a)\wp_z, v(t - \vartheta_a)\wp_r), & \vartheta_a < t < \vartheta_a + n\lim_{z_i\to\infty}(\vartheta), \\ \begin{pmatrix} z_b + \left(t - \vartheta_a + n\lim_{z_i\to\infty}(\vartheta)\right)\cos\theta, \\ r_b + \left(t - \vartheta_a + n\lim_{z_i\to\infty}(\vartheta)\right)\sin\theta \end{pmatrix}, & t > \vartheta_a + n\lim_{z_i\to\infty}(\vartheta) \end{cases}}, \tag{5.57}$$

where $v = c/n$ is the speed of light inside the lens and c is the speed of light in vacuum (outside the lens, we set $n_o = n_i = 1$). The case in which $n_o \neq n_i \neq 1$ is left as an exercise for the reader.

5.8 Mathematica code

This section presents `Mathematica` code for the aspheric wavefront generator lens, which is partially based on equations (5.54), (5.55), and (5.57).

```
In[7]:= w[x_,y_]:=50

e=10^(-5);

wdx[x_,y_]:=(w[x+e,y]-w[x,y])/e

wdy[x_,y_]:=(w[x,y+e]-w[x,y])/e

wx[x_,y_]:= -wdx[x,y] / Sqrt[wdx[x,y]^2+wdy[x,y]^2+1]
```

$$\text{wy}[\text{x\_},\text{y\_}]:=\frac{-\text{wdy}[x,y]}{\sqrt{\text{wdx}[x,y]^2+\text{wdy}[x,y]^2+1}}$$

$$\text{wz}[\text{x\_},\text{y\_}]:=\frac{1}{\sqrt{\text{wdx}[x,y]^2+\text{wdy}[x,y]^2+1}}$$

$$\boldsymbol{\vartheta}[\text{x\_},\text{y\_}]:=-\frac{\text{w}[x,y]+p\ x+s\ y-\text{zo}-\sqrt{1+p^2+s^2}\ \text{zo}}{\sqrt{1+p^2+s^2}+p\ \text{wx}[x,y]+s\ \text{wy}[x,y]+\text{wz}[x,y]}$$

```
xa[x_,y_]:=x+𝛝[x,y]wx[x,y]

ya[x_,y_]:=y+𝛝[x,y]wy[x,y]

za[x_,y_]:=w[x,y]+𝛝[x,y]wz[x,y]

dxxa[x_,y_]:=(xa[x+e,y]-xa[x,y])/e
dyxa[x_,y_]:=(xa[x,y+e]-xa[x,y])/e
dxya[x_,y_]:=(ya[x+e,y]-ya[x,y])/e
dyya[x_,y_]:=(ya[x,y+e]-ya[x,y])/e
dxza[x_,y_]:=(za[x+e,y]-za[x,y])/e
dyza[x_,y_]:=(za[x,y+e]-za[x,y])/e

Φx[x_,y_]:=(-dxza[x,y] dyya[x,y]+dxya[x,y] dyza[x,y])/
   (√((-dxya[x,y] dyxa[x,y]+dxxa[x,y] dyya[x,y])^2+
    (dxza[x,y] dyxa[x,y]-dxxa[x,y] dyza[x,y])^2+
    (-dxza[x,y] dyya[x,y]+dxya[x,y] dyza[x,y])^2))

Φy[x_,y_]:=(dxza[x,y] dyxa[x,y]-dxxa[x,y] dyza[x,y])/
   (√((-dxya[x,y] dyxa[x,y]+dxxa[x,y] dyya[x,y])^2+
    (dxza[x,y] dyxa[x,y]-dxxa[x,y] dyza[x,y])^2+
    (-dxza[x,y] dyya[x,y]+dxya[x,y] dyza[x,y])^2))

Φz[x_,y_]:=(-dxya[x,y] dyxa[x,y]+dxxa[x,y] dyya[x,y])/
   (√((-dxya[x,y] dyxa[x,y]+dxxa[x,y] dyya[x,y])^2+
    (dxza[x,y] dyxa[x,y]-dxxa[x,y] dyza[x,y])^2+
    (-dxza[x,y] dyya[x,y]+dxya[x,y] dyza[x,y])^2))
```

$$\varrho x[x\_,y\_]=-\frac{1}{ni}no\ (wy[x,y]\ \mathbf{\Phi}x[x,y]\ \mathbf{\Phi}y[x,y]+wz[x,y]\ \mathbf{\Phi}x[x,y]\ \mathbf{\Phi}z[x,y]-wx[x,y]\ (\mathbf{\Phi}y[x,y]^2+\mathbf{\Phi}z[x,y]^2))+\mathbf{\Phi}x[x,y]\ \surd(1-\frac{1}{ni^2}no^2\ ((wy[x,y]\ \mathbf{\Phi}x[x,y]-wx[x,y]\ \mathbf{\Phi}y[x,y])^2+(wz[x,y]\ \mathbf{\Phi}x[x,y]-wx[x,y]\ \mathbf{\Phi}z[x,y])^2+(wz[x,y]\ \mathbf{\Phi}y[x,y]-wy[x,y]\ \mathbf{\Phi}z[x,y])^2));$$

$$\varrho y[x\_,y\_]=\frac{1}{ni}no\ (-\mathbf{\Phi}y[x,y]\ (wx[x,y]\ \mathbf{\Phi}x[x,y]+wz[x,y]\ \mathbf{\Phi}z[x,y])+wy[x,y]\ (\mathbf{\Phi}x[x,y]^2+\mathbf{\Phi}z[x,y]^2))+\mathbf{\Phi}y[x,y]\ \surd(1-\frac{1}{ni^2}no^2\ ((wy[x,y]\ \mathbf{\Phi}x[x,y]-wx[x,y]\ \mathbf{\Phi}y[x,y])^2+(wz[x,y]\ \mathbf{\Phi}x[x,y]-wx[x,y]\ \mathbf{\Phi}z[x,y])^2+(wz[x,y]\ \mathbf{\Phi}y[x,y]-wy[x,y]\ \mathbf{\Phi}z[x,y])^2));$$

$$\varrho z[x\_,y\_]=\frac{1}{ni}no\ (wz[x,y]\ (\mathbf{\Phi}x[x,y]^2+\mathbf{\Phi}y[x,y]^2)-(wx[x,y]\ \mathbf{\Phi}x[x,y]+wy[x,y]\ \mathbf{\Phi}y[x,y])\ \mathbf{\Phi}z[x,y])+\mathbf{\Phi}z[x,y]\ \surd(1-\frac{1}{ni^2}no^2\ ((wy[x,y]\ \mathbf{\Phi}x[x,y]-wx[x,y]\ \mathbf{\Phi}y[x,y])^2+(wz[x,y]\ \mathbf{\Phi}x[x,y]-wx[x,y]\ \mathbf{\Phi}z[x,y])^2+(wz[x,y]\ \mathbf{\Phi}y[x,y]-wy[x,y]\ \mathbf{\Phi}z[x,y])^2));$$

$$wx2=\frac{p}{\sqrt{1+p^2+s^2}};$$

$$wy2=\frac{s}{\sqrt{1+p^2+s^2}};$$

$$wz2=\frac{1}{\sqrt{1+p^2+s^2}};$$

```
c=1;
v=c/ni;
no=1;
ni=-1;
rmax=10;
zo=60;
s=0.75;
p=0;
```

```
L[x_,y_]:=60-ϑ[x,y];

 Homotopy={
{Piecewise[{
{wx[x,y]tt+x,tt<ϑ[x,y]},{- wx2(tt-ϑ[x,y])+xa[x,y],tt>ϑ[x,y]}}],
Piecewise[{
{wy[x,y]tt+y,tt<ϑ[x,y]},{ -wy2(tt-ϑ[x,y])+ya[x,y],tt>ϑ[x,y]}}],
Piecewise[{
{wz[x,y]tt+w[x,y],tt<ϑ[x,y]},{-wz2(tt-ϑ[x,y])+za[x,y],tt>ϑ[x,y]}}]}};


Waves:=Table[{ParametricPlot3D[Homotopy,{x,-rmax,rmax},{y,-rmax,rmax},
PlotRange→All,PlotStyle→{Blue,Opacity[.5]}, Mesh→True]},{tt,0,60,10}]
l[x_,y_]:=Graphics3D[{Purple,Line[{{x,y,w[x,y]},{xa[x,y],ya[x,y],za[x,y]}}]}]


wn[x_,y_]:=Graphics3D[{Red,Thick,
          Line[{{x,y,w[x,y]},{x+wx[x,y],y+wy[x,y],w[x,y]+wz[x,y]}}]}]


zan[x_,y_]:=Graphics3D[{Red,Thick, Line[{{xa[x,y],ya[x,y],za[x,y]},
           {xa[x,y]-Φx[x,y],ya[x,y]-Φy[x,y],za[x,y]-Φz[x,y] }}]}]


v2[x_,y_]:=Graphics3D[{Purple,Line[{{xa[x,y],ya[x,y],za[x,y]},{xa[x,y]-
          L[x,y] ϱx[x,y] ,ya[x,y]-L[x,y] ϱy[x,y] ,za[x,y]-L[x,y] ϱz[x,y] }}]}]


v2a[x_,y_]:=Graphics3D[{Purple,Line[{{xa[x,y],ya[x,y],za[x,y]},{xa[x,y]-
           L[x,y] wx2 ,ya[x,y]-L[x,y] wy2 ,za[x,y]-L[x,y]wz2 }}]}]


g0=Plot3D[{w[x,y]},{x,-rmax,rmax},{y,-rmax,rmax},
      Axes→True,PlotStyle→{Blue,Opacity[.5]},AxesLabel→{x,y,z},
      AspectRatio→Automatic,Boxed→False];


g1=ParametricPlot3D[{xa[x,y],ya[x,y],za[x,y]},{x,-rmax,rmax},{y,-rmax,rmax},
      Axes→True,PlotStyle→LightYellow,Boxed→True,AxesLabel→{x,y,z},
      AspectRatio→Automatic];
```

```
g2=ParametricPlot3D[{xa[x,y]-L[x,y] wx2 ,ya[x,y]-L[x,y] wy2 ,za[x,y]-L[x,y]wz2 },
{x,-rmax,rmax},{y,-rmax,rmax},
PlotRange→All,PlotStyle→Blue, Mesh→True,Boxed→True];
```

```
Show[g1,g0,Waves,Table[l[x,y],{x,-rmax,rmax,3},{y,-rmax,rmax,3}],
Table[v2a[x,y],{x,-rmax,rmax,3},{y,-rmax,rmax,3}],PlotRange→All]
```

5.9 Examples

Examples of the output generated by the Mathematica code for equations (5.54), (5.55), and (5.57) are presented in this subsection in figures 5.7, 5.8, 5.9, and 5.10.

The results of equations (5.54) and (5.55) are very precise. The reason that equations (5.54) and (5.55) have such outstanding performance is that they are fully analytical and closed form. Their deduction was performed without recurring to paraxial optics, optimization, or numerical approximation. There is only one solution for a given set of input parameters and this solution is given by equations (5.54) and (5.55), which demonstrates its its uniqueness.

The limitation of equations (5.54) and (5.55) is that two rays may not touch each other at any point of the first or second surfaces. Equations (5.54) and (5.55) are

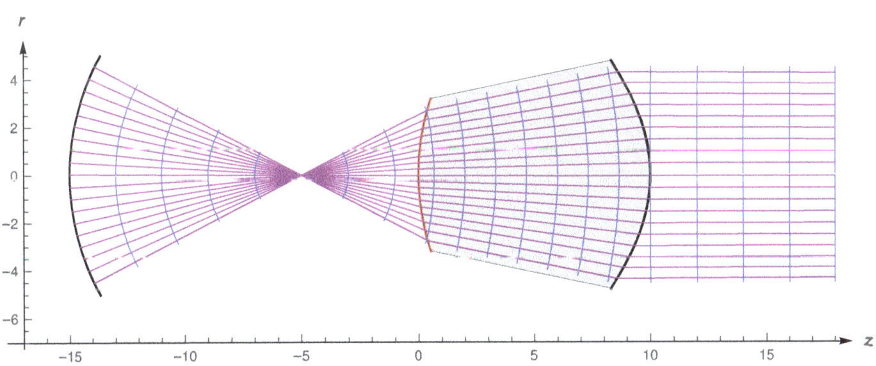

Figure 5.7. Ray-traced graph of a stigmatic lens; the incoming wavefront is a circle and the image is at an infinite distance. Observe the natural spherical aberration that the incoming wave presents, while the output delivered by the system collimates the rays, thus guaranteeing the desired stigmatic behaviour. The calculations were made using equations (5.54), (5.55), and (5.57) and by substituting the corresponding ϑ_a equation for the incoming wavefront. The parameters of the lens design's input wavefront are: $R - \sqrt{R^2 - r^2}$, $\vartheta_a = \frac{-r\ wr(r) + R wz(r) - w(r)\ wz(r)}{-\sqrt{R^2\ wz(r)^2 + 2R\ wr(r)\ (w(r)\ wr(r) - r\ wz(r)) - (w(r)\ wr(r) - r\ wz(r))^2}}$, refractive index $n = 1.5$, object distance: $zo = -15$ mm, thickness: $\tau = 10$ mm, radius: $R = 10$ mm, $r_{max} = 5$ mm.

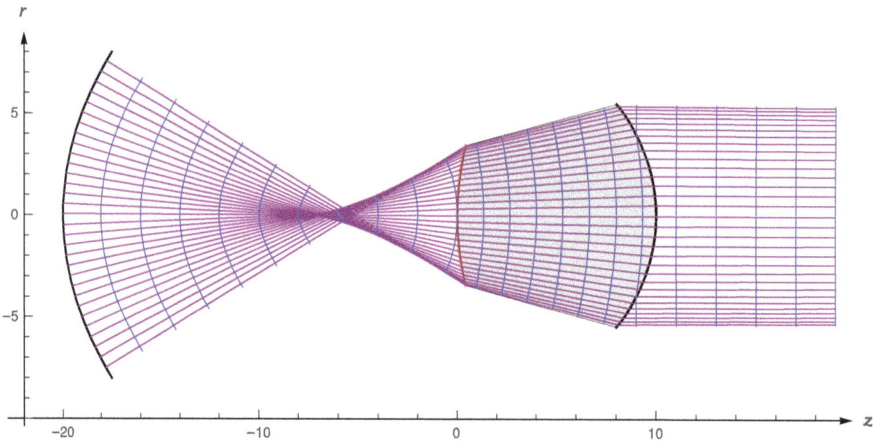

Figure 5.8. Ray-traced graph of a stigmatic lens; the incoming wavefront is a parabola and the image is at an infinite distance. Observe the natural spherical aberration that the incoming wave presents, while the output delivered by the system collimates the rays, thus guaranteeing the desired stigmatic behaviour. The calculations were made using equations (5.54), (5.55), and (5.57) and by substituting the corresponding ϑ_a equation for the incoming wavefront. The parameters of the lens design's input wavefront are: $(r^2/25) + zo$, $\vartheta_a = \frac{\left(-r\, wr(r) - 2f_a wz(r) + \sqrt{f_a}\sqrt{-w(r)wr(r)^2 + rwr(r)wz(r) + f_a wz(r)^2}\right)}{wr(r)^2}$, refractive index : $n = 1.5$, object distance: $zo = -20$ mm, thickness: $\tau = 10$ mm; the parameters of the parabola are: $f_a = -7$ mm, $r_{max} = 8$ mm.

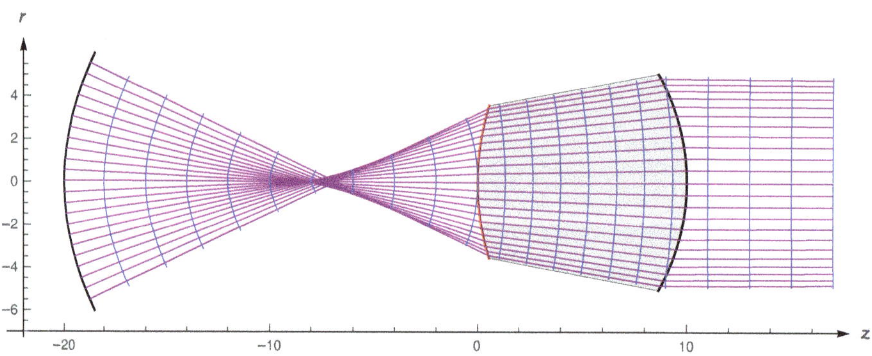

Figure 5.9. Ray-traced graph of a stigmatic lens; the incoming wavefront is an ellipse and the image is at an infinite distance. Observe the natural spherical aberration that the incoming wave presents, while the output delivered by the system collimates the rays, thus guaranteeing the desired stigmatic behaviour. The calculations were made using equations (5.54), (5.55), and (5.57) and by substituting the corresponding ϑ_a equation for the incoming wavefront. The parameters of the lens design's input wavefront are: $-b + \sqrt{(-a^2 * b^2 + b^2 r^2)/ - a^2} + zo$,

$$\vartheta_a = -\frac{1}{(b^2 wr(r)^2 - a^2 wz(r)^2)} \times \Big(-b^2 rwr(r) + a^2 bwz(r) + a^2 w(r)wz(r) - \sqrt{a^2 b^2 (w(r)^2 wr(r)^2 - 2rw(r)wr(r)wz(r) + (a^2 + r^2)wz(r)^2 + 2bwr(r)(w(r)wr(r) - rwz(r)))}\Big),$$

refractive index : $n = 1.5$, object distance: $zo = -20$ mm, thickness: $\tau = 10$ mm, constant : $a = 20$ mm, constant : $b = 35$ mm, $r_{max} = 6$ mm.

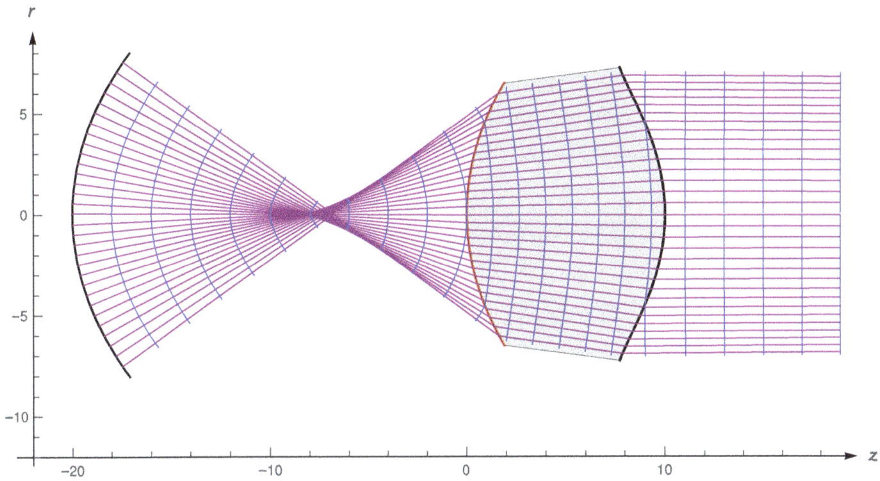

Figure 5.10. Ray-traced graph of a stigmatic lens; the incoming wavefront is a hyperbola and the image is at an infinite distance. Observe the natural spherical aberration that the incoming wave presents, while the output delivered by the system collimates the rays, thus guaranteeing the desired stigmatic behaviour. The calculations were made using equations (5.54), (5.55), and (5.57) and by substituting the corresponding ϑ_a equation for the incoming wavefront. The parameters of the lens design's input wavefront are: $b - \sqrt{(-a^2b^2 - b^2r^2)/ - a^2} + zo$,

$$\vartheta_a = -\frac{1}{b^2 wr(r)^2 + a^2 wz(r)^2} \times \left(b^2 r wr(r) - a^2 b wz(r) + a^2 w(r) wz(r) + \sqrt{(a^2b^2(-w(r)^2 wr(r)^2 + 2rw(r)wr(r)wz(r) + (a^2 - r^2)wz(r)^2 + 2bwr(r)(w(r)wr(r) - rwz(r))))}\right),$$

refractive index : $n = 1.5$, object distance: $zo = -20$ mm, image distance: $zi = 50$ mm, thickness: $\tau = 10$ mm, constant : $a = 20$ mm, constant : $b = 35$ mm, $r_{max} = 8$ mm.

deduced by comparing the OPLs of the axial and non-axial rays. The OPLs of the neighbouring rays are not considered. Therefore, there is a one-to-one relationship between an input ray, its path, and the points where it strikes the first and second surfaces. Thus, if two rays strike the same point of the second surface, then the second surface overlaps with itself.

The lenses presented in section (5.2) and (5.6) can be seen as null lenses for the input w. This type of lens is called a null lens, which is why we called this chapter 'the general equation for the aspheric null lens', i.e. because we consider any aspheric w and the main application of equations (5.54) and (5.55) is the creation of null lenses.

5.10 End notes

In this chapter, we have obtained general equations, equations (5.54) and (5.55), that allow one to design a lens that produces a stigmatic image point given an input wavefront. These general equations are fully analytical and closed form. Equations (5.54) and (5.55) have been tested using ray-tracing techniques, and their performance was as expected. Their properties, implications, and limitations have been discussed. Finally, the lenses proposed here can be easily generalized to freeform surfaces using freeform input wavefronts, and this generalization will be presented in chapter 9.

Part III

Freeform optical systems and the path of light

Chapter 6

General mirror for adaptive optical systems

In this chapter, we present the general formula that allows a mirror to be designed in three dimensions such that it reflects the light of a given arbitrary wavefront as a spherical wave or a plane. The formula is fully analytical and closed form. We test our equations using ray-tracing techniques. We do not use any paraxial concepts or numerical approximations during their derivation; we only use Fermat's principle.

6.1 The crux of adaptive optics

We are about to begin the chapter that we think is the most important chapter of this treatise. We began this treatise with an introductory chapter, chapter 1, and then worked with stigmatic mirrors in two dimensions in chapter 2. The mirrors of chapter 2 were very interesting because their input wavefronts were not necessarily flat or spherical, and their output wavefronts were indeed flat or spherical. The mathematical equations presented in chapter 2 were very useful as a pedagogical resource, because it is very easy to see dimensions and parameters in $\mathbb{R}^2$. Unfortunately, since the model presented in chapter 2 is not radially symmetric, it cannot be implemented in $\mathbb{R}^3$. Therefore, we need to deduce $\mathbb{R}^3$ versions of all the equations of chapter 2. The visualization of the parameters in $\mathbb{R}^3$ will be clearer as a result of the experience gained in chapter 2.

We start by recalling the basic diagram of an adaptive optical system (figure 1.12). After passing through the primary and secondary mirrors of the telescope shown, the light has a non-flat wavefront. The deformable mirror has a key role in converting the aberrated wavefront into a flat wavefront. The shape of that mirror obtained via the analytical closed-form equation is the main topic of this chapter. So, in this chapter, we are going to design mirrors such that the input wavefront is not necessarily flat or spherical, and the output wavefront is indeed flat or spherical, again only using Fermat's principle. The equation used to describe such mirrors is the general equation for the abovementioned problem in $\mathbb{R}^3$. Thus, this equation is the crux of adaptive optics, because this mirror performs the phase conjugation in an

doi:10.1088/978-0-7503-4705-1ch6

adaptive optical system. Page 29 of the classical adaptive optics book *Introduction to Adaptive Optics* by Robert K Tyson says:

...The crux of adaptive optics is phase conjugation. If we can measure the aberrations, then all we have to do is put the opposite aberration on the beam, at the right time, to cancel the aberrations. The concept of phase conjugation is fairly simple, even as it applies in nonlinear optics. The propagating electric field (the beam of light) is an electromagnetic wave represented by an amplitude and a phase, If we apply a change that flattens the phase, we multiply it by the complex conjugate of the field. To do this physically, we add an optical aberration to the beam at precisely the right time, using a deformable mirror...

The very interesting thing about the approach presented in this chapter is that we do not mention concepts such as the electric field or the complex conjugate of the field. We only use Fermat's principle, which dictates the path of light and governs the optical path theory. The solution presented in this chapter is analytical and closed form and the only premise needed is Fermat's principle; thus, this solution is a natural implication of the path of light. This chapter is organized as follows: first we obtain the general formula for freeform deformable mirrors for finite images and their wavefronts in sections 6.2 and 6.3. The `Mathematica code` and examples are presented in sections 6.4 and 6.5, respectively. We then focus on the general formula for freeform deformable mirrors for images at infinity in sections 6.6 and 6.7. Finally, conclusions are presented in section 6.10.

6.2 The general formula for freeform deformable mirrors for images at finite distances

We begin this section with the diagram presented in figure 6.1. It shows the geometry of the problem that we presented in the introduction.

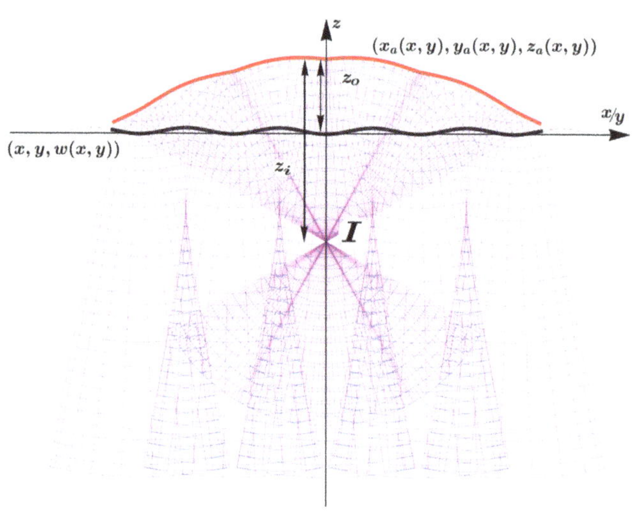

Figure 6.1. Schematic diagram of the configuration of an adaptive mirror with an image at a finite distance.

The origin of the Cartesian coordinate system is at the vertex of the input wavefront w. The input w is a continuous function that depends only on one independent variable r; w is given and it is not necessarily flat or spherical. The mirror's surface is described by (x_a, y_a, z_a), where x_a is the x component of the mirror's surface, y_a is the y component of the mirror's surface, and z_a is the profile of the mirror. Let z_o be the distance between the input wavefront w and the mirror's surface. So, mathematically speaking, we want to find (x_a, y_a, z_a) such that the mirror reflects the input wave w as a spherical wavefront.

We will follow the strategy of chapter 2, as we mentioned in the introduction. We are going to obtain the phase conjugation without evoking the concepts of the electric field or complex conjugate of the field. We only use Fermat's principle, from which we know that the rays start at w, then move towards to (x_a, y_a, z_a), and are reflected as the rays of a spherical wavefront.

The optical path length (OPL) of any ray can be described by

$$\mathrm{OPL} = \sqrt{x_a^2 + y_a^2 + (z_a + z_i - z_o)^2} + \sqrt{(x_a - x)^2 + (y_a - y)^2 + (z_a - w)^2}, \quad (6.1)$$

and for the axial ray, the OPL is

$$\mathrm{OPL} = z_o + z_i \equiv k, \quad (6.2)$$

where k is a constant with the value of the OPL. The OPLs of all rays must be constant, and they are the (x_a, y_a, z_a) that we are interested in, so

$$\sqrt{x_a^2 + y_a^2 + (z_a + z_i - z_o)^2} + \sqrt{(x_a - x)^2 + (y_a - y)^2 + (z_a - w)^2} = z_o - z_i \equiv k. \quad (6.3)$$

Since equation (6.3) considers the paths of all rays, we need to follow them by cosine directors and express (x_a, y_a, z_a) in terms of the input rays. Then, using the Malus–Dupin theorem, we can express the input rays of the wavefront w in the three-dimensional model:

$$w_x = \frac{-\partial_x w}{\sqrt{\partial_x w^2 + \partial_y w^2 + 1}}, \quad w_y = \frac{-\partial_y w}{\sqrt{\partial_x w^2 + \partial_y w^2 + 1}}, \quad w_z = \frac{1}{\sqrt{\partial_x w^2 + \partial_y w^2 + 1}}, \quad (6.4)$$

where $\partial_x \equiv \partial w / \partial x$ is the derivative of w to x, $\partial_y w \equiv \partial w / \partial y$ is the derivative of w to y, and w_x, w_y and w_z are the cosine directors in the x, y, and z directions, respectively. The surface of the mirror in three dimensions expressed by these input cosine directors is then given by

$$x_a = x + \vartheta w_x, \quad y_a = y + \vartheta w_y, \quad z_a = w + \vartheta w_z, \quad (6.5)$$

where, in the three-dimensional model, ϑ changes to

$$\vartheta = \sqrt{(x_a - x)^2 + (y_a - y)^2 + (z_a - w)^2}. \quad (6.6)$$

Here, ϑ is the distance each ray travels from w to (x_a, y_a, z_a). We have the mirror in terms of the input rays; we now need to express the OPL in terms of the input rays as well, so that we can define ϑ without using (x_a, y_a, z_a). Therefore, we substitute equations (6.5) and (6.6) into equation (6.3):

$$\sqrt{(x + \vartheta w_x)^2 + \left(y + \vartheta w_y\right)^2 + [(w + \vartheta w_z) - z_i - z_o]^2} + \vartheta = k. \tag{6.7}$$

Let τ be given by

$$\tau = w + z_i - z_o; \tag{6.8}$$

we now replace τ in equation (6.7) and square both sides:

$$(w_x\vartheta + x)^2 + \left(w_y\vartheta + y\right)^2 + (\tau + w_z\vartheta)^2 = (k - \vartheta)^2. \tag{6.9}$$

Expanding the result yields

$$\begin{aligned} &\tau^2 + w_x^2\vartheta^2 + 2w_x x\vartheta + w_y^2\vartheta^2 + 2w_y y\vartheta + w_z^2\vartheta^2 + 2\tau w_z\vartheta + x^2 + y^2 \\ &= k^2 - 2k\vartheta + \vartheta^2, \end{aligned} \tag{6.10}$$

and solving for ϑ gives

$$\vartheta = \frac{k^2 - \tau^2 - x^2 - y^2}{2(k + w_x x + w_y y + \tau w_z)}. \tag{6.11}$$

Now that we know ϑ, we can express (x_a, y_a, z_a) correctly in terms of the given parameters, substituting equation (6.11) into equation (6.5). The x component of the mirror is

$$\boxed{x_a = x - \frac{\partial_x w[-(w + z_i - z_o)^2 + k^2 - x^2 - y^2]}{2\left(k\sqrt{\partial_x w^2 + \partial_y w^2 + 1} + w - x\partial_x w - y\partial_y w + z_i - z_o\right)}}, \tag{6.12}$$

the y component of the mirror is

$$\boxed{y_a = y - \frac{\partial_y w[-(w + z_i - z_o)^2 + k^2 - x^2 - y^2]}{2\left(k\sqrt{\partial_x w^2 + \partial_y w^2 + 1} + w - x\partial_x w - y\partial_y w + z_i - z_o\right)}}, \tag{6.13}$$

and finally the profile of the mirror is

$$\boxed{z_a = w - \frac{w(w + 2z_i - 2z_o) - k^2 + x^2 + y^2 + (z_i - z_o)^2}{2\left(k\sqrt{\partial_x w^2 + \partial_y w^2 + 1} + w - x\partial_x w - y\partial_y w + z_i - z_o\right)}}. \tag{6.14}$$

Equations (6.12), (6.13), and (6.14) express the shape the mirror must be so that for a given input wavefront, we obtain a spherical wavefront at the output centred on z_i.

Note that from equation (6.3) to equations (6.12), (6.13), and (6.14), we do not use the complex conjugate of any field or any other related concept. Also observe that we do not use any paraxial concepts, numerical approach, or iteration. Equations (6.12), (6.13), and (6.14) are the general analytical closed form of the problem presented in the introduction. As with other similar results presented in previous chapters, equations (6.12), (6.13), and (6.14) give a unique answer for a unique set of values. The derivation of these equations implicitly confirms that they are a uniqueness theorem.

6.3 The wavefront for finite images

The next step is to obtain the wavefront at all points and all times for the mirrors described by equations (6.12), (6.13), and (6.14). At the moment, we know the input rays and their cosine directors (equation (6.4)). We are now interested in the output rays, so we apply Snell's law to (x_a, y_a, z_a):

$$\boxed{\vec{\mathbf{v}}_2 = [\vec{\mathbf{v}}_1 - (\vec{\mathbf{n}}_a \cdot \vec{\mathbf{v}}_1)\vec{\mathbf{n}}_a] - \vec{\mathbf{n}}_a\sqrt{1 - (\vec{\mathbf{n}}_a \times \vec{\mathbf{v}}_1)^2} \quad \text{for} \quad \vec{\mathbf{v}}_2, \vec{\mathbf{v}}_1, \vec{\mathbf{n}}_a \in \mathbb{R}^3,} \tag{6.15}$$

where $\vec{\mathbf{v}}_1$ is the unit vector of the incident ray, $\vec{\mathbf{v}}_2$ is unit vector of the refracted ray, and $\vec{\mathbf{n}}_a$ is the normal vector of the first surface (see figure 6.1):

$$\begin{aligned} \vec{\mathbf{v}}_1 &= w_x \vec{\mathbf{e}}_1 + w_y \vec{\mathbf{e}}_2 + w_z \vec{\mathbf{e}}_3, \\ \vec{\mathbf{v}}_2 &= \wp_x \vec{\mathbf{e}}_1 + \wp_y \vec{\mathbf{e}}_2 + \wp_z \vec{\mathbf{e}}_3, \\ \vec{\mathbf{n}}_a &= \Phi_x \vec{\mathbf{e}}_1 + \Phi_y \vec{\mathbf{e}}_2 + \Phi_z \vec{\mathbf{e}}_3, \end{aligned} \tag{6.16}$$

where $\vec{\mathbf{e}}_1$ denotes the x direction, $\vec{\mathbf{e}}_2$ denotes the y direction and $\vec{\mathbf{e}}_3$ denotes the z direction. The components of the normal vector are given by the well-known expression for the normal vector of a surface in $\mathbb{R}^3$:

$$\vec{\mathbf{n}}_a = \Phi_x \vec{\mathbf{e}}_1 + \Phi_y \vec{\mathbf{e}}_2 + \Phi_z \vec{\mathbf{e}}_3 - \pm \frac{\dfrac{\partial(x_a \vec{\mathbf{e}}_1 + y_a \vec{\mathbf{e}}_2 + z_a \vec{\mathbf{e}}_3)}{\partial x} \times \dfrac{\partial(x_a \vec{\mathbf{e}}_1 + y_a \vec{\mathbf{e}}_2 + z_a \vec{\mathbf{e}}_3)}{\partial y}}{\left| \dfrac{\partial(x_a \vec{\mathbf{e}}_1 + y_a \vec{\mathbf{e}}_2 + z_a \vec{\mathbf{e}}_3)}{\partial x} \times \dfrac{\partial(x_a \vec{\mathbf{e}}_1 + y_a \vec{\mathbf{e}}_2 + z_a \vec{\mathbf{e}}_3)}{\partial y} \right|}. \tag{6.17}$$

By computing the partial derivatives and cross-products in equation (6.17), we get the x component of the normal vector of equation (6.17), as follows:

$$\Phi_x = \frac{\partial_x y_a \partial_y z_a - \partial_x z_a \partial_y y_a}{\sqrt{(\partial_x x_a \partial_y y_a - \partial_x y_a \partial_y x_a)^2 + (\partial_x z_a \partial_y x_a - \partial_x x_a \partial_y z_a)^2 + (\partial_x y_a \partial_y z_a - \partial_x z_a \partial_y y_a)^2}}; \tag{6.18}$$

the y component of the normal is

$$\Phi_y = \frac{\partial_x z_a \partial_y x_a - \partial_x x_a \partial_y z_a}{\sqrt{(\partial_x x_a \partial_y y_a - \partial_x y_a \partial_y x_a)^2 + (\partial_x z_a \partial_y x_a - \partial_x x_a \partial_y z_a)^2 + (\partial_x y_a \partial_y z_a - \partial_x z_a \partial_y y_a)^2}}; \tag{6.19}$$

and the z component is

$$\Phi_z = \frac{\partial_x x_a \partial_y y_a - \partial_x y_a \partial_y x_a}{\sqrt{(\partial_x x_a \partial_y y_a - \partial_x y_a \partial_y x_a)^2 + (\partial_x z_a \partial_y x_a - \partial_x x_a \partial_y z_a)^2 + (\partial_x y_a \partial_y z_a - \partial_x z_a \partial_y y_a)^2}}. \tag{6.20}$$

Now, that we know the normal vector of the mirror, we proceed to compute Snell's law at the surface of the mirror; we begin with the term $[\vec{\mathbf{v}}_1 - (\vec{\mathbf{n}}_a \cdot \vec{\mathbf{v}}_1)\vec{\mathbf{n}}_a]$ of equation (6.15):

$$\begin{aligned}[\vec{\mathbf{v}}_1 - (\vec{\mathbf{n}}_a \cdot \vec{\mathbf{v}}_1)\vec{\mathbf{n}}_a] &= \left[-w_x\left(\Phi_y^2 + \Phi_z^2\right) + w_y\Phi_x\Phi_y + w_z\Phi_x\Phi_z\right]\vec{\mathbf{e}}_1 \\ &+ \left[w_y(\Phi_x^2 + \Phi_z^2) - \Phi_y(w_x\Phi_x + w_z\Phi_z)\right]\vec{\mathbf{e}}_2 \\ &+ \left[w_z\left(\Phi_x^2 + \Phi_y^2\right) - \Phi_z(w_x\Phi_x + w_y\Phi_y)\right]\vec{\mathbf{e}}_3;\end{aligned} \tag{6.21}$$

we then compute the squared root term of equation (6.15):

$$\begin{aligned}&- \vec{\mathbf{n}}_a\sqrt{1 - (\vec{\mathbf{n}}_a \times \vec{\mathbf{v}}_1)^2} = (\Phi_x\vec{\mathbf{e}}_1 + \Phi_y\vec{\mathbf{e}}_2 + \Phi_z\vec{\mathbf{e}}_3) \\ &\times \sqrt{1 - \left[\left(w_y\Phi_x - w_x\Phi_y\right)^2 + (w_z\Phi_x - w_x\Phi_z)^2 + \left(w_z\Phi_y - w_y\Phi_z\right)^2\right]}.\end{aligned} \tag{6.22}$$

We obtain the cosine directors from equations (6.21) and (6.22). The direction cosine in the x direction is

$$\begin{aligned}\wp_x =& \left[- w_x\left(\Phi_y^2 + \Phi_z^2\right) + w_y\Phi_x\Phi_y + w_z\Phi_x\Phi_z\right] \\ &+ \Phi_x\sqrt{1 - \left[\left(w_y\Phi_x - w_x\Phi_y\right)^2 + (w_z\Phi_x - w_x\Phi_z)^2 + \left(w_z\Phi_y - w_y\Phi_z\right)^2\right]},\end{aligned} \tag{6.23}$$

in the y direction, it is

$$\begin{aligned}\wp_y =& \left[w_y(\Phi_x^2 + \Phi_z^2) - \Phi_y(w_x\Phi_x + w_z\Phi_z)\right] \\ &+ \Phi_y\sqrt{1 - \left[\left(w_y\Phi_x - w_x\Phi_y\right)^2 + (w_z\Phi_x - w_x\Phi_z)^2 + \left(w_z\Phi_y - w_y\Phi_z\right)^2\right]},\end{aligned} \tag{6.24}$$

and in the z direction, it is

$$\begin{aligned}\wp_z =& \left[w_z\left(\Phi_x^2 + \Phi_y^2\right) - \Phi_z(w_x\Phi_x + w_y\Phi_y)\right] \\ &+ \Phi_z\sqrt{1 - \left[\left(w_y\Phi_x - w_x\Phi_y\right)^2 + (w_z\Phi_x - w_x\Phi_z)^2 + \left(w_z\Phi_y - w_y\Phi_z\right)^2\right]}.\end{aligned} \tag{6.25}$$

Equation (6.4) gives the input cosine directors, and equations (6.23), (6.24), and (6.25) give the output cosine directors. Thus, we can obtain the eikonal using the piecewise function given below, which depends on t and the independent variables x and y:

$$\boxed{H(t,\, r_a) = \begin{cases} [x + tw_x,\, y + tw_y,\, w + tw_z], & t < \vartheta, \\ \left[x_a + \wp_x(t - \vartheta),\, y_a + \wp_y(t - \vartheta),\, z_a + \wp_z(t - \vartheta)\right], & t > \vartheta, \end{cases}} \tag{6.26}$$

Naturally, ϑ (given by equation (6.11)) is the dividing condition in the piecewise function, since it is the distance that the input ray travels from the input wavefront to the surface of the mirror.

6.4 Mathematica code

In this section we present an example of a mirror designed using equations (6.12), (6.13), and (6.14) and its wavefronts, given by equation (6.26).

```
In[8]:=   w[x_,y_]:=Cos[x/3]+Cos[y/5]

          e=10^(-5);

          wdx[x_,y_]:=(w[x+e,y]-w[x,y])/e

          wdy[x_,y_]:=(w[x,y+e]-w[x,y])/e
```

$$wx[x\_,y\_]:=\frac{-wdx[x,y]}{\sqrt{wdx[x,y]^2+wdy[x,y]^2+1}}$$

$$wy[x\_,y\_]:=\frac{-wdy[x,y]}{\sqrt{wdx[x,y]^2+wdy[x,y]^2+1}}$$

$$wz[x\_,y\_]:=\frac{1}{\sqrt{wdx[x,y]^2+wdy[x,y]^2+1}}$$

```
          τ[x_,y_]:=w[x,y]+zi-zo
```

$$\vartheta[\text{x\_},\text{y\_}]:=\frac{\text{k}^2\text{-x}^2\text{-y}^2\text{-}\boldsymbol{\tau}[\text{x,y}]^2}{2\ (\text{k+wx[x,y] x+wy[x,y] y+wz[x,y]}\ \boldsymbol{\tau}[\text{x,y}])};$$

xa[x_,y_]:=x+ϑ[x,y]wx[x,y]

ya[x_,y_]:=y+ϑ[x,y]wy[x,y]

za[x_,y_]:=w[x,y]+ϑ[x,y]wz[x,y]

k=zo+zi;

rmax=10;

zi=20;

zo=60;

$$\text{wz2[x\_,y\_]:=}\frac{\text{-(za[x,y]+zi-zo)}}{\sqrt{\text{(za[x,y]+zi-zo)}^2\text{+(xa[x,y])}^2\text{+(ya[x,y])}^2}}$$

$$\text{wx2[x\_,y\_]:=}\frac{\text{-(xa[x,y])}}{\sqrt{\text{(za[x,y]+zi-zo)}^2\text{+(xa[x,y])}^2\text{+(ya[x,y])}^2}}$$

$$\text{wy2[x\_,y\_]:=}\frac{\text{-(ya[x,y])}}{\sqrt{\text{(za[x,y]+zi-zo)}^2\text{+(xa[x,y])}^2\text{+(ya[x,y])}^2}}$$

```
Homotopy={{
  Piecewise[{
   {wx[x,y]tt+x,tt<ϑ[x,y]},{  wx2[x,y](tt-ϑ[x,y])+xa[x,y],tt>ϑ[x,y]}}],
  Piecewise[{
   {wy[x,y]tt+y,tt<ϑ[x,y]},{  wy2[x,y](tt-ϑ[x,y])+ya[x,y],tt>ϑ[x,y]}}],
  Piecewise[{
  {wz[x,y]tt+w[x,y],tt<ϑ[x,y]},{ wz2[x,y](tt-ϑ[x,y])+za[x,y],tt>ϑ[x,y]}}]}};
```

```
Waves:=Table[{ParametricPlot3D[Homotopy,{x,-rmax,rmax},{y,-rmax,rmax},
PlotRange→All,PlotStyle→{Blue,Opacity[.5]},PlotPoints→10,Mesh→False]},
{tt,5,zi+zo,10}]

Rays:=Table[{ParametricPlot3D[Homotopy,{tt,0,zi+zo},
PlotRange→All,PlotStyle→{Purple,Opacity[.5]},PlotPoints→10,Mesh→False]},
{x,-rmax,rmax,rmax},{y,-rmax,rmax,rmax}]

g0=Plot3D[{w[x,y]},{x,-rmax,rmax},{y,-rmax,rmax},Axes→True,
PlotStyle→{Blue,Opacity[.5]},AxesLabel→{x,y,z},AspectRatio→Automatic,
Boxed→True,Mesh→False];

g1=ParametricPlot3D[{xa[x,y],ya[x,y],za[x,y]},{x,-rmax,rmax},{y,-rmax,rmax},
Axes→True,PlotStyle→{Red,Opacity[.95]},Boxed→True,AxesLabel→{x,y,z},
AspectRatio→Automatic,Mesh→False,PlotPoints→10];

Show[g1,g0,Waves,Rays,PlotRange→All]
```

6.5 Examples

Figures 6.2, 6.3, 6.4, 6.5, 6.6, and 6.7 display the outputs of the algorithm presented in the previous section. Based on the figures shown in this gallery, we can confirm once again the robustness of the optical path theory behind the shapes of the mirrors in these examples.

The performance should not be surprising, since we mentioned previously that equations (6.12), (6.13), and (6.14) were obtained without using any approximations and only assume Fermat's principle.

6.6 The crux of adaptive optics

For reasons of didacticism and clarity, we presented the equations of chapter 2 in two dimensions. In this chapter, we follow the same strategy, but in $\mathbb{R}^3$. In this section, we now turn to the mirror that converts an input wavefront w into a flat wavefront. The variables used are the same as in section 4.2, except for the definitions of the following new parameters: x_i is the distance from the vertex of the mirror to the image in the x direction and y_i is the distance from the vertex of the mirror to the image in the y direction (see figure 6.8). The other parameters presented in the figure are the same as those described in section 4.2.1.

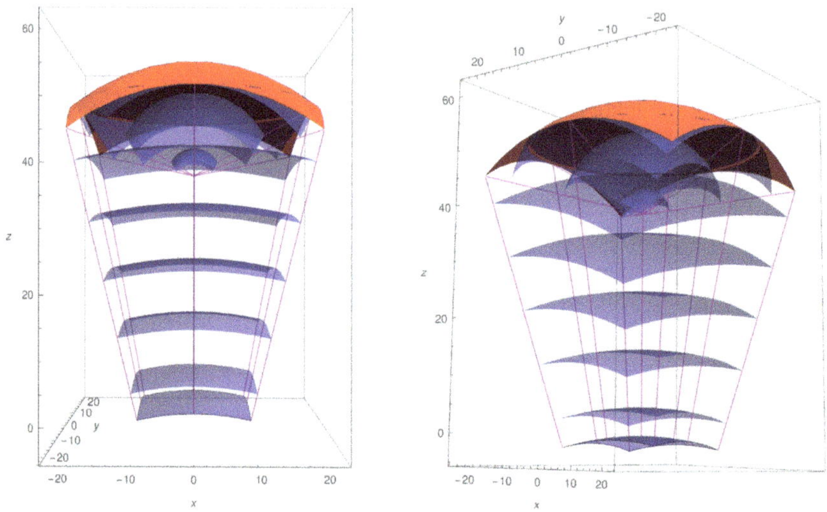

(a) A frontal view point of the adaptative mirror system. (b) A frontal perspective view point of the adaptative mirror system.

Figure 6.2. Mirror design specifications for a parabolic input wavefront; the system produces a stigmatic output wavefront: (a) the input wavefront is $-\frac{(x)^2+(y^2)}{12}$, refractive index $n = 1.5$, object distance $z_o = 20$ mm, image distance $z_i = 60$ mm, $\mathrm{r}_{\max} = 10$ mm, $(r_a, z_a) =$ equations (6.12), (6.13), and (6.14), $H(t, r_a) =$ equation (6.26); (b) a frontal perspective of the same configuration.

Our first step is the obtain the OPL of an axial ray, which is the one that goes from the vertex of the input wavefront w to the vertex of the mirror (x_a, y_a, z_a) and then to the image. Note that in this step, the image is at a finite distance; later, we will set it at infinity. With this in mind, we can obtain the OPL of the axial ray using

$$\mathrm{OPL} = \sqrt{x_i^2 + y_i^2 + z_i^2} + z_o, \tag{6.27}$$

and the OPL of the non-axial ray from

$$\begin{aligned}\mathrm{OPL} = &\sqrt{(x_a - x_i)^2 + \left(y_a - y_i\right)^2 + (z_a - z_i - z_o)^2} \\ &+ \sqrt{(x_a - x)^2 + \left(y_a - y\right)^2 + (z_a - w)^2} = \sqrt{x_i^2 + y_i^2 + z_i^2} + z_o.\end{aligned} \tag{6.28}$$

Both must be equal because of the shape of (x_a, y_a, z_a); therefore, we can write

$$\begin{aligned}&\sqrt{(x_a - x_i)^2 + \left(y_a - y_i\right)^2 + (z_a - z_i - z_o)^2} \\ &+ \sqrt{(x_a - x)^2 + \left(y_a - y\right)^2 + (z_a - w)^2} = \sqrt{x_i^2 + y_i^2 + z_i^2} + z_o.\end{aligned} \tag{6.29}$$

The next step is to set the image at infinity. First, we compute the limit $x_i \to pz_i$ in equation (6.29):

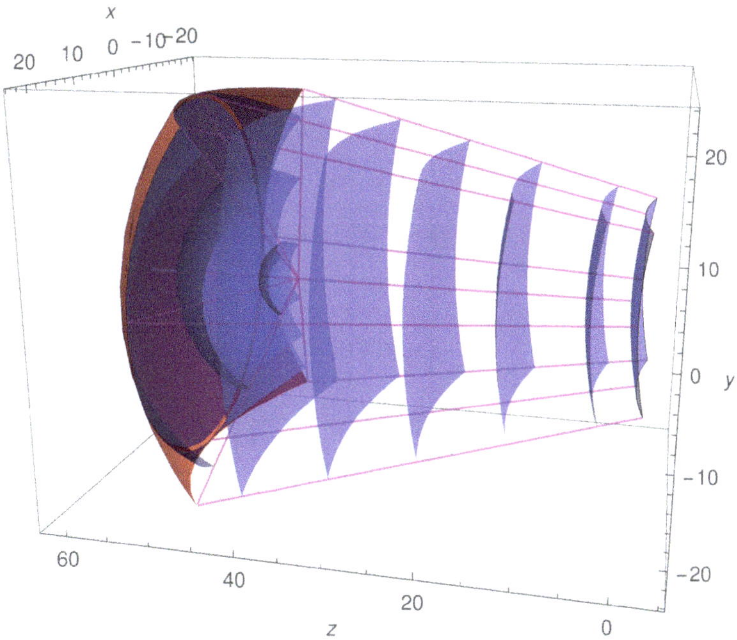

(a) A frontal left perspective view point of the adaptative mirror system.

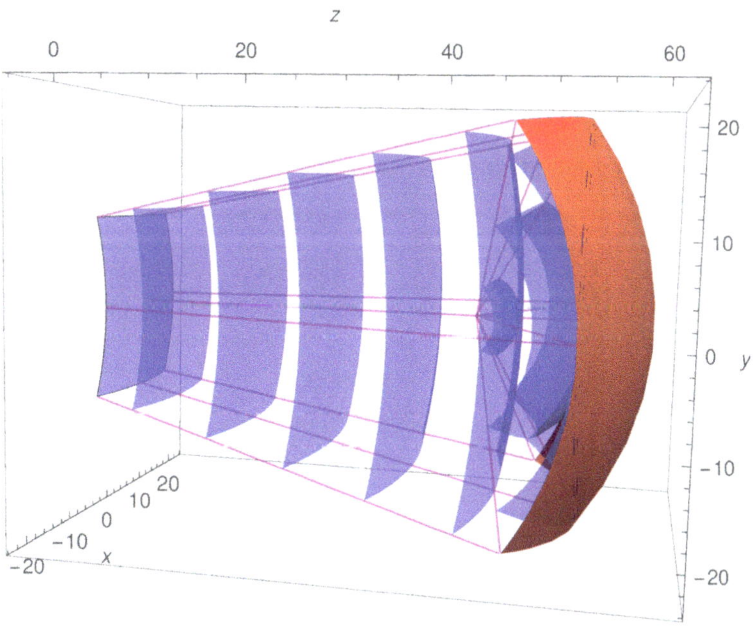

(b) A frontal right perspective view point of the adaptative mirror system.

Figure 6.3. Mirror design specifications for a parabolic input wavefront; the system produces a stigmatic output wavefront: (a) a frontal left perspective of the same configuration as that shown in figure 6.2; (b) a frontal right perspective of the same configuration as that shown in figure 6.2.

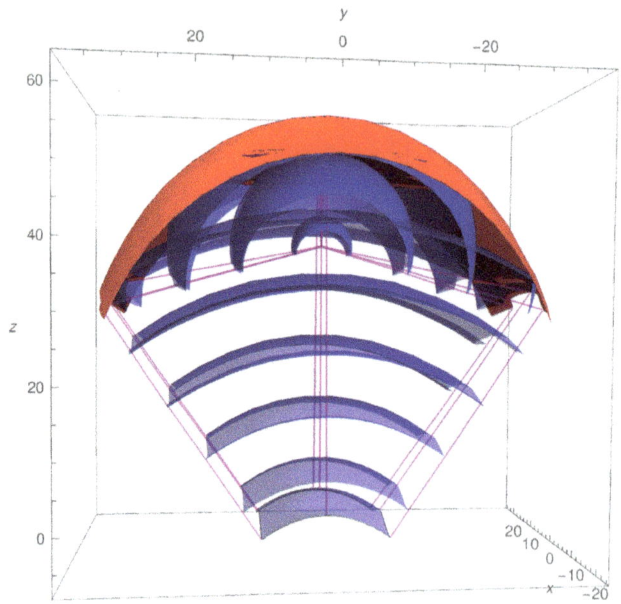

(a) A frontal view point of the adaptative mirror system.

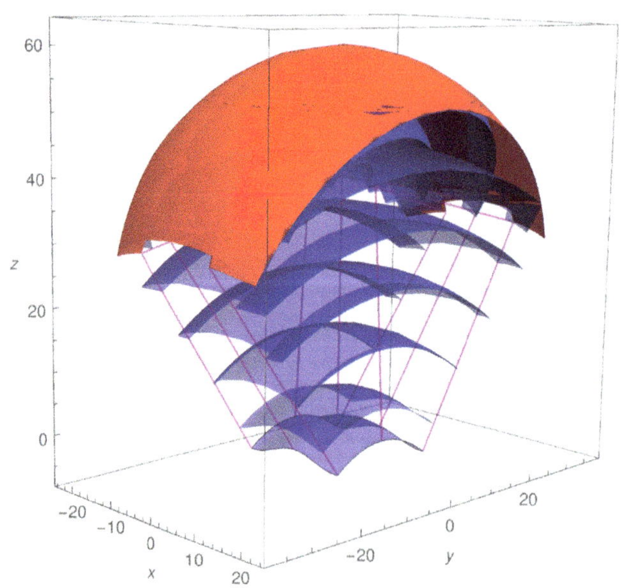

(b) A frontal perspective view point of the adaptative mirror system.

Figure 6.4. Mirror design specifications for an arbitrary input wavefront; the system produces a stigmatic output wavefront: (a) the input wavefront is $\cos(x/3) - \frac{(y^2)}{28}$, refractive index $n = 1.5$, object distance $z_o = 20$ mm, image distance $z_i = 60$ mm, $r_{max} = 10$ mm, $(r_a, z_a) =$ equations (6.12), (6.13), and (6.14), $H(t, r_a) =$ equation (6.26); (b) a frontal perspective of the same configuration.

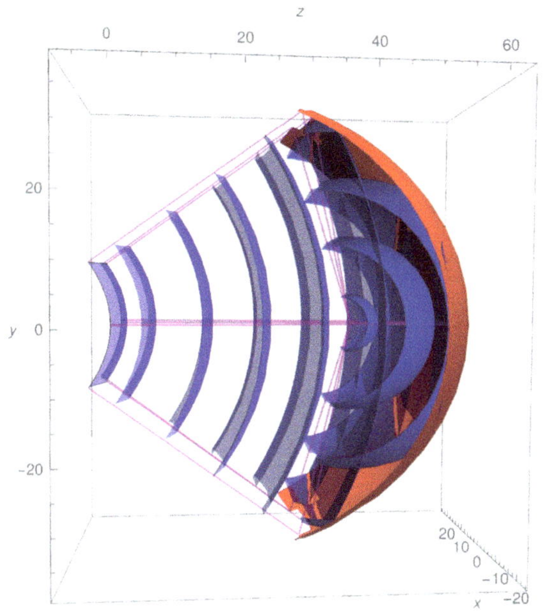

(a) A frontal left perspective view point of the adaptative mirror system.

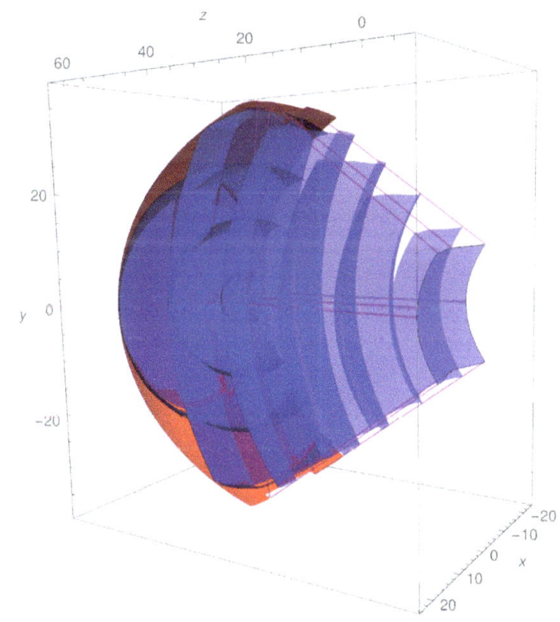

(b) A frontal right perspective view point of the adaptative mirror system.

Figure 6.5. Mirror design specifications for a parabolic input wavefront; the system produces a stigmatic output wavefront: (a) a frontal left perspective of the same configuration as that shown in figure 6.4; (b) a frontal right perspective of the same configuration as that shown in figure 6.4.

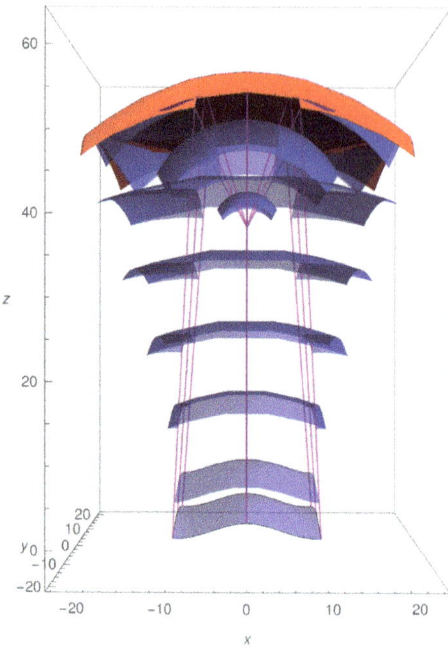

(a) A frontal view point of the adaptative mirror system.

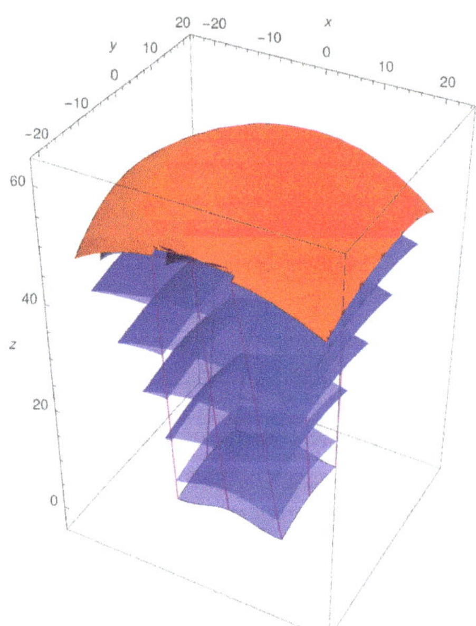

(b) A frontal perspective view point of the adaptative mirror system.

Figure 6.6. Mirror design specifications for an arbitrary input wavefront; the system produces a stigmatic output wavefront: (a) the input wavefront is $\cos(x/3) - \cos(y/5)$, refractive index $n = 1.5$, object distance $z_o = 20$ mm, image distance $z_i = 60$ mm, $r_{max} = 10$ mm, $(r_a, z_a) =$ equations (6.12), (6.13), and (6.14), $H(t, r_a) =$ equation (6.26); (b) a frontal perspective of the same configuration.

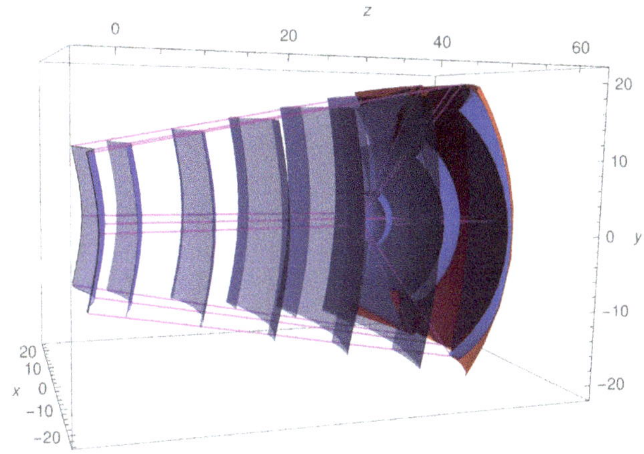

(a) A frontal left perspective view point of the adaptative mirror system.

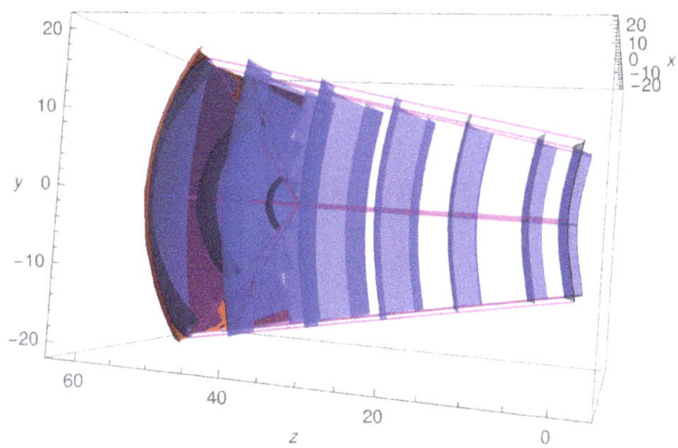

(b) A frontal right perspective view point of the adaptative mirror system.

Figure 6.7. Mirror design specifications for a parabolic input wavefront; the system produces a stigmatic output wavefront; (a) a frontal left perspective of the same configuration as that shown in figure 6.6; (b) a frontal right perspective of the same configuration as that shown in figure 6.6.

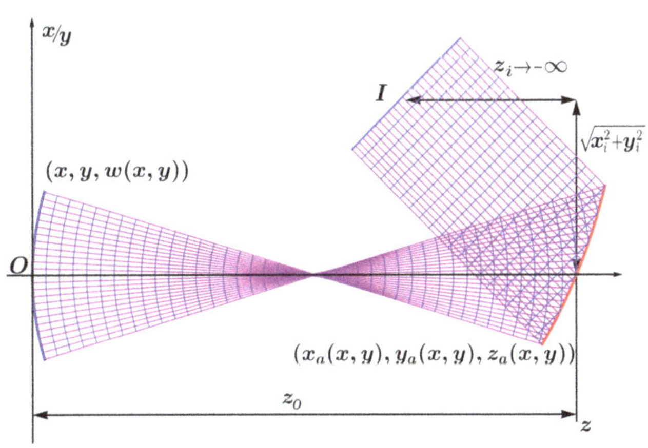

Figure 6.8. Schematic diagram of the configuration of an adaptive mirror with an image at infinity.

$$\lim_{x_i \to pz_i}\left[\sqrt{(x_a - x_i)^2 + (y_a - y_i)^2 + (z_a - z_i - z_o)^2}\right.$$
$$\left.+ \sqrt{(x_a - x)^2 + (y_a - y)^2 + (z_a - w)^2} - \sqrt{x_i^2 + y_i^2 + z_i^2} - z_o\right]$$
$$= -\sqrt{y_i^2 + (p^2 + 1)z_i^2} + \sqrt{(y_i - y_a)^2 + (x_a - pz_i)^2 + (-z_a + z_i + z_o)^2}$$
$$+ \sqrt{(w - z_a)^2 + (x - x_a)^2 + (y - y_a)^2} - z_o. \tag{6.30}$$

Second, we compute the limit $y_i \to sz_i$ in equation (6.30):

$$\lim_{y_i \to sz_i}\left[-\sqrt{y_i^2 + (p^2 + 1)z_i^2} + \sqrt{(y_i - y_a)^2 + (x_a - pz_i)^2 + (-z_a + z_i + z_o)^2}\right.$$
$$\left.+ \sqrt{(w - z_a)^2 + (x - x_a)^2 + (y - y_a)^2} - z_o\right]$$
$$= -\sqrt{z_i^2(p^2 + s^2 + 1)} + \sqrt{(x_a - pz_i)^2 + (y_a - sz_i)^2 + (-z_a + z_i + z_o)^2}$$
$$+ \sqrt{(w - z_a)^2 + (x - x_a)^2 + (y - y_a)^2} - z_o, \tag{6.31}$$

We then need to express the shape of the mirror in terms of the wavefront w and the independent variables x and y. From the last section we know that the cosine directors of the wavefront w in the three-dimensional model are

$$w_x = \frac{-\partial_x w}{\sqrt{\partial_x w^2 + \partial_y w^2 + 1}}, \quad w_y = \frac{-\partial_y w}{\sqrt{\partial_x w^2 + \partial_y w^2 + 1}}, \quad w_z = \frac{1}{\sqrt{\partial_x w^2 + \partial_y w^2 + 1}}, \tag{6.32}$$

where $\partial_x \equiv \partial w/\partial x$ is the derivative of w to x, $\partial_y w \equiv \partial w/\partial y$ is the derivative of w to y, and w_x, w_y, and w_z are the cosine directors in the x, y, and z directions, respectively. Thus, the surface of the mirror in three dimensions is given by equation (6.5), which we recall here:

$$x_a = x + \vartheta w_x, \quad y_a = y + \vartheta w_y, \quad z_a = w + \vartheta w_z, \tag{6.33}$$

where ϑ is given by equation (6.6), which we recall below:

$$\vartheta = \sqrt{(x_a - x)^2 + (y_a - y)^2 + (z_a - w)^2}. \tag{6.34}$$

We substitute equations (6.5) and (6.6) into equation (6.31) and compute the limit $z_i \to \infty$ in order to get a plane wave at the output of the system:

$$\lim_{z_i \to \infty}\left[-\sqrt{z_i^2(p^2 + s^2 + 1)} + \vartheta - z_o\right.$$
$$\left.+ \sqrt{\left(x + \vartheta - pz_i\right)^2 + (y + \vartheta - sz_i)^2 + [-(w + \vartheta w_z) + z_i + z_o]^2}\right]$$
$$= \frac{\lim_{z_i \to \infty}(\vartheta)(pw_x + sw_y + w_z) + px + sy + w - z_o}{\sqrt{p^2 + s^2 + 1}} - z_o + \lim_{z_i \to \infty}(\vartheta). \tag{6.35}$$

Note that the result of the limit $z_i \to \infty$ over equation (6.35) must be zero, since it is the computation over the OPD:

$$\frac{\lim_{z_i\to\infty}(\vartheta)(pw_x + sw_y + w_z) + px + sy + w - z_o}{\sqrt{p^2 + s^2 + 1}} - z_o + \lim_{z_i\to\infty}(\vartheta) = 0. \tag{6.36}$$

We can solve the last expression for $\lim_{z_i\to\infty}(\vartheta)$:

$$\lim_{z_i\to\infty}(\vartheta) = -\frac{-z_o\sqrt{p^2 + s^2 + 1} + px + sy + w - z_o}{\sqrt{p^2 + s^2 + 1} + pw_x + sw_y + w_z}, \tag{6.37}$$

and using ϑ, we can express the mirror surface properly. We substitute equation (6.37) into equation (6.5), obtaining the following in x:

$$\boxed{\lim_{z_i\to\infty}(x_a) = x - \frac{\partial_x w\left(w - z_o\left(\sqrt{p^2 + s^2 + 1} + 1\right) + px + sy\right)}{-\sqrt{p^2 + s^2 + 1}\sqrt{\partial_x w^2 + \partial_y w^2 + 1} + p\partial_x w + s\partial_y w - 1}}, \tag{6.38}$$

the following in y:

$$\boxed{\lim_{z_i\to\infty}(y_a) = y - \frac{\partial_y w\left(w - z_o\left(\sqrt{p^2 + s^2 + 1} + 1\right) + px + sy\right)}{-\sqrt{p^2 + s^2 + 1}\sqrt{\partial_x w^2 + \partial_y w^2 + 1} + p\partial_x w + s\partial_y w - 1}}, \tag{6.39}$$

and the following in z,

$$\boxed{\lim_{z_i\to\infty}(z_a) = w + \frac{w - z_o\left(\sqrt{p^2 + s^2 + 1} + 1\right) + px + sy}{-\sqrt{p^2 + s^2 + 1}\sqrt{\partial_x w^2 + \partial_y w^2 + 1} + p\partial_x w + s\partial_y w - 1}}. \tag{6.40}$$

Equations (6.38), (6.39), and (6.40) are the most important equations of the chapter and also the most important equations of the treatise.

Observe the similarities between the equations given in chapter 2 and the equations of this section. Since equations (6.38), (6.39), and (6.40) are expressed in three dimensions, they are the ones needed for applications. Since equations (2.37) and (2.38) are expressed in two dimensions, they are best suited for visualisations. Each set of equations is fully analytical and closed form and is derived from Fermat's principle. Thus, both share the same properties and limitations discussed in chapter 2. We recall them below.

We used neither the paraxial concept, nor numerical approximations, nor optimisations to obtain equations (6.38), (6.39), and (6.40). Equations (6.38), (6.39), and (6.40) are analytic closed-form solutions to the proposed problem. Thus, we can confirm that equations (6.38), (6.39), and (6.40) are natural implications of Fermat's principle. They are *a priori* predicted by the optical path theory. Based on the procedure used to obtain equations (6.38), (6.39), and (6.40), we can conclude that they implicitly represent a uniqueness theorem. This means that for given input values, equations (6.38), (6.38), and (6.40) compute a single mirror that transforms an input wavefront w into a perfect plane wave. This implies that the set of mirrors designed using equations (6.38), (6.39), and 6.40 contains all the possible mirrors that produce a plane wave from the input wavefront w.

6.7 The eikonal of the crux of adaptive optics

We are interested in obtaining the wavefront at all times t and at all positions. To get this wavefront we only need to know the cosine directors of the input and output vectors. The former are given by equation (6.4). For the cosine directors of the output vectors, we can compute the limit $z_i \to \infty$ over the normal vector of equation (6.3). To compute this task, we can, without loss of generality, take the forms of expression of equations (6.23), (6.24), and (6.25) and replace the terms x_a, y_a, and z_a with equations (6.38), (6.39), and (6.40). Therefore, the direction cosine of the reflected ray in x is

$$\begin{aligned}
\lim_{z_i\to\infty}(\wp_x) = &\left\{-w_x\left[\lim_{z_i\to\infty}(\Phi_y^2) + \lim_{z_i\to\infty}(\Phi_z^2)\right]\right.\\
&\left.+ w_y \lim_{z_i\to\infty}(\Phi_x)\lim_{z_i\to\infty}(\Phi_y) + w_z \lim_{z_i\to\infty}(\Phi_x)\lim_{z_i\to\infty}(\Phi_z)\right\}\\
&+ \lim_{z_i\to\infty}(\Phi_x)\left(1 - \left\{\left[w_y \lim_{z_i\to\infty}(\Phi_x) - w_x \lim_{z_i\to\infty}(\Phi_y)\right]^2\right.\right.\\
&+ \left[w_z \lim_{z_i\to\infty}(\Phi_x) - w_x \lim_{z_i\to\infty}(\Phi_z)\right]^2\\
&\left.\left.+ \left[w_z \lim_{z_i\to\infty}(\Phi_y) - w_y \lim_{z_i\to\infty}(\Phi_z)\right]^2\right\}\right)^{\frac{1}{2}};
\end{aligned} \tag{6.41}$$

in y, equation (6.24) changes to

$$\begin{aligned}
\lim_{z_i\to\infty}(\wp_y) = &\left\{\left[w_y\left[\lim_{z_i\to\infty}(\Phi_x^2) + \lim_{z_i\to\infty}(\Phi_z^2)\right]\right.\right.\\
&\left.\left.- \lim_{z_i\to\infty}(\Phi_y)\left[w_x \lim_{z_i\to\infty}(\Phi_x) + w_z \lim_{z_i\to\infty}(\Phi_z)\right]\right]\right\}\\
&+ \lim_{z_i\to\infty}(\Phi_y)\left(1 - \left\{\left[w_y \lim_{z_i\to\infty}(\Phi_x) - w_x \lim_{z_i\to\infty}(\Phi_y)\right]^2\right.\right.\\
&+ \left[w_z \lim_{z_i\to\infty}(\Phi_x) - w_x \lim_{z_i\to\infty}(\Phi_z)\right]^2\\
&\left.\left.+ \left[w_z \lim_{z_i\to\infty}(\Phi_y) - w_y \lim_{z_i\to\infty}(\Phi_z)\right]^2\right\}\right)^{\frac{1}{2}};
\end{aligned} \tag{6.42}$$

and in z, equation (6.25) changes to

$$
\begin{aligned}
\lim_{z_i\to\infty}(\wp_z) = &\left\{\left[w_z\left[\lim_{z_i\to\infty}(\Phi_x^2) + \lim_{z_i\to\infty}(\Phi_y^2)\right]\right.\right.\\
&\left.\left. - \lim_{z_i\to\infty}(\Phi_z)\left[w_x \lim_{z_i\to\infty}(\Phi_x) + w_y \lim_{z_i\to\infty}(\Phi_y)\right]\right]\right\}\\
&+ \lim_{z_i\to\infty}(\Phi_y)\left(1 - \left\{\left[w_y \lim_{z_i\to\infty}(\Phi_x) - w_x \lim_{z_i\to\infty}(\Phi_y)\right]^2\right.\right.\\
&+\left[w_z \lim_{z_i\to\infty}(\Phi_x) - w_x \lim_{z_i\to\infty}(\Phi_z)\right]^2\\
&\left.\left.+\left[w_z \lim_{z_i\to\infty}(\Phi_y) - w_y \lim_{z_i\to\infty}(\Phi_z)\right]^2\right\}\right)^{\frac{1}{2}},
\end{aligned}
\tag{6.43}
$$

where $\lim_{z_i\to\infty}(\Phi_x)$ is the x component of the normal vector of the surface of the mirror designed using equations (6.17), (6.18), (6.19), and (6.20). In these equations, we just replace (x_a, y_a, z_a) with equations (6.38), (6.39), and (6.40), so equation (6.18) changes to

$$
\begin{aligned}
\lim_{z_i\to\infty}(\Phi_x) = &\left[\partial_x \lim_{z_i\to\infty}(y_a)\partial_y \lim_{z_i\to\infty}(z_a) - \partial_x \lim_{z_i\to\infty}(z_a)\partial_y \lim_{z_i\to\infty}(y_a)\right]\\
&\times\left\{\left[\partial_x \lim_{z_i\to\infty}(x_a)\partial_y \lim_{z_i\to\infty}(y_a) - \partial_x \lim_{z_i\to\infty}(y_a)\partial_y \lim_{z_i\to\infty}(x_a)\right]^2\right.\\
&+\left[\partial_x \lim_{z_i\to\infty}(z_a)\partial_y \lim_{z_i\to\infty}(x_a) - \partial_x \lim_{z_i\to\infty}(x_a)\partial_y \lim_{z_i\to\infty}(z_a)\right]^2\\
&\left.+\left[\partial_x \lim_{z_i\to\infty}(y_a)\partial_y \lim_{z_i\to\infty}(z_a) - \partial_x \lim_{z_i\to\infty}(z_a)\partial_y \lim_{z_i\to\infty}(y_a)\right]^2\right\}^{\frac{1}{2}};
\end{aligned}
\tag{6.44}
$$

$\lim_{z_i\to\infty}(\Phi_y)$ is the y component of the normal vector of the surface of the mirror designed by substituting (6.38), (6.39), and (6.40) into equation (6.19), so equation (6.19) changes to

$$
\begin{aligned}
\lim_{z_i\to\infty}(\Phi_y) = &\left[\partial_x \lim_{z_i\to\infty}(z_a)\partial_y \lim_{z_i\to\infty}(x_a) - \partial_x \lim_{z_i\to\infty}(x_a)\partial_y \lim_{z_i\to\infty}(z_a)\right]\\
&\times\left\{\left[\partial_x \lim_{z_i\to\infty}(x_a)\partial_y \lim_{z_i\to\infty}(y_a) - \partial_x \lim_{z_i\to\infty}(y_a)\partial_y \lim_{z_i\to\infty}(x_a)\right]^2\right.\\
&+\left[\partial_x \lim_{z_i\to\infty}(z_a)\partial_y \lim_{z_i\to\infty}(x_a) - \partial_x \lim_{z_i\to\infty}(x_a)\partial_y \lim_{z_i\to\infty}(z_a)\right]^2\\
&\left.+\left[\partial_x \lim_{z_i\to\infty}(y_a)\partial_y \lim_{z_i\to\infty}(z_a) - \partial_x \lim_{z_i\to\infty}(z_a)\partial_y \lim_{z_i\to\infty}(y_a)\right]^2\right\}^{\frac{1}{2}};
\end{aligned}
\tag{6.45}
$$

$\lim_{z_i \to \infty}(\Phi_y)$ is the z component of the normal vector of the surface of the mirror designed by substituting (6.38), (6.39), and (6.40) into equation (6.20), so equation (6.20) changes to

$$\begin{aligned}\lim_{z_i\to\infty}(\Phi_z) = &\left[\partial_x \lim_{z_i\to\infty}(x_a)\partial_y \lim_{z_i\to\infty}(y_a) - \partial_x \lim_{z_i\to\infty}(y_a)\partial_y \lim_{z_i\to\infty}(x_a)\right] \\ &\times\left\{\left[\partial_x \lim_{z_i\to\infty}(x_a)\partial_y \lim_{z_i\to\infty}(y_a) - \partial_x \lim_{z_i\to\infty}(y_a)\partial_y \lim_{z_i\to\infty}(x_a)\right]^2\right. \\ &+\left[\partial_x \lim_{z_i\to\infty}(z_a)\partial_y \lim_{z_i\to\infty}(x_a) - \partial_x \lim_{z_i\to\infty}(x_a)\partial_y \lim_{z_i\to\infty}(z_a)\right]^2 \\ &\left.+\left[\partial_x \lim_{z_i\to\infty}(y_a)\partial_y \lim_{z_i\to\infty}(z_a) - \partial_x \lim_{z_i\to\infty}(z_a)\partial_y \lim_{z_i\to\infty}(y_a)\right]^2\right\}^{\frac{1}{2}}.\end{aligned} \tag{6.46}$$

Note that we just replace the function that describes the mirror in equations (6.12), (6.13), and (6.14) by equations (6.38), (6.39), and (6.40), which are the equations for the mirror such that the output wavefront is flat rather than spheric. In equations (6.44), (6.45), and (6.46), there are several parameters of interest related to the partial deviation of the mirror surface: $\partial_x \lim_{z_i\to\infty}(x_a)$ and $\partial_y \lim_{z_i\to\infty}(x_a)$ are the partial derivatives with respect to x and y of equation (6.38). $\partial_x \lim_{z_i\to\infty}(x_a)$ is given by

$$\begin{aligned}\partial_x \lim_{z_i\to\infty}(x_a) = 1 &+ \frac{\partial_x w(\partial_x w + p)}{\sqrt{p^2+s^2+1}\sqrt{\partial_x w^2 + \partial_y w^2 + 1} - p\partial_x w - s\partial_y w + 1} - \\ &\frac{\partial_{xx} w\left(w - z_o\left(\sqrt{p^2+s^2+1}+1\right) + px + sy\right)}{-\sqrt{p^2+s^2+1}\sqrt{\partial_x w^2 + \partial_y w^2 + 1} + p\partial_x w + s\partial_y w - 1} \times \\ &\left\{\partial_x w\left(\partial_x w\partial_{xx} w + \partial_y w\partial_{yx} w\right)\left(w - z_o\left(\sqrt{p^2+s^2+1}+1\right) + px + sy\right)\right\} \times \\ &\left\{\left(\partial_x w^2 + \partial_y w^2 + 1\right)\left[-\sqrt{p^2+s^2+1}\sqrt{\partial_x w^2 + \partial_y w^2 + 1} + p\partial_x w + s\partial_y w - 1\right]\right\}^{-1} \\ &+ \left\{\partial_x w\left(w - z_o\left(\sqrt{p^2+s^2+1}+1\right) + px + sy\right)\right. \\ &\left.\left(\partial_{xx} w\left(p\partial_y w^2 - s\partial_x w\partial_y w + \partial_x w + p\right) + \partial_{yx} w\left(-p\partial_x w\partial_y w + s\partial_x w^2 + \partial_y w + s\right)\right)\right\} \times \\ &\left\{\left(\partial_x w^2 + \partial_y w^2 + 1\right) \times \left(-\sqrt{p^2+s^2+1}\sqrt{\partial_x w^2 + \partial_y w^2 + 1} + p\partial_x w + s\partial_y w - 1\right)^2\right\}^{-1},\end{aligned} \tag{6.47}$$

and $\partial_y \lim_{z_i\to\infty}(x_a)$ is given by

$$
\begin{aligned}
\partial_y \lim_{z_i \to \infty} (x_a) = {} & - \Big\{ \partial_x w \big(\partial_y w + s \big) \\
& \times \Big(-\sqrt{p^2 + s^2 + 1} \sqrt{\partial_x w^2 + \partial_y w^2 + 1} + p \partial_x w + s \partial_y w - 1 \Big) \\
& - \Big[\partial_x w \big(\partial_x w \partial_{xy} w + \partial_y w \partial_{yy} w \big) \Big(w - z_o \Big(\sqrt{p^2 + s^2 + 1} + 1 \Big) \\
& + px + sy \Big) \Big(-\sqrt{p^2 + s^2 + 1} \sqrt{\partial_x w^2 + \partial_y w^2 + 1} + p \partial_x w + s \partial_y w - 1 \Big) \Big] \\
& \times \Big[\partial_x w^2 + \partial_y w^2 + 1 \Big]^{-1} - \Big[\partial_x w \Big(w - z_o \Big(\sqrt{p^2 + s^2 + 1} + 1 \Big) + px + sy \Big) \\
& \times \Big(\partial_{xy} w \big(p \partial_y w^2 - s \partial_x w \partial_y w + \partial_x w + p \big) \\
& \quad + \partial_{yy} w \big(-p \partial_x w \partial_y w + s \partial_x w^2 + \partial_y w + s \big) \Big) \Big] \\
& \times \Big[\partial_x w^2 + \partial_y w^2 + 1 \Big]^{-1} \Big\} \\
& \quad \Big\{ \Big(-\sqrt{p^2 + s^2 + 1} \sqrt{\partial_x w^2 + \partial_y w^2 + 1} + p \partial_x w + s \partial_y w - 1 \Big)^2 \Big\}^{-2}.
\end{aligned}
\tag{6.48}
$$

$\partial_x \lim_{z_i \to \infty}(y_a)$ and $\partial_y \lim_{z_i \to \infty}(y_a)$ are the partial derivatives with respect to x and y of the y component of the mirror (see equation (6.39)). $\partial_x \lim_{z_i \to \infty}(y_a)$ is given by

$$
\begin{aligned}
\partial_x \lim_{z_i \to \infty} (y_a) = {} & \Big\{ \partial_y w (\partial_x w + p) \Big(\sqrt{p^2 + s^2 + 1} \sqrt{\partial_x w^2 + \partial_y w^2 + 1} - p \partial_x w - s \partial_y w + 1 \Big) \\
& - \partial_{yx} w \Big(-\sqrt{p^2 + s^2 + 1} \sqrt{\partial_x w^2 + \partial_y w^2 + 1} + p \partial_x w + s \partial_y w - 1 \Big) \\
& \times \Big(w - z_o \Big(\sqrt{p^2 + s^2 + 1} + 1 \Big) + px + sy \Big) + \Big[\partial_x w^2 + \partial_y w^2 + 1 \Big]^{-1} \\
& \times \Big[\partial_y w \Big(w - z_o \Big(\sqrt{p^2 + s^2 + 1} + 1 \Big) + px + sy \Big) \times \big(\partial_x w \partial_{xx} w + \partial_y w \partial_{yx} w \big) \\
& \times \Big(-\sqrt{p^2 + s^2 + 1} \sqrt{\partial_x w^2 + \partial_y w^2 + 1} + p \partial_x w + s \partial_y w - 1 \Big) \Big] \\
& \quad + \Big[\partial_x w^2 + \partial_y w^2 + 1 \Big]^{-1} \\
& \times \Big[\Big(\partial_{xx} w \big(p \partial_y w^2 - s \partial_x w \partial_y w + \partial_x w + p \big) \partial_{yx} w \\
& \quad + \big(-p \partial_x w \partial_y w + s \partial_x w^2 + \partial_y w + s \big) \Big) \Big] \Big\} \\
& \times \Big\{ -\sqrt{p^2 + s^2 + 1} \sqrt{\partial_x w^2 + \partial_y w^2 + 1} + p \partial_x w + s \partial_y w - 1 \Big\}^{-2},
\end{aligned}
\tag{6.49}
$$

and $\partial_y \lim_{z_i \to \infty}(y_a)$ is given by

$$
\begin{aligned}
\partial_y \lim_{z_i\to\infty}(y_a) = {} & 1 + \frac{\partial_y w\left(\partial_y w + s\right)}{\sqrt{p^2 + s^2 + 1}\sqrt{\partial_x w^2 + \partial_y w^2 + 1} - p\partial_x w - s\partial_y w + 1} \\
& - \frac{\partial_{yy} w\left(w - z_o\left(\sqrt{p^2 + s^2 + 1} + 1\right) + px + sy\right)}{-\sqrt{p^2 + s^2 + 1}\sqrt{\partial_x w^2 + \partial_y w^2 + 1} + p\partial_x w + s\partial_y w - 1} \\
& + \left\{\partial_y w\left(\partial_x w \partial_{xy} w + \partial_y w \partial_{yy} w\right)\left(w - z_o\left(\sqrt{p^2 + s^2 + 1} + 1\right) + px + sy\right)\right\} \\
& \times \left\{\left(\partial_x w^2 + \partial_y w^2 + 1\right)\left(-\sqrt{p^2 + s^2 + 1}\sqrt{\partial_x w^2 + \partial_y w^2 + 1} + p\partial_x w + s\partial_y w - 1\right)\right\}^{-1} \\
& + \left\{\partial_y w\left(w - z_o\left(\sqrt{p^2 + s^2 + 1} + 1\right) + px + sy\right)\right. \\
& \times \left(\partial_{xy} w\left(p\partial_y w^2 - s\partial_x w\partial_y w + \partial_x w + p\right) + \partial_{yy} w\left(-p\partial_x w\partial_y w + s\partial_x w^2 + \partial_y w + s\right)\right)\Big\} \\
& \times \left\{\left(\partial_x w^2 + \partial_y w^2 + 1\right)\left(-\sqrt{p^2 + s^2 + 1}\sqrt{\partial_x w^2 + \partial_y w^2 + 1} + p\partial_x w + s\partial_y w - 1\right)^2\right\}^{-1}.
\end{aligned}
\tag{6.50}
$$

Finally, $\partial_x \lim\limits_{z_i\to\infty}(y_a)$ and $\partial_y \lim\limits_{z_i\to\infty}(y_a)$ are the partial derivatives with respect to x and y of the z component of the mirror (the profile of the mirror), derived from equation (6.40) with respect to x and y. $\partial_y \lim\limits_{z_i\to\infty}(z_a)$ is

$$
\begin{aligned}
\partial_x \lim_{z_i\to\infty}(z_a) = {} & \partial_x w + \frac{\partial_x w + p}{-\sqrt{p^2 + s^2 + 1}\sqrt{\partial_x w^2 + \partial_y w^2 + 1} + p\partial_x w + s\partial_y w - 1} \\
& + \left\{\left(\partial_x w \partial_{xx} w + \partial_y w \partial_{yx} w\right)\left(w - z_o\left(\sqrt{p^2 + s^2 + 1} + 1\right) + px + sy\right)\right\} \\
& \times \left\{\left(\partial_x w^2 + \partial_y w^2 + 1\right)\left(\sqrt{p^2 + s^2 + 1}\sqrt{\partial_x w^2 + \partial_y w^2 + 1} - p\partial_x w - s\partial_y w + 1\right)\right\}^{-1} \\
& + \left\{\left(w - z_o\left(\sqrt{p^2 + s^2 + 1} + 1\right) + px + sy\right)\right. \\
& \times \left(\partial_{xx} w\left(p\partial_y w^2 - s\partial_x w\partial_y w + \partial_x w + p\right) + \partial_{yx} w\left(-p\partial_x w\partial_y w + s\partial_x w^2 + \partial_y w + s\right)\right)\Big\} \\
& \times \left\{\left(\partial_x w^2 + \partial_y w^2 + 1\right)\left(-\sqrt{p^2 + s^2 + 1}\sqrt{\partial_x w^2 + \partial_y w^2 + 1} + p\partial_x w + s\partial_y w - 1\right)^2\right\}^{-1},
\end{aligned}
\tag{6.51}
$$

and $\partial_y \lim\limits_{z_i\to\infty}(z_a)$ is

$$
\begin{aligned}
\partial_y \lim_{z_i\to\infty}(z_a) = {} & \partial_y w + \frac{\partial_y w + s}{-\sqrt{p^2 + s^2 + 1}\sqrt{\partial_x w^2 + \partial_y w^2 + 1} + p\partial_x w + s\partial_y w - 1} \\
& + \left\{\left(\partial_x w \partial_{xy} w + \partial_y w \partial_{yy} w\right)\left(w - z_o\left(\sqrt{p^2 + s^2 + 1} + 1\right) + px + sy\right)\right\} \\
& \times \left\{\left(\partial_x w^2 + \partial_y w^2 + 1\right)\left(\sqrt{p^2 + s^2 + 1}\sqrt{\partial_x w^2 + \partial_y w^2 + 1} - p\partial_x w - s\partial_y w + 1\right)\right\}^{-1} \\
& + \left\{\left(w - z_o\left(\sqrt{p^2 + s^2 + 1} + 1\right) + px + sy\right)\right. \\
& \times \left(\partial_{xy} w\left(p\partial_y w^2 - s\partial_x w\partial_y w + \partial_x w + p\right) + \partial_{yy} w\left(-p\partial_x w\partial_y w + s\partial_x w^2 + \partial_y w + s\right)\right)\Big\} \\
& \times \left\{\left(\partial_x w^2 + \partial_y w^2 + 1\right)\left(-\sqrt{p^2 + s^2 + 1}\sqrt{\partial_x w^2 + \partial_y w^2 + 1} + p\partial_x w + s\partial_y w - 1\right)^2\right\}.
\end{aligned}
\tag{6.52}
$$

It is important to remark that to calculate $z_i \to \infty$ over equations (6.12), (6.13), and (6.14), we set the output wavefront to a plane wavefront with a known direction related to the constants p and s. As a consequence we obtain equations (6.38), (6.39), and (6.40) and equations (6.41), (6.43), and (6.42), but in fact they can be drastically reduced. Equation (6.41) changes to

$$\lim_{z_i \to \infty} (\wp_x) = \frac{p}{\sqrt{p^2 + s^2 + 1}}, \tag{6.53}$$

equation (6.42) becomes

$$\lim_{z_i \to \infty} (\wp_y) = \frac{s}{\sqrt{p^2 + s^2 + 1}}, \tag{6.54}$$

and equation (6.43) changes to

$$\lim_{z_i \to \infty} (\wp_z) = \frac{1}{\sqrt{p^2 + s^2 + 1}}. \tag{6.55}$$

Using equations (6.4), (6.53), (6.55), and (6.54), we construct the eikonal as a piecewise function using the condition designed by ϑ from equation (6.37); the eikonal can therefore be written as follows:

$$\boxed{\lim_{z_i \to \infty} H(t, r_a) = \begin{cases} [x + tw_x,\ y + tw_y,\ w + tw_z], & t < \lim\limits_{z_i \to \infty} (\vartheta), \\ \begin{bmatrix} \lim\limits_{z_i \to \infty} (x_a) + \lim\limits_{z_i \to \infty} (\wp_x) \left\{ t - \lim\limits_{z_i \to \infty} (\vartheta) \right\}, \\ \lim\limits_{z_i \to \infty} (y_a) + \lim\limits_{z_i \to \infty} (\wp_y) \left\{ t - \lim\limits_{z_i \to \infty} (\vartheta) \right\}, \\ \lim\limits_{z_i \to \infty} (z_a) + \lim\limits_{z_i \to \infty} (\wp_z) \left\{ t - \lim\limits_{z_i \to \infty} (\vartheta) \right\} \end{bmatrix}, & t > \lim\limits_{z_i \to \infty} (\vartheta), \end{cases}} . \tag{6.56}$$

Equation (6.56) along with equations (6.38), (6.39), and (6.40) contain the physical information most relevant to the problem proposed in this section.

6.8 Mathematica code

This section presents Mathematica code for a freeform deformable mirror based on equations (6.38), (6.39), (6.40), and (6.56).

```
In[9]:= w[x_,y_]:=50

e=10^(-5);
wdx[x_,y_]:=(w[x+e,y]-w[x,y])/e

wdy[x_,y_]:=(w[x,y+e]-w[x,y])/e
```

$$wx[x\_,y\_]:=\frac{-wdx[x,y]}{\sqrt{wdx[x,y]^2+wdy[x,y]^2+1}}$$

$$wy[x\_,y\_]:=\frac{-wdy[x,y]}{\sqrt{wdx[x,y]^2+wdy[x,y]^2+1}}$$

$$wz[x\_,y\_]:=\frac{1}{\sqrt{wdx[x,y]^2+wdy[x,y]^2+1}}$$

$$\vartheta[x\_,y\_]:=-\frac{w[x,y]+p\ x+s\ y-zo-\sqrt{1+p^2+s^2}\ zo}{\sqrt{1+p^2+s^2}+p\ wx[x,y]+s\ wy[x,y]+wz[x,y]}$$

```
xa[x_,y_]:=x+ϑ[x,y]wx[x,y]

ya[x_,y_]:=y+ϑ[x,y]wy[x,y]

za[x_,y_]:=w[x,y]+ϑ[x,y]wz[x,y]

dxxa[x_,y_]:=(xa[x+e,y]-xa[x,y])/e

dyxa[x_,y_]:=(xa[x,y+e]-xa[x,y])/e

dxya[x_,y_]:=(ya[x+e,y]-ya[x,y])/e

dyya[x_,y_]:=(ya[x,y+e]-ya[x,y])/e

dxza[x_,y_]:=(za[x+e,y]-za[x,y])/e
```

```
dyza[x_,y_]:=(za[x,y+e]-za[x,y])/e
```

```
Φx[x_,y_]:=(-dxza[x,y] dyya[x,y]+dxya[x,y] dyza[x,y])/
  (√((-dxya[x,y] dyxa[x,y]+dxxa[x,y] dyya[x,y])^2+
   (dxza[x,y] dyxa[x,y]-dxxa[x,y] dyza[x,y])^2+
   (-dxza[x,y] dyya[x,y]+dxya[x,y] dyza[x,y])^2))
```

```
Φy[x_,y_]:=(dxza[x,y] dyxa[x,y]-dxxa[x,y] dyza[x,y])/
  (√((-dxya[x,y] dyxa[x,y]+dxxa[x,y] dyya[x,y])^2+
   (dxza[x,y] dyxa[x,y]-dxxa[x,y] dyza[x,y])^2+
   (-dxza[x,y] dyya[x,y]+dxya[x,y] dyza[x,y])^2))
```

```
Φz[x_,y_]:=(-dxya[x,y] dyxa[x,y]+dxxa[x,y] dyya[x,y])/
  (√((-dxya[x,y] dyxa[x,y]+dxxa[x,y] dyya[x,y])^2+
   (dxza[x,y] dyxa[x,y]-dxxa[x,y] dyza[x,y])^2+
   (-dxza[x,y] dyya[x,y]+dxya[x,y] dyza[x,y])^2))
```

```
ϱx[x_,y_]=-(1/ni) no (wy[x,y] Φx[x,y] Φy[x,y]+wz[x,y] Φx[x,y] Φz[x,y]
     -wx[x,y] (Φy[x,y]^2+Φz[x,y]^2))+Φx[x,y] √(1-(1/ni^2) no^2
     ((wy[x,y] Φx[x,y]-wx[x,y] Φy[x,y])^2+(wz[x,y] Φx[x,y]-wx[x,y] Φz[x,y])^2
     +(wz[x,y] Φy[x,y]-wy[x,y] Φz[x,y])^2));
```

```
ϱy[x_,y_]=(1/ni) no (-Φy[x,y] (wx[x,y] Φx[x,y]+wz[x,y] Φz[x,y])
     +wy[x,y] (Φx[x,y]^2+Φz[x,y]^2))+Φy[x,y] √(1-(1/ni^2) no^2
     ((wy[x,y] Φx[x,y]-wx[x,y] Φy[x,y])^2+(wz[x,y] Φx[x,y]-wx[x,y] Φz[x,y])^2
     +(wz[x,y] Φy[x,y]-wy[x,y] Φz[x,y])^2));
```

```
ϱz[x_,y_]=(1/ni) no (wz[x,y] (Φx[x,y]^2+Φy[x,y]^2)-(wx[x,y] Φx[x,y]+wy[x,y] Φy[x,y])
    Φz[x,y])+Φz[x,y] √(1-(1/ni^2) no^2 ((wy[x,y] Φx[x,y] wx[x,y] Φy[x,y])^2
    +(wz[x,y] Φx[x,y]-wx[x,y] Φz[x,y])^2+(wz[x,y] Φy[x,y]-wy[x,y] Φz[x,y])^2));
```

```
wx2 = p/Sqrt[1+p^2+s^2];

wy2 = s/Sqrt[1+p^2+s^2];

wz2 = 1/Sqrt[1+p^2+s^2];


c=1;
v=c/ni;
no=1;
ni=-1;
rmax=10;
zo=60;
s=0.75;
p=0;


L[x_,y_]:=60-ϑ[x,y];


Homotopy={
{Piecewise[{
{wx[x,y]tt+x,tt<ϑ[x,y]},{- wx2(tt-ϑ[x,y])+xa[x,y],tt>ϑ[x,y]}}],
Piecewise[{
{wy[x,y]tt+y,tt<ϑ[x,y]},{ -wy2(tt-ϑ[x,y])+ya[x,y],tt>ϑ[x,y]}}],
Piecewise[{
{wz[x,y]tt+w[x,y],tt<ϑ[x,y]},{-wz2(tt-ϑ[x,y])+za[x,y],tt>ϑ[x,y]}}]}};


Waves:=Table[{ParametricPlot3D[Homotopy,{x,-rmax,rmax},{y,-rmax,rmax},
PlotRange→All,PlotStyle→{Blue,Opacity[.5]}, Mesh→True]},{tt,0,60,10}]
l[x_,y_]:=Graphics3D[{Purple,Line[{{x,y,w[x,y]},{xa[x,y],ya[x,y],za[x,y]}}]}]


wn[x_,y_]:=Graphics3D[{Red,Thick,
          Line[{{x,y,w[x,y]},{x+wx[x,y],y+wy[x,y],w[x,y]+wz[x,y]}}]}]


zan[x_,y_]:=Graphics3D[{Red,Thick, Line[{{xa[x,y],ya[x,y],za[x,y]},
          {xa[x,y]-Φx[x,y],ya[x,y]-Φy[x,y],za[x,y]-Φz[x,y] }}]}]
```

```
v2[x_,y_]:=Graphics3D[{Purple,Line[{{xa[x,y],ya[x,y],za[x,y]},{xa[x,y]-
        L[x,y] ϱx[x,y] ,ya[x,y]-L[x,y] ϱy[x,y] ,za[x,y]-L[x,y] ϱz[x,y] }}]}]

v2a[x_,y_]:=Graphics3D[{Purple,Line[{{xa[x,y],ya[x,y],za[x,y]},{xa[x,y]-
        L[x,y] wx2 ,ya[x,y]-L[x,y] wy2 ,za[x,y]-L[x,y]wz2 }}]}]

g0=Plot3D[{w[x,y]},{x,-rmax,rmax},{y,-rmax,rmax},
    Axes→True,PlotStyle→{Blue,Opacity[.5]},AxesLabel→{x,y,z},
    AspectRatio→Automatic,Boxed→False];

g1=ParametricPlot3D[{xa[x,y],ya[x,y],za[x,y]},{x,-rmax,rmax},{y,-rmax,rmax},
    Axes→True,PlotStyle→LightYellow,Boxed→True,AxesLabel→{x,y,z},
    AspectRatio→Automatic];

g2=ParametricPlot3D[{xa[x,y]-L[x,y] wx2 ,ya[x,y]-L[x,y] wy2 ,za[x,y]-L[x,y]wz2 },
{x,-rmax,rmax},{y,-rmax,rmax},
PlotRange→All,PlotStyle→Blue, Mesh→True,Boxed→True];

Show[g1,g0,Waves,Table[l[x,y],{x,-rmax,rmax,3},{y,-rmax,rmax,3}],
Table[v2a[x,y],{x,-rmax,rmax,3},{y,-rmax,rmax,3}],PlotRange→All]
```

6.9 Examples

In this subsection, we present Mathematica examples of the mirrors of the previous section in figures 6.9, 6.10, 6.11, 6.12, 6.13, and 6.14. It is very important to remark that the solutions provided by equations (6.38), (6.39), and (6.40) can work perfectly for any wavefront w that depends on time t. We mean that we can still use equations (6.38), (6.39), and 6.40 to simulate the design of an adaptive optics mirror for every w at time t. The equations are robust enough that at all times, the mirror will be such that it reflects the light of a given wavefront w as a plane wave. The deduction of equations (6.38), (6.39), and (6.40) reveals that equations (6.38), (6.39), and (6.40) are not only a general solution to this problem, but the solution is also unique. Thus, for given input values there is only one mirror that can accomplish the task mentioned in the introduction of this chapter. All of this is achieved without using complex conjugates; everything is ultimately related to Fermat's principle and its optical path theory.

Finally, the limitations of the mirror in chapter 3 are similar to those of the mirror in this section. If no more than two rays strike the same location on the mirror's surface, there can be no overlap on the surface. This occurs because the OPLs of

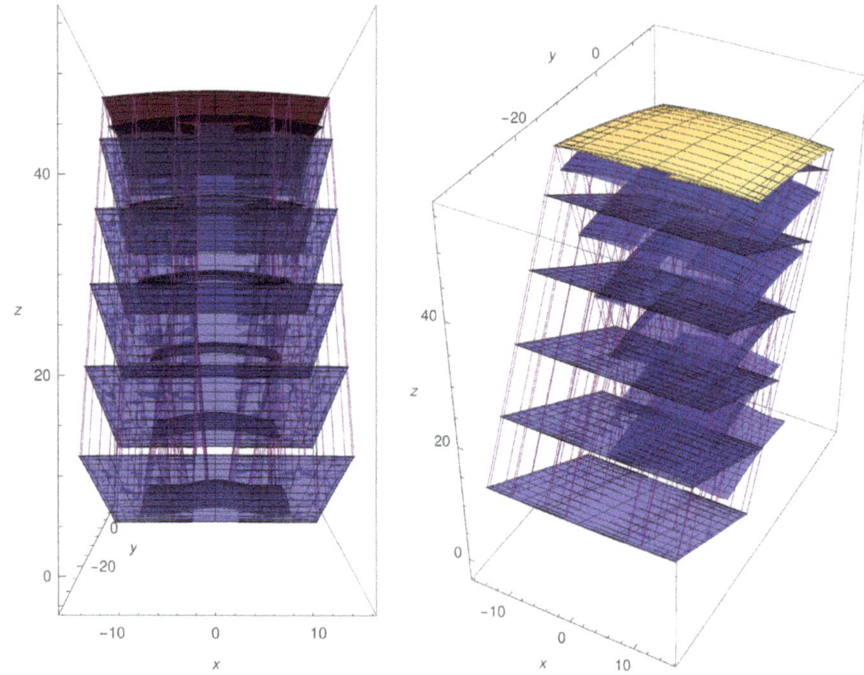

(a) A frontal view point of the adaptative mirror system.

(b) A frontal perspective view point of the adaptative mirror system.

Figure 6.9. Mirror design specifications for a parabolic input wavefront; the system produces a flat output wavefront: (a) the input wavefront is $-\cos(x/5)^2 + \cos(y/8)^2$, refractive index $n = 1.5$, object distance $z_o = 50$ mm, $\mathrm{r}_{\max} = 10$ mm, parameter $s = 0.5$; parameter $p = 0$; $(r_a, z_a) =$ equations (6.38), (6.39), and (6.40), $H(t, r_a) =$ equation (6.56); (b) a frontal perspective of the same configuration.

each ray must be the same, and we can conclude from equations (6.38), (6.39), and (6.40) that two rays cannot touch the same point of the mirror's surface. This occurs because equations (6.38), (6.39), and (6.40) describe the relationships between the OPLs of axial and non-axial rays, but they do not take into account the case in which a non-axial ray and its neighbour strike the same point.

The implication of equations (6.38), (6.39), and (6.40) is therefore that a non-axial ray and its neighbouring ray have the same OPL, and they should not strike the same point on the mirror's surface.

6.10 End notcs

In this chapter, we found out how to design a mirror that can reflect a wavefront's light as a spherical wave or a plane wave. To obtain the design of this mirror, we used Fermat's principle. The equations obtained are fully analytical and closed form. There was no need for paraxial calculations, numerical approaches, or optimization in the deduction. We tested the behaviour of the resulting mirrors in several scenarios, and in all cases, the results were satisfactory and as predicted.

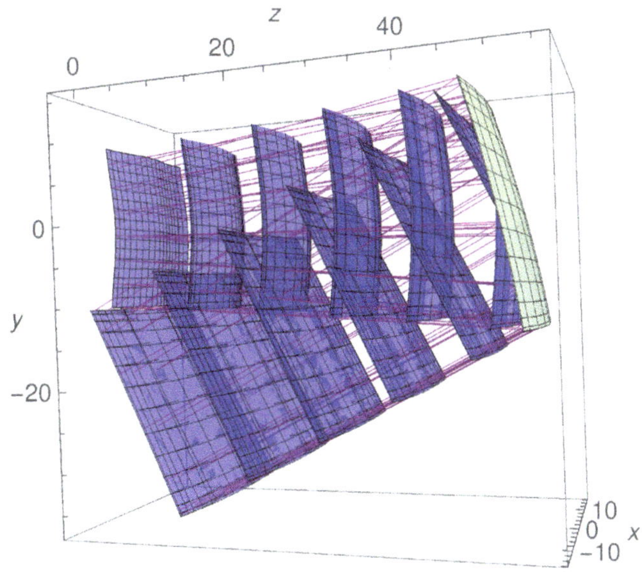

(a) A frontal left view point of the adaptative mirror system.

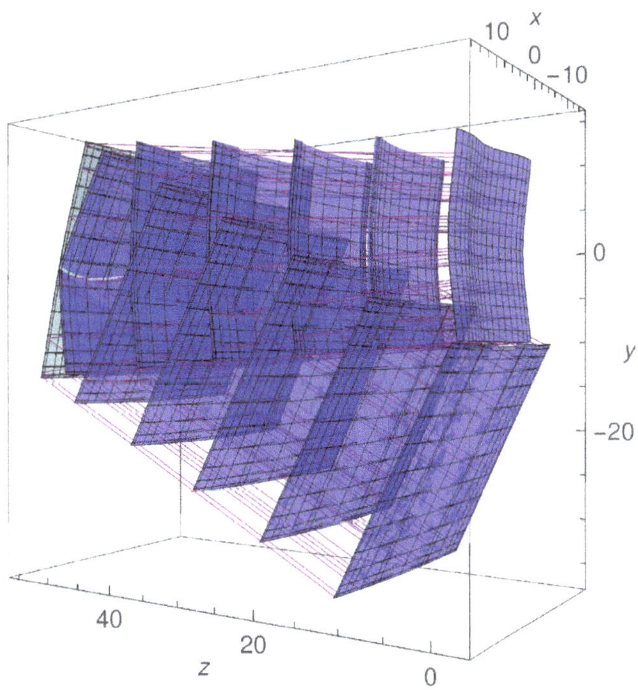

(b) A frontal right view point of the adaptative mirror system.

Figure 6.10. Mirror design specifications for a parabolic input wavefront; the system produces a flat output wavefront: (a) a frontal left perspective of the same configuration as that shown in figure 6.9; (b) a frontal right perspective of the same configuration as that shown in figure 6.9.

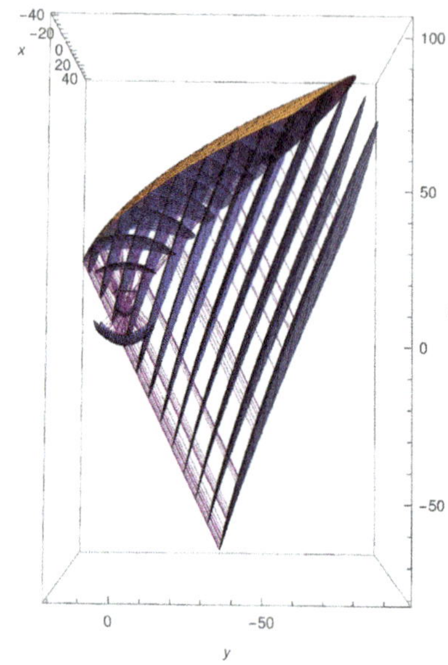

(a) A frontal view point of the adaptative mirror system.

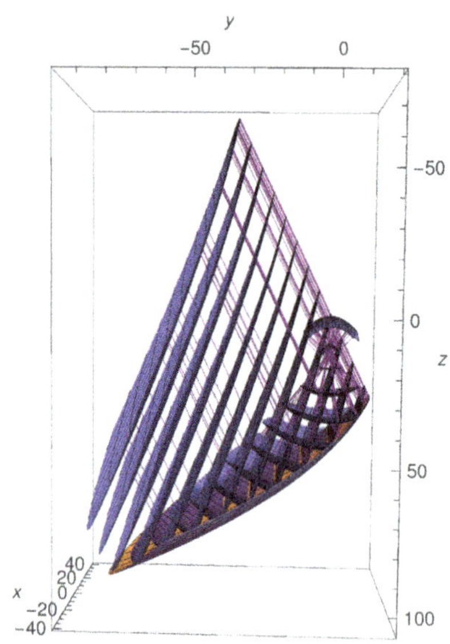

(b) A frontal perspective view point of the adaptative mirror system.

Figure 6.11. Mirror design specifications for an arbitrary input wavefront; the system produces a flat output wavefront: (a) the input wavefront is $\frac{(x)^2+(1.2y)^2}{28}$, refractive index $n = 1.5$, object distance $z_o = 50$ mm, $r_{max} = 10$ mm, parameter $s = 20$; parameter $p = 0$; (r_a, z_a) = equations (6.12), (6.13), and (6.14), $H(t, r_a)$ = equation (6.26); (b) a frontal perspective of the same configuration.

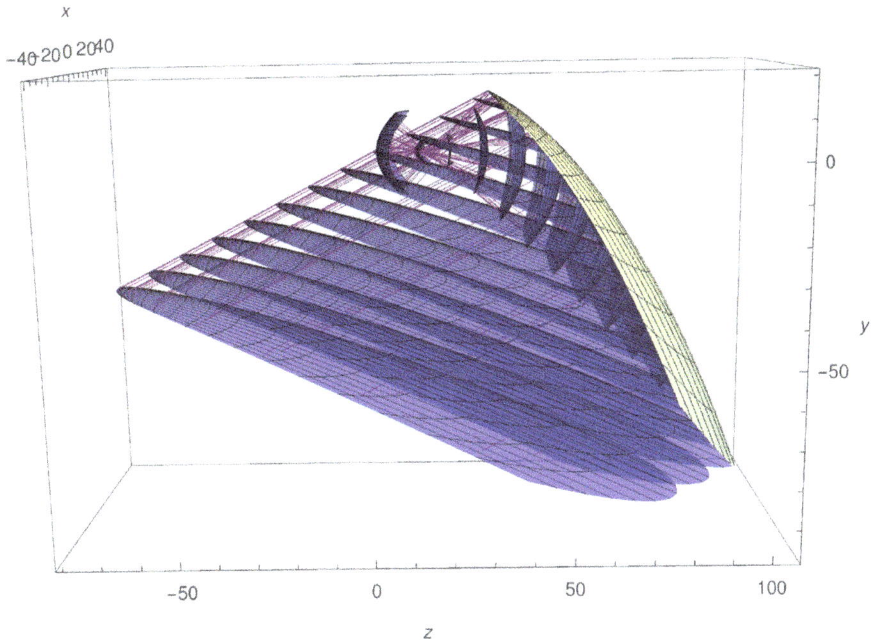

(a) A frontal left view point of the adaptative mirror system.

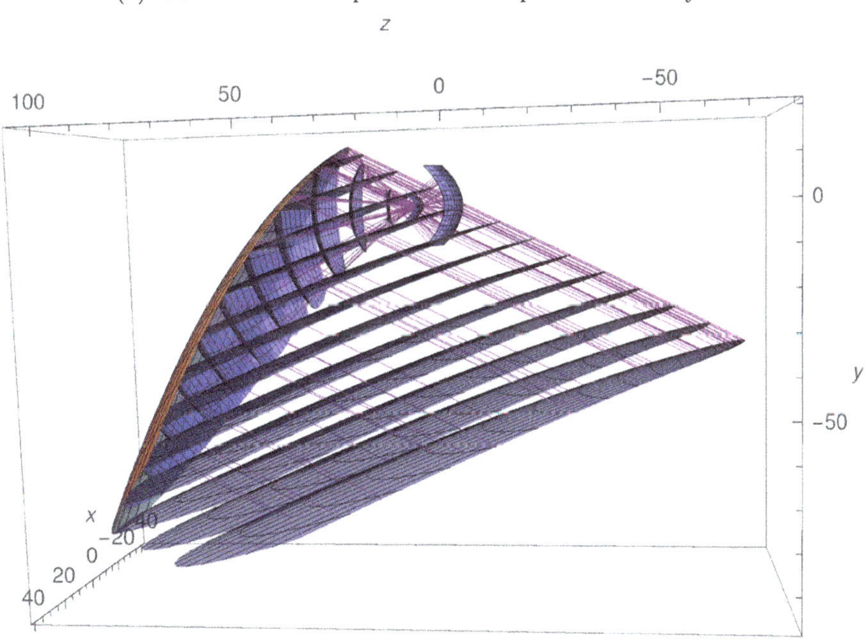

(b) A frontal right view point of the adaptative mirror system.

Figure 6.12. Mirror design specifications for a parabolic input wavefront; the system produces a stigmatic output wavefront: (a) a frontal left perspective of the same configuration as that shown in figure 6.11; (b) a frontal right perspective of the same configuration as that shown in figure 6.11.

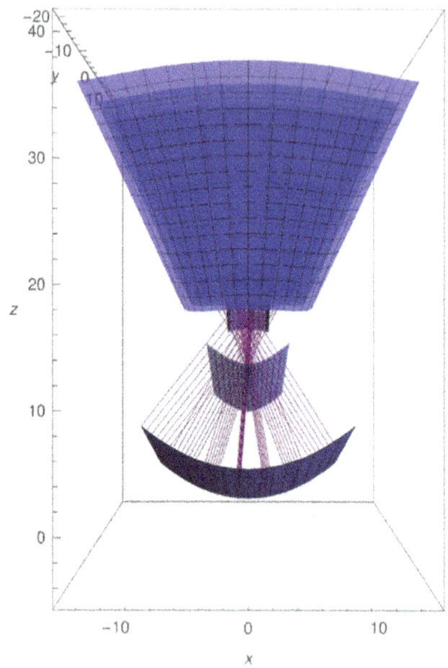

(a) A frontal view point of the adaptative mirror system.

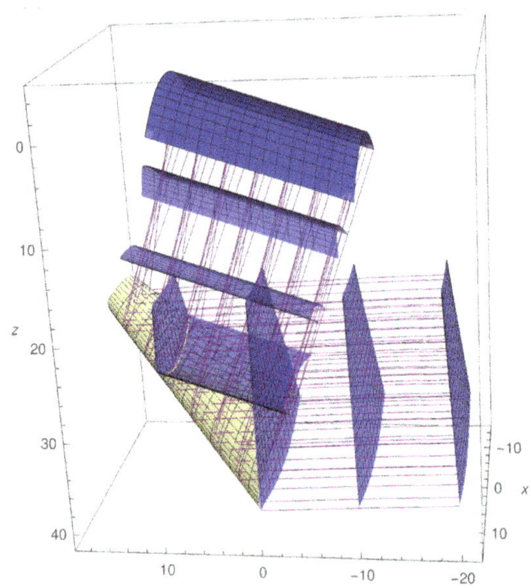

(b) A frontal perspective view point of the adaptative mirror system.

Figure 6.13. Mirror design specifications for an arbitrary input wavefront; the system produces a flat output wavefront: (a) the input wavefront is $(x/5)^2 - (y/3)$, refractive index $n = 1.5$, object distance $z_o = 40$ mm, $r_{max} = 10$ mm, parameter $s = 10$; parameter $p = 0$; (r_a, z_a) = equations (6.12), (6.13), and (6.14), $H(t, r_a)$ = equation (6.26); (b) a frontal perspective of the same configuration.

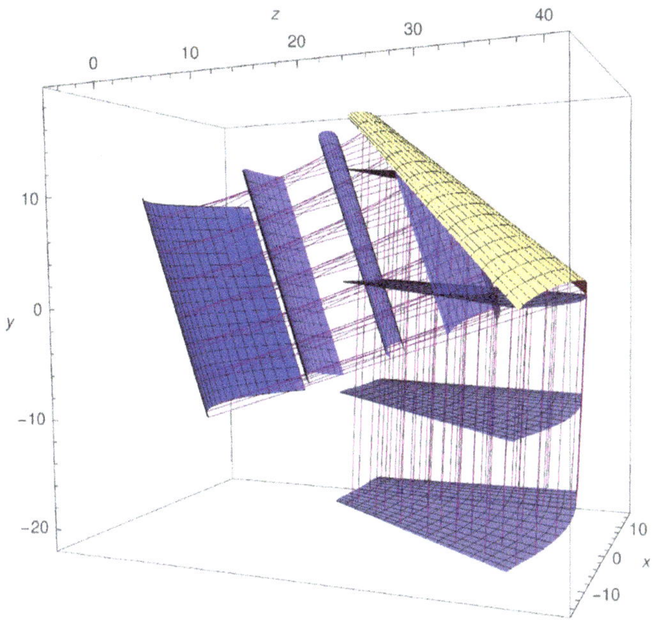

(a) A frontal left view point of the adaptative mirror system.

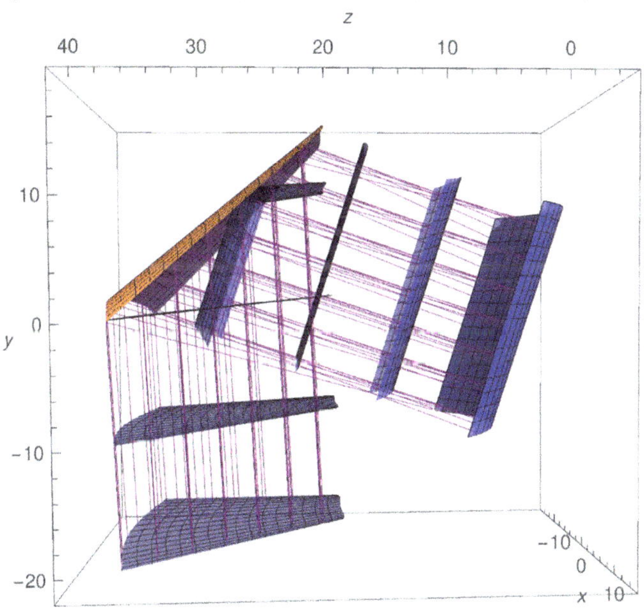

(b) A frontal right view point of the adaptative mirror system.

Figure 6.14. Mirror design specifications for a parabolic input wavefront; the system produces a flat output wavefront: (a) a frontal left perspective of the same configuration as that shown in figure 6.13; (b) a frontal right perspective of the same configuration as that shown in figure 6.13.

IOP Publishing

Optical Path Theory
Fundamentals to freeform adaptive optics
Rafael G González-Acuña and Héctor A Chaparro-Romo

Chapter 7

General freeform dioptric stigmatic surfaces

In this chapter, we derive the analytical surface that refracts a given arbitrary freeform input wavefront into a single image point or a wave with a flat wavefront. The derivation of the surface is fully analytical; it does not use paraxial or numerical approximations, and it only assumes Fermat's principle. The performance of the resulting surfaces is as predicted by the equations that described them.

7.1 Introduction

In chapter 3, we studied surfaces that we called general dioptric stigmatic surfaces. The main property of these surfaces is that they transform an arbitrary wavefront into a perfect point by means of refraction. The equations presented in chapter 3 only considered aspherical input wavefronts, and the image points were placed along the optical axis.

The model of chapter 3 was very interesting for pedagogical purposes, since all the parameters lay in $\mathbb{R}^2$. Their visualization is easier than in $\mathbb{R}^3$. Using the experience gained in chapter 3, generalization to $\mathbb{R}^3$ is straightforward. In this chapter, our objective is to generalize the model of chapter 3. This generalization consists of allowing the input wavefront to have a freeform geometry and removing the requirement for the image to be placed on the optical axis. The generalization of these stigmatic surfaces contains all the results presented in chapter 3; thus, the Cartesian oval can be considered to be a special case of general dioptric stigmatic surfaces in which the input and output wavefronts are spherical.

This chapter is organized as follows: we first discuss the generalization for freeform input wavefronts and spherical output wavefronts in sections 7.2, 7.3, 7.4, and 7.5. The generalization for freeform input wavefronts and flat output wavefronts is then presented in sections 7.6, 7.7, 7.8, and 7.9. Finally, conclusions are given in 7.10.

doi:10.1088/978-0-7503-4705-1ch7

7.2 Mathematical model of freeform stigmatic surfaces for images at finite distances

In the introduction we mentioned the goal of this chapter, which is that given a freeform wavefront w, we need to find a refractive surface (x_a, y_a, z_a) that refracts the input wavefront into a spherical wavefront centred at (z_i, x_i, y_i), where z_i, x_i, and y_i are the distances in the z, x, and y directions, respectively, as can be seen in figure 7.1. The distances z_i, x_i, and y_i are measured from the origin of the coordinate system. The vertex of w is placed at the origin of the coordinate system. The distance from the vertex of w to the surface is z_o.

Let n_o, n_i be the refractive indexes of the input and output spaces, respectively. Finally, the only independent variables of the system are x and y, so (x_a, y_a, z_a) is expressed in terms of x and y. In order for a surface to be a refractive stigmatic freeform surface, from Fermat's principle, we know that the optical path lengths (OPLs) of all the rays that are refracted by the refractive surface must be equal to a constant, k. Thus, the OPL of the axial ray and the OPL of a non-axial ray must be equal to k. The OPL of the axial ray is given by

$$k \equiv n_o z_o + n_i\sqrt{z_i^2 + x_i^2 + y_i^2}, \tag{7.1}$$

and the OPL of the no-axial ray is given by

$$\begin{aligned} k = n_i&\sqrt{(x_a - x_i)^2 + (y_a - y_i)^2 + (z_a - z_i - z_o)^2} \\ &+ n_o\sqrt{(z_a - w)^2 + (x_a - x)^2 + (y_a - y)^2}. \end{aligned} \tag{7.2}$$

According to Fermat's principle, we can equate equations (7.1) and (7.2), which gives:

$$\begin{aligned} k &\equiv n_o z_o + n_i\sqrt{z_i^2 + x_i^2 + y_i^2} \\ &= n_i\sqrt{(x_a - x_i)^2 + (y_a - y_i)^2 + (z_a - z_i - z_o)^2} \\ &\quad + n_o\sqrt{(z_a - w)^2 + (x_a - x)^2 + (y_a - y)^2}. \end{aligned} \tag{7.3}$$

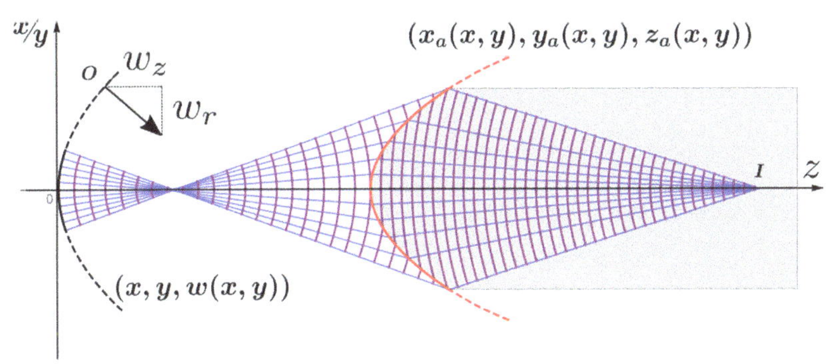

Figure 7.1. Schematic diagram of the configuration of a freeform surface with an image at a finite distance.

The next step is to describe each ray path of the input w using the Malus–Dupin theorem. According to this theorem, rays are perpendicular to the wavefront. Thus, the rays have the same direction as the normal vector of the input w, so

$$w_x = -\frac{\partial_x w}{\sqrt{\partial_x w^2 + \partial_y w^2 + 1}}, \quad w_y = -\frac{\partial_y w}{\sqrt{\partial_x w^2 + \partial_y w^2 + 1}}, \\ w_z = \frac{1}{\sqrt{\partial_x w^2 + \partial_y w^2 + 1}}, \tag{7.4}$$

where w_x, w_y, and w_z, are the cosine directors of the input rays in the x, y, and z directions, respectively. The partial derivatives of w are given by $\partial_x w \equiv \partial w/\partial x$ and $\partial_y w \equiv \partial w/\partial y$. The next step is to define the distance between the input wavefront w and the refractive surface (x_a, y_a, z_a):

$$\vartheta \equiv \sqrt{(z_a - w)^2 + (x_a - x)^2 + (y_a - y)^2}. \tag{7.5}$$

Using equations (7.4) and (7.5), we can express the refractive surface as

$$x_a = x + \vartheta w_x, \quad y_a = y + \vartheta w_y, \quad z_a = w + \vartheta w_z. \tag{7.6}$$

Observe that equation (7.6) has the structure of the solution: it express the shape of the refractive surface, but it depends on ϑ, which in turn depends on (x_a, y_a, z_a). Therefore, we need to express ϑ in terms of the known parameters, in other words, in terms of w_x, w_y, and w_z, the distances z_o, z_i, x_i, and y_i, and the refractive indexes n_o and n_i. We therefore substitute equations (7.5) and (7.6) into equation (7.3):

$$n_o z_o + n_i\sqrt{z_i^2 + x_i^2 + y_i^2} = n_i\sqrt{(x_a - x_i)^2 + (y_a - y_i)^2 + (z_a - z_i - z_o)^2} + n_o\vartheta. \tag{7.7}$$

By squaring both sides, we obtain

$$(k - n_o\vartheta)^2 = n_i^2\left[(x_a - x_i)^2 + (y_a - y_i)^2 + (z_a - z_i - z_o)^2\right], \tag{7.8}$$

which we expand, giving

$$n_i^2\left(\tau_x^2 + \tau_y^2 + \tau_z^2\right) + \vartheta^2 n_i^2 + \vartheta\left(2n_i^2\tau_x w_x + 2n_i^2\tau_y w_y + 2n_i^2\tau_z w_z\right) \\ = k^2 - 2kn_o\vartheta + n_o^2\vartheta^2. \tag{7.9}$$

In order to decrease the size of equation (7.9), let us define the following parameters:

$$\tau_x = x - x_i, \quad \tau_y = y - y_i, \quad \tau_z = w - z_i - z_o; \tag{7.10}$$

we now substitute equation (7.10) into equation (7.9):

$$n_i^2\left(\tau_x^2 + \tau_y^2 + \tau_z^2\right) - k^2 + \vartheta^2\left(n_o^2 - n_i^2\right) + 2\vartheta\left[n_i^2(\tau_x w_x + \tau_y w_y + \tau_z w_z) + kn_o\right] = 0, \tag{7.11}$$

multiply both sides by $4(n_o^2 - n_i^2)$:

$$\begin{aligned}4\left(n_o^2 - n_i^2\right)^2\vartheta^2 + 8\left(n_o^2 - n_i^2\right)\left[n_i^2(\tau_x w_x + \tau_y w_y + \tau_z w_z)\right] + 4\vartheta^2\left[n_i^2(\tau_x w_x + \tau_y w_y + \tau_z w_z)\right]\\ = 4\vartheta^2\left[n_i^2(\tau_x w_x + \tau_y w_y + \tau_z w_z) + kn_o\right]^2 - 4(n_o^2 - n_i^2)[n_i^2(\tau_x^2 + \tau_y^2 + \tau_z^2) - k^2],\end{aligned} \tag{7.12}$$

and complete the squared binomial:

$$\begin{aligned}\left\{2\left(n_o^2 - n_i^2\right)\vartheta + 2\left[n_i^2(\tau_x w_x + \tau_y w_y + \tau_z w_z) + kn_o\right]\right\}^2\\ = 4\vartheta^2\left[n_i^2(\tau_x w_x + \tau_y w_y + \tau_z w_z) + kn_o\right]^2 - 4\left(n_o^2 - n_i^2\right)\left[n_i^2\left(\tau_x^2 + \tau_y^2 + \tau_z^2\right) - k^2\right].\end{aligned} \tag{7.13}$$

Finally, we solve for ϑ,

$$\begin{aligned}\vartheta = \frac{kn_o + n_i^2\left(\tau_x w_x + \tau_y w_y + \tau_z w_z\right)}{n_o^2 - n_i^2}\\ - \frac{\sqrt{\left(kn_o + n_i^2\left(\tau_x w_x + \tau_y w_y + \tau_z w_z\right)\right)^2 - (n_i^2 - n_o^2)\left(n_i^2\left(\tau_x^2 + \tau_y^2 + \tau_z^2\right) - k^2\right)}}{n_o^2 - n_i^2}.\end{aligned} \tag{7.14}$$

Now that we know ϑ from equation (7.14), we can substitute it into equation (7.6) to properly describe the stigmatic surface. The x component of the stigmatic surface is

$$\boxed{\begin{aligned}x_a = x - &\frac{\partial_x w}{\left(n_o^2 - n_i^2\right)\sqrt{\partial_x w^2 + \partial_y w^2 + 1}}\\ &\times\left\{\frac{n_i^2\left(x_i\partial_x w + y_i\partial_y w + w - x\partial_x w - y\partial_y w - z_i - z_o\right)}{\sqrt{\partial_x w^2 + \partial_y w^2 + 1}} + \left(n_i\sqrt{x_i^2 + y_i^2 + z_i^2} + n_o z_o\right)n_o\right.\\ &+\left[\left(\frac{n_i^2\left(x_i\partial_x w + y_i\partial_y w + w - x\partial_x w - y\partial_y w - z_i - z_o\right)}{\sqrt{\partial_x w^2 + \partial_y w^2 + 1}} + \left(n_i\sqrt{x_i^2 + y_i^2 + z_i^2} + n_o z_o\right)n_o\right)^2\right.\\ &-\left(n_i^2 - n_o^2\right)\\ &\left.\left.\left(n_i^2\left((w - z_i - z_o)^2 + (x - x_i)^2 + (y - y_i)^2\right) - \left(n_i\sqrt{x_i^2 + y_i^2 + z_i^2} + n_o z_o\right)^2\right)\right]^{\frac{1}{2}}\right\}\end{aligned}}, \tag{7.15}$$

the y component of the stigmatic surface is

$$
\begin{aligned}
y_a = {} & y - \frac{\partial_y w}{\left(n_o^2 - n_i^2\right)\sqrt{\partial_x w^2 + \partial_y w^2 + 1}} \\
& \times \left\{ \frac{n_i^2\left(x_i \partial_x w + y_i \partial_y w + w - x \partial_x w - y \partial_y w - z_i - z_o\right)}{\sqrt{\partial_x w^2 + \partial_y w^2 + 1}} + \left(n_i\sqrt{x_i^2 + y_i^2 + z_i^2} + n_o z_o\right) n_o \right. \\
& + \left[\left(\frac{n_i^2\left(x_i \partial_x w + y_i \partial_y w + w - x \partial_x w - y \partial_y w - z_i - z_o\right)}{\sqrt{\partial_x w^2 + \partial_y w^2 + 1}} + \left(n_i\sqrt{x_i^2 + y_i^2 + z_i^2} + n_o z_o\right) n_o \right)^2 \right. \\
& - \left(n_i^2 - n_o^2\right) \\
& \left. \left. \left(n_i^2\left((w - z_i - z_o)^2 + (x - x_i)^2 + (y - y_i)^2\right) - \left(n_i\sqrt{x_i^2 + y_i^2 + z_i^2} + n_o z_o\right)^2 \right) \right]^{\frac{1}{2}} \right\},
\end{aligned}
\tag{7.16}
$$

and the z component of the stigmatic surface is

$$
\begin{aligned}
z_a = {} & w + \frac{1}{\left(n_o^2 - n_i^2\right)\sqrt{\partial_x w^2 + \partial_y w^2 + 1}} \\
& \times \left\{ \frac{n_i^2\left(x_i \partial_x w + y_i \partial_y w + w - x \partial_x w - y \partial_y w - z_i - z_o\right)}{\sqrt{\partial_x w^2 + \partial_y w^2 + 1}} + \left(n_i\sqrt{x_i^2 + y_i^2 + z_i^2} + n_o z_o\right) n_o \right. \\
& + \left[\left(\frac{n_i^2\left(x_i \partial_x w + y_i \partial_y w + w - x \partial_x w - y \partial_y w - z_i - z_o\right)}{\sqrt{\partial_x w^2 + \partial_y w^2 + 1}} + \left(n_i\sqrt{x_i^2 + y_i^2 + z_i^2} + n_o z_o\right) n_o \right)^2 \right. \\
& - \left(n_i^2 - n_o^2\right) \\
& \left. \left. \times \left(n_i^2\left((w - z_i - z_o)^2 + (x - x_i)^2 + (y - y_i)^2\right) - \left(n_i\sqrt{x_i^2 + y_i^2 + z_i^2} + n_o z_o\right)^2 \right) \right]^{\frac{1}{2}} \right\}.
\end{aligned}
\tag{7.17}
$$

This chapter's most important equations are equations (7.15), (7.16), and (7.17); they accurately describe the refractive surface that we discussed in the introduction to this chapter. Equations (7.15), (7.16), and (7.17) provide a closed-form solution to the problem discussed in the introduction to this chapter. Observe that these equations are not only the solution, but also the general equation, since w is supplied by the user along with the other design parameters. From the derivation of equations (7.15), (7.16), and (7.17) it is clear that they are the unique solution.

7.3 The wavefronts of images at finite distances

Once we know the shape (x_a, y_a, z_a), our next objective is to examine the exact ray tracing generated by (x_a, y_a, z_a). So, we compute Snell's law for the surface of (x_a, y_a, z_a):

$$\boxed{\vec{\mathbf{v}}_2 = \frac{n_o}{n_i}[\vec{\mathbf{v}}_1 - (\vec{\mathbf{n}}_a \cdot \vec{\mathbf{v}}_1)\vec{\mathbf{n}}_a] - \vec{\mathbf{n}}_a\sqrt{1 - \frac{n_o^2}{n_i^2}(\vec{\mathbf{n}}_a \times \vec{\mathbf{v}}_1)^2} \quad \text{for} \quad \vec{\mathbf{v}}_2, \vec{\mathbf{v}}_1, \vec{\mathbf{n}}_a \in \mathbb{R}^3,} \tag{7.18}$$

where $\vec{\mathbf{v}}_1$ is the unit vector of the incident ray, $\vec{\mathbf{v}}_2$ is the unit vector of the refracted ray, and $\vec{\mathbf{n}}_a$ is the normal vector of the first surface:

$$\begin{aligned} \vec{\mathbf{v}}_1 &= w_x\vec{\mathbf{e}}_1 + w_y\vec{\mathbf{e}}_2 + w_z\vec{\mathbf{e}}_3, \\ \vec{\mathbf{v}}_2 &= \wp_x\vec{\mathbf{e}}_1 + \wp_y\vec{\mathbf{e}}_2 + \wp_z\vec{\mathbf{e}}_3, \\ \vec{\mathbf{n}}_a &= \Phi_x\vec{\mathbf{e}}_1 + \Phi_y\vec{\mathbf{e}}_2 + \Phi_z\vec{\mathbf{e}}_3, \end{aligned} \tag{7.19}$$

where $\vec{\mathbf{e}}_1$ denotes the x direction, $\vec{\mathbf{e}}_2$ denotes the y direction and $\vec{\mathbf{e}}_3$ denotes the z direction. The normal vector of (x_a, y_a, z_a) is not at all trivial; since (x_a, y_a, z_a) is in $\mathbb{R}^3$, it is given by the following expression:

$$\vec{\mathbf{n}}_a = \Phi_x\vec{\mathbf{e}}_1 + \Phi_y\vec{\mathbf{e}}_2 + \Phi_z\vec{\mathbf{e}}_3 = \pm\frac{\dfrac{\partial(x_a\vec{\mathbf{e}}_1 + y_a\vec{\mathbf{e}}_2 + z_a\vec{\mathbf{e}}_3)}{\partial x} \times \dfrac{\partial(x_a\vec{\mathbf{e}}_1 + y_a\vec{\mathbf{e}}_2 + z_a\vec{\mathbf{e}}_3)}{\partial y}}{\left|\dfrac{\partial(x_a\vec{\mathbf{e}}_1 + y_a\vec{\mathbf{e}}_2 + z_a\vec{\mathbf{e}}_3)}{\partial x} \times \dfrac{\partial(x_a\vec{\mathbf{e}}_1 + y_a\vec{\mathbf{e}}_2 + z_a\vec{\mathbf{e}}_3)}{\partial y}\right|}. \tag{7.20}$$

Using the last equation, we can compute the cross-products and separate the components; in x, we have

$$\Phi_x = \frac{\partial_x y_a \partial_y z_a - \partial_x z_a \partial_y y_a}{\sqrt{(\partial_x x_a \partial_y y_a - \partial_x y_a \partial_y x_a)^2 + (\partial_x z_a \partial_y x_a - \partial_x x_a \partial_y z_a)^2 + (\partial_x y_a \partial_y z_a - \partial_x z_a \partial_y y_a)^2}}, \tag{7.21}$$

in y, we get

$$\Phi_y = \frac{\partial_x z_a \partial_y x_a - \partial_x x_a \partial_y z_a}{\sqrt{(\partial_x x_a \partial_y y_a - \partial_x y_a \partial_y x_a)^2 + (\partial_x z_a \partial_y x_a - \partial_x x_a \partial_y z_a)^2 + (\partial_x y_a \partial_y z_a - \partial_x z_a \partial_y y_a)^2}}, \tag{7.22}$$

and finally in z, we obtain

$$\Phi_z = \frac{\partial_x x_a \partial_y y_a - \partial_x y_a \partial_y x_a}{\sqrt{(\partial_x x_a \partial_y y_a - \partial_x y_a \partial_y x_a)^2 + (\partial_x z_a \partial_y x_a - \partial_x x_a \partial_y z_a)^2 + (\partial_x y_a \partial_y z_a - \partial_x z_a \partial_y y_a)^2}}. \tag{7.23}$$

Once we have the components of the normal vector of (x_a, y_a, z_a), we can proceed to compute Snell's law. We shall begin with the term $[\vec{\mathbf{v}}_1 - (\vec{\mathbf{n}}_a \cdot \vec{\mathbf{v}}_1)\vec{\mathbf{n}}_a]$ of equation (7.18):

$$\left[\vec{\mathbf{v}}_1 - \left(\vec{\mathbf{n}}_a \cdot \vec{\mathbf{v}}_1\right)\vec{\mathbf{n}}_a\right] = \left[-w_x\left(\Phi_y^2 + \Phi_z^2\right) + w_y\Phi_x\Phi_y + w_z\Phi_x\Phi_z\right]\vec{\mathbf{e}}_1 + \left[w_y(\Phi_x^2 + \Phi_z^2) - \Phi_y\left(w_x\Phi_x + w_z\Phi_z\right)\right]\vec{\mathbf{e}}_2 + \left[w_z\left(\Phi_x^2 + \Phi_y^2\right) - \Phi_z\left(w_x\Phi_x + w_y\Phi_y\right)\right]\vec{\mathbf{e}}_3. \tag{7.24}$$

We then compute the term containing the square root in equation (7.18),

$$-\vec{\mathbf{n}}_a\sqrt{1 - \frac{n_o^2}{n_i^2}(\vec{\mathbf{n}}_a \times \vec{\mathbf{v}}_1)^2} = \left(\Phi_x\vec{\mathbf{e}}_1 + \Phi_y\vec{\mathbf{e}}_2 + \Phi_z\vec{\mathbf{e}}_3\right) \times \sqrt{1 - \frac{n_o^2}{n_i^2}\left[\left(w_y\Phi_x - w_x\Phi_y\right)^2 + (w_z\Phi_x - w_x\Phi_z)^2 + \left(w_z\Phi_y - w_y\Phi_z\right)^2\right]}. \tag{7.25}$$

We sum equations (7.24) and (7.25) and separate the components. We conclude that the cosine directors are, in x:

$$\wp_x = \frac{n_o}{n_i}\left[-w_x\left(\Phi_y^2 + \Phi_z^2\right) + w_y\Phi_x\Phi_y + w_z\Phi_x\Phi_z\right] + \Phi_x\sqrt{1 - \frac{n_o^2}{n_i^2}\left[\left(w_y\Phi_x - w_x\Phi_y\right)^2 + (w_z\Phi_x - w_x\Phi_z)^2 + \left(w_z\Phi_y - w_y\Phi_z\right)^2\right]}, \tag{7.26}$$

in y:

$$\wp_y = \frac{n_o}{n_i}\left[w_y(\Phi_x^2 + \Phi_z^2) - \Phi_y\left(w_x\Phi_x + w_z\Phi_z\right)\right] + \Phi_y\sqrt{1 - \frac{n_o^2}{n_i^2}\left[\left(w_y\Phi_x - w_x\Phi_y\right)^2 + (w_z\Phi_x - w_x\Phi_z)^2 + \left(w_z\Phi_y - w_y, \Phi_z\right)^2\right]}, \tag{7.27}$$

and finally in z:

$$\wp_z = \frac{n_o}{n_i}\left[w_z\left(\Phi_x^2 + \Phi_y^2\right) - \Phi_z\left(w_x\Phi_x + w_y\Phi_y\right)\right] + \Phi_z\sqrt{1 - \frac{n_o^2}{n_i^2}\left[\left(w_y\Phi_x - w_x\Phi_y\right)^2 + (w_z\Phi_x - w_x\Phi_z)^2 + \left(w_z\Phi_y - w_y\Phi_z\right)^2\right]}. \tag{7.28}$$

Using equation (7.4), we can find the cosine directors of the input rays and in equations (7.26), (7.27), and (7.28), we have the cosine directors of the output. Therefore, there is nothing to stop us from defining the eikonal of the system. If we set $n_o = 1$, the eikonal $H(t, x, y)$ is given by

$$\boxed{H\left(t, x, y\right) = \begin{cases} [x + tw_x, y + tw_y, w + tw_z], & t < \vartheta, \\ \left[x_a + v\wp_x\left(t - \vartheta\right), y_a + v\wp_y\left(t - \vartheta\right), z_a + v\wp_z\left(t - \vartheta\right)\right], & t > \vartheta, \end{cases}} \tag{7.29}$$

where v is the velocity of the light inside the medium of the refractive surface. $H(t, x, y)$ has the information for the wavefront at all points and at all times. Finally, ϑ is defined as the distance rays travel until they strike the refractive surface, because ϑ is the natural condition for the piecewise function $H(t, x, y)$.

7.4 Mathematica

In the last section we obtained some interesting equations, equations (7.15), (7.16), (7.17), and (7.29). We now explore their shapes and performance using `Mathematica`. Here, we present `Mathematica` code to compute the refractive surfaces of the previous section.

```
In[10]:=  w[x_,y_]:=4 BesselJ[0,3Sqrt[0.5(x)^2+0.5(y )^2]/5] / 10

e=10^(-5);

wdx[x_,y_]:=(w[x+e,y]-w[x,y])/e

wdy[x_,y_]:=(w[x,y+e]-w[x,y])/e

wx[x_,y_]:= -wdx[x,y] / Sqrt[wdx[x,y]^2+wdy[x,y]^2+1]

wy[x_,y_]:= -wdy[x,y] / Sqrt[wdx[x,y]^2+wdy[x,y]^2+1]

wz[x_,y_]:= 1 / Sqrt[wdx[x,y]^2+wdy[x,y]^2+1]
```

```
τz[x_,y_]:=w[x,y]-zi-zo

τx[x_,y_]:=x-hi

τy[x_,y_]:=y-ki
```

$$\vartheta[x\_,y\_]:=\frac{1}{(1-\frac{no^2}{ni^2})}\Big(-\big(wx[x,y]\,\tau x[x,y]+wy[x,y]\,\tau y[x,y]+wz[x,y]\,\tau z[x,y]+\frac{k\,no}{ni^2}\big)$$
$$-\sqrt{}\Big(\big(wx[x,y]\,\tau x[x,y]+wy[x,y]\,\tau y[x,y]+wz[x,y]\,\tau z[x,y]+\frac{k\,no}{ni^2}\big)^2$$
$$-\big(\tau x[x,y]^2+\tau y[x,y]^2+\tau z[x,y]^2-\frac{k^2}{ni^2}\big)\big(1-\frac{no^2}{ni^2}\big)\Big)\Big)$$

```
xa[x_,y_]:=x+ϑ[x,y]wx[x,y]

ya[x_,y_]:=y+ϑ[x,y]wy[x,y]

za[x_,y_]:=w[x,y]+ϑ[x,y]wz[x,y]
```

```
dxxa[x_,y_]:=(xa[x+e,y]-xa[x,y])/e
dyxa[x_,y_]:=(xa[x,y+e]-xa[x,y])/e
dxya[x_,y_]:=(ya[x+e,y]-ya[x,y])/e
dyya[x_,y_]:=(ya[x,y+e]-ya[x,y])/e
dxza[x_,y_]:=(za[x+e,y]-za[x,y])/e
dyza[x_,y_]:=(za[x,y+e]-za[x,y])/e
```

```
Φx[x_,y_]:=(-dxza[x,y] dyya[x,y]+dxya[x,y] dyza[x,y])/
  (√((-dxya[x,y] dyxa[x,y]+dxxa[x,y] dyya[x,y])^2
    +(dxza[x,y] dyxa[x,y]-dxxa[x,y] dyza[x,y])^2
    +(-dxza[x,y] dyya[x,y]+dxya[x,y] dyza[x,y])^2))
```

```
Φy[x_,y_]:=(dxza[x,y] dyxa[x,y]-dxxa[x,y] dyza[x,y])/
  (√((-dxya[x,y] dyxa[x,y]+dxxa[x,y] dyya[x,y])^2
    +(dxza[x,y] dyxa[x,y]-dxxa[x,y] dyza[x,y])^2
```

$$+(\text{-dxza}[x,y]\ \text{dyya}[x,y]+\text{dxya}[x,y]\ \text{dyza}[x,y])^2))$$

$$\begin{aligned}&\Phi z[x\_,y\_]:=(\text{-dxya}[x,y]\ \text{dyxa}[x,y]+\text{dxxa}[x,y]\ \text{dyya}[x,y])/\\&\quad(\surd((\text{-dxya}[x,y]\ \text{dyxa}[x,y]+\text{dxxa}[x,y]\ \text{dyya}[x,y])^2\\&\qquad+(\text{dxza}[x,y]\ \text{dyxa}[x,y]\text{-dxxa}[x,y]\ \text{dyza}[x,y])^2\\&\qquad+(\text{-dxza}[x,y]\ \text{dyya}[x,y]+\text{dxya}[x,y]\ \text{dyza}[x,y])^2))\end{aligned}$$

$$\begin{aligned}&\varrho x[x\_,y\_]:=-\frac{1}{\text{ni}}\text{no}\ (\text{wy}[x,y]\ \Phi x[x,y]\quad \Phi y[x,y]+\text{wz}[x,y]\ \Phi x[x,y]\\&\quad\Phi z[x,y]\text{-wx}[x,y]\ (\Phi y[x,y]^2+\Phi z[x,y]^2))+\Phi x[x,y]\\&\quad\surd(1-\frac{1}{\text{ni}^2}\text{no}^2\ ((\text{wy}[x,y]\ \Phi x[x,y]\text{-wx}[x,y]\ \Phi y[x,y])^2\\&\qquad+\ (\text{wz}[x,y]\ \Phi x[x,y]\text{-wx}[x,y]\ \Phi z[x,y])^2\\&\qquad+(\text{wz}[x,y]\ \Phi y[x,y]\text{-wy}[x,y]\ \Phi z[x,y])^2));\end{aligned}$$

$$\begin{aligned}&\varrho y[x\_,y\_]:=\frac{1}{\text{ni}}\text{no}\ (-\Phi y[x,y](\text{wx}[x,y]\Phi x[x,y]+\text{wz}[x,y]\ \Phi z[x,y])\\&\quad+\text{wy}[x,y](\Phi x[x,y]^2\ +\Phi z[x,y]^2))+\Phi y[x,y]\\&\quad\surd(1-\frac{1}{\text{ni}^2}\text{no}^2\ ((\text{wy}[x,y]\ \Phi x[x,y]\text{-wx}[x,y]\ \Phi y[x,y])^2\\&\qquad+(\text{wz}[x,y]\ \Phi x[x,y]\text{-wx}[x,y]\ \Phi z[x,y])^2\\&\qquad+(\text{wz}[x,y]\ \Phi y[x,y]\text{-wy}[x,y]\ \Phi z[x,y])^2));\end{aligned}$$

$$\begin{aligned}&\varrho z[x\_,y\_]:=\frac{1}{\text{ni}}\text{no}\ (\text{wz}[x,y]\ (\Phi x[x,y]^2+\Phi y[x,y]^2)\\&\quad-(\text{wx}[x,y]\ \Phi x[x,y]+\text{wy}[x,y]\ \Phi y[x,y])\ \Phi z[x,y])+\Phi z[x,y]\\&\quad\surd(1-\frac{1}{\text{ni}^2}\text{no}^2\ ((\text{wy}[x,y]\ \Phi x[x,y]\text{-wx}[x,y]\ \Phi y[x,y])^2\\&\qquad+(\text{wz}[x,y]\ \Phi x[x,y]\text{-wx}[x,y]\ \Phi z[x,y])^2\\&\qquad+(\text{wz}[x,y]\ \Phi y[x,y]\text{-wy}[x,y]\ \Phi z[x,y])^2));\end{aligned}$$

$$\text{wz2}[x\_,y\_]:=\frac{-(\text{za}[x,y]\text{-zi-zo})}{\sqrt{(\text{za}[x,y]\text{-zi-zo})^2+(\text{xa}[x,y]\text{-hi})^2+(\text{ya}[x,y]\text{-ki})^2}}$$

$$\text{wx2}[x\_,y\_]:=\frac{-(\text{xa}[x,y]\text{-hi})}{\sqrt{(\text{za}[x,y]\text{-zi-zo})^2+(\text{xa}[x,y]\text{-hi})^2+(\text{ya}[x,y]\text{-ki})^2}}$$

```
wy2[x_,y_]:= -(ya[x,y]-ki) / Sqrt[(za[x,y]-zi-zo)^2+(xa[x,y]-hi)^2+(ya[x,y]-ki)^2]

c=1;
v=c/ni;
no=1;
ni=1.5;
rmax=5;
zo=4;
hi=1;
ki=-1;
zi=30;
k=no zo+ni Sqrt[zi^2+hi^2+ki^2];


Homotopy={{
  Piecewise[{
   {wx[x,y]tt+x,tt<ϑ[x,y]},{ v wx2[x,y](tt-ϑ[x,y])+xa[x,y],tt>ϑ[x,y]}}],
  Piecewise[{
   {wy[x,y]tt+y,tt<ϑ[x,y]},{ v wy2[x,y](tt-ϑ[x,y])+ya[x,y],tt>ϑ[x,y]}}],
  Piecewise[{
   {wz[x,y]tt+w[x,y],tt<ϑ[x,y]},{v wz2[x,y](tt-ϑ[x,y])+za[x,y],tt>ϑ[x,y]}}]}};


Waves:=Table[{ParametricPlot3D[Homotopy,{x,-rmax,rmax},{y,-rmax,rmax},
  PlotRange→All,PlotStyle→{Blue,Opacity[.5]},PlotPoints→10,
  Mesh→False]},{tt,5,k,5}]

Rays:=Table[{ParametricPlot3D[Homotopy,{tt,0,k},PlotRange→All,
  PlotStyle→{Purple,Opacity[.5]}, PlotPoints→10,Mesh→False]},
  {x,-rmax,rmax,rmax},{y,-rmax,rmax,rmax}]

g0=Plot3D[{w[x,y]},{x,-rmax,rmax},{y,-rmax,rmax},Axes→True,
  PlotStyle→{Blue,Opacity[.5]},AxesLabel→{x,y,z},AspectRatio→Automatic,
  Boxed→True,Mesh→False];


g1=ParametricPlot3D[{xa[x,y],ya[x,y],za[x,y]},{x,-rmax,rmax},{y,-rmax,rmax},
  Axes→True,PlotStyle→{Red,Opacity[.75]},Boxed→True,AxesLabel→{x,y,z},
  AspectRatio→Automatic,Mesh→False,PlotPoints→10];


Show[g1,g0, Waves,Rays,PlotRange→All]
```

7.5 Examples

Example outputs generated by the `Mathematica` code shown in the last section are presented here as a gallery consisting of figures 7.2, 7.3, 7.4, 7.5, 7.6, and 7.7; the calculations used equations (7.15), (7.16), (7.17), and (7.29).

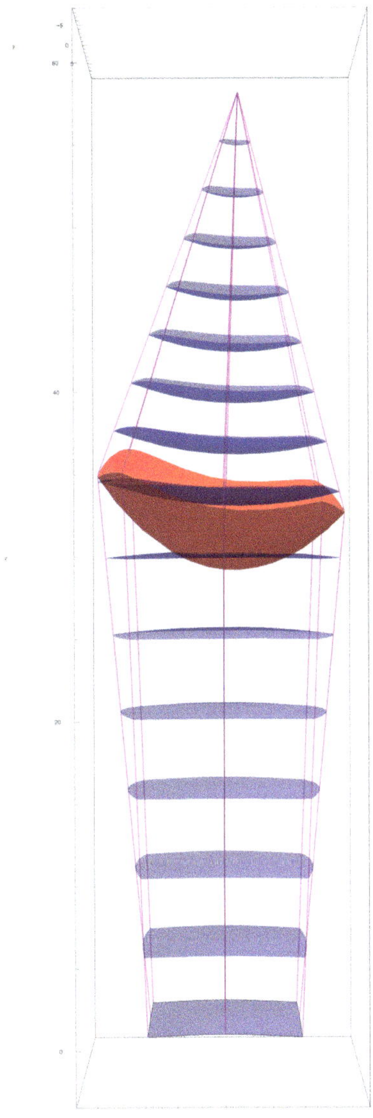

(a) A frontal view point of the adaptative mirror system.

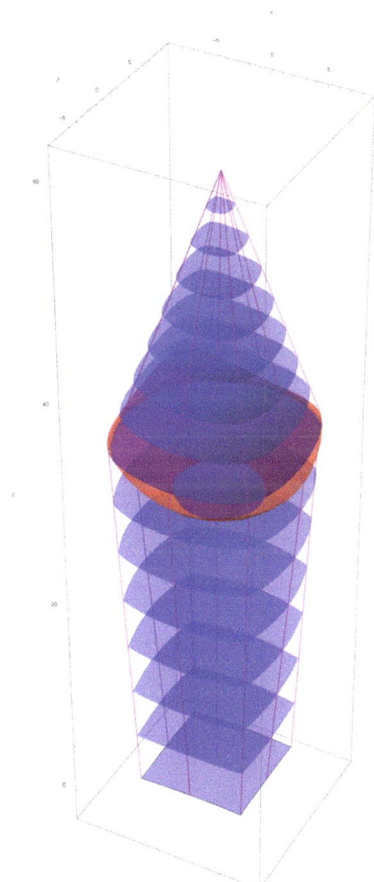

(b) A frontal perspective view point of the adaptative mirror system.

Figure 7.2. Surface specifications for a parabolic input wavefront, the system produces a flat output wavefront: (a) the input wavefront is $-\cos(x/5)^2 + \cos(y/8)^2$, refractive index $n = 1.5$, object distance $z_o = 50$ mm, $r_{\max} = 10$ mm, parameter $s = 0.5$; parameter $p = 0$; $(r_a, z_a) =$ equations (7.15), (7.16), and (7.17), $H(t, r_a) =$ equation (7.29); (b) a frontal perspective of the same configuration.

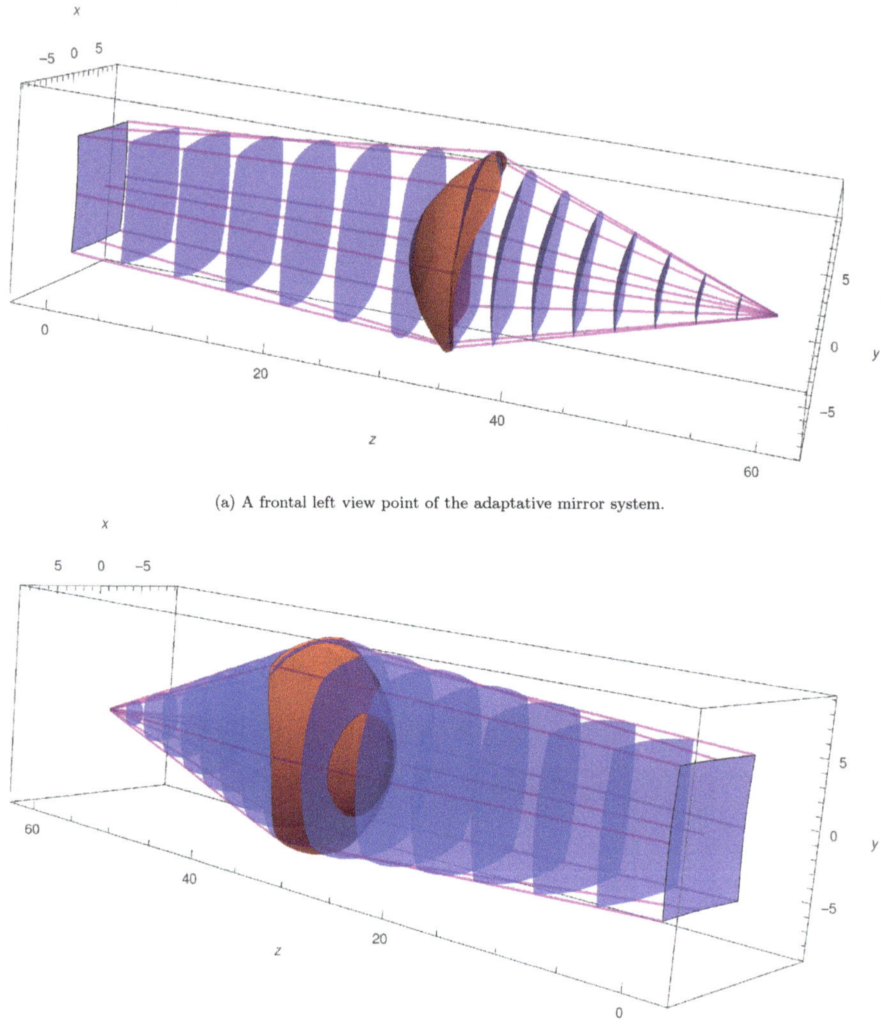

(a) A frontal left view point of the adaptative mirror system.

(b) A frontal right view point of the adaptative mirror system.

Figure 7.3. Surface specifications for a parabolic input wavefront; the system produces a flat output wavefront: (a) a frontal left perspective of the same configuration as that shown in figure 7.2; (b) a frontal right perspective of the same configuration as that shown in figure 7.2.

7.6 Mathematical model of freeform stigmatic surfaces for images at infinity

In this section, the objective is to obtain freeform stigmatic surfaces for images at infinity. This problem can be solved using the experience gained in section 7.3, in which we solved the same problem but for finite-distance images; see figure 7.8.

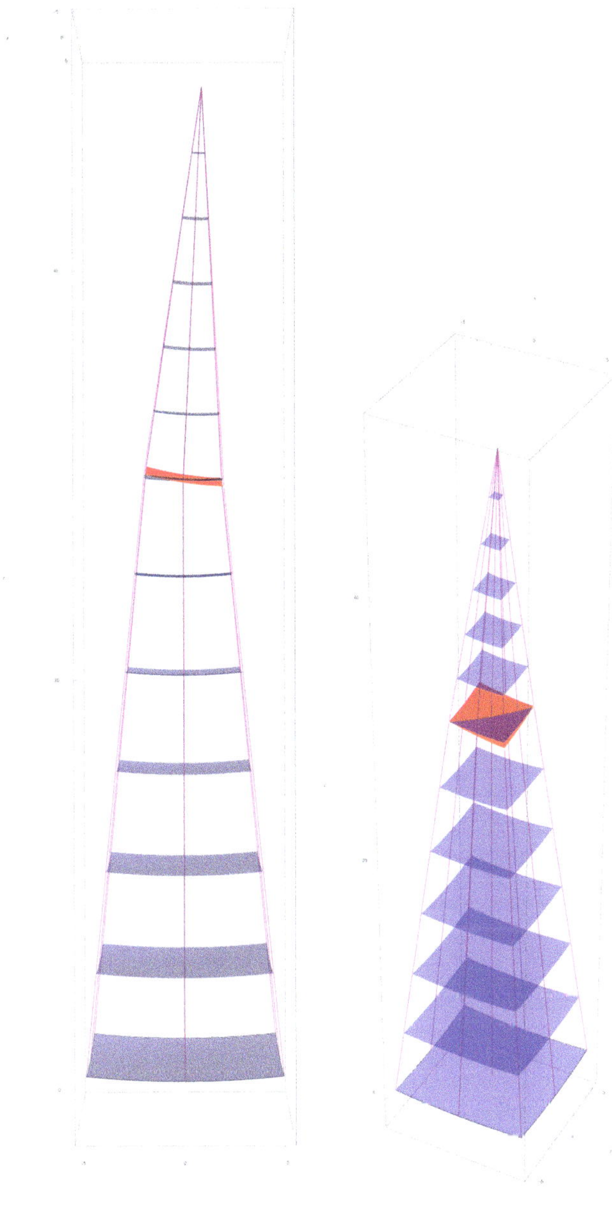

(a) A frontal view point of the adaptative mirror system.

(b) A frontal perspective view point of the adaptative mirror system.

Figure 7.4. Surface specifications for an arbitrary input wavefront; the system produces a flat output wavefront: (a) the input wavefront is $\frac{(x)^2+(1.2y)^2}{28}$, refractive index $n = 1.5$, object distance $z_o = 50$ mm, $r_{max} = 10$ mm, parameter $s = 20$; parameter $p = 0$; (r_a, z_a) = equations (7.15), (7.16), and (7.17), $H(t, r_a)$ = equation (7.29); (b) a frontal perspective of the same configuration.

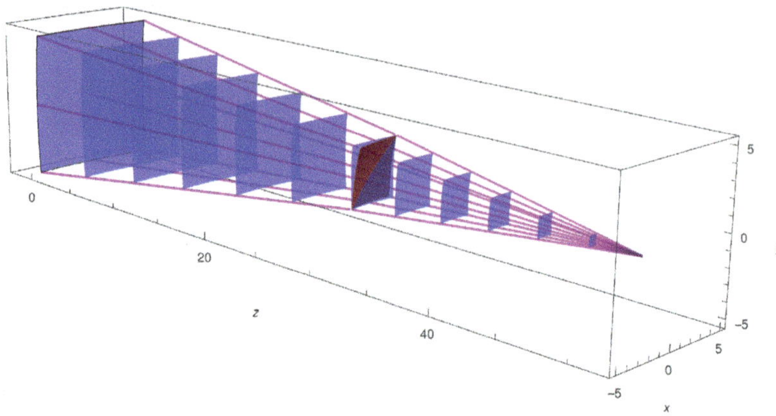

(a) A frontal left view point of the adaptative mirror system.

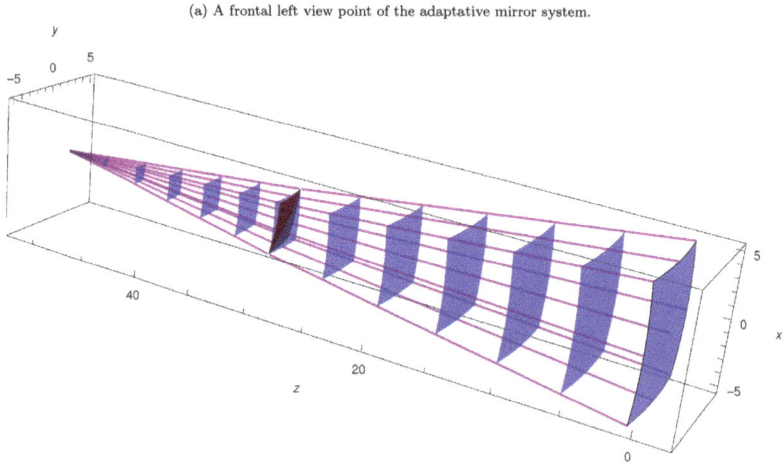

(b) A frontal right view point of the adaptative mirror system.

Figure 7.5. Surface specifications for a parabolic input wavefront; the system produces a stigmatic output wavefront: (a) a frontal left perspective of the same configuration as that shown in figure 7.4; (b) a frontal right perspective of the same configuration as that shown in figure 7.4.

The key equation implemented in section 7.3 was Fermat's principle, equation (7.3); we recall it below:

$$\begin{aligned} n_o z_o + n_i\sqrt{z_i^2 + x_i^2 + y_i^2} &= \\ n_i\sqrt{(x_a - x_i)^2 + (y_a - y_i)^2 + (z_a - z_i - z_o)^2} & \\ + n_o\sqrt{(z_a - w)^2 + (x_a - x)^2 + (y_a - y)^2}. & \end{aligned} \tag{7.30}$$

In order to get an image at infinity, but not necessarily along the optical axis, we first compute the limits $x_i \to sz_i$ and $y_i \to pz_i$ over equation (7.3). Therefore, we apply $x_i \to sz_i$ over equation (7.3):

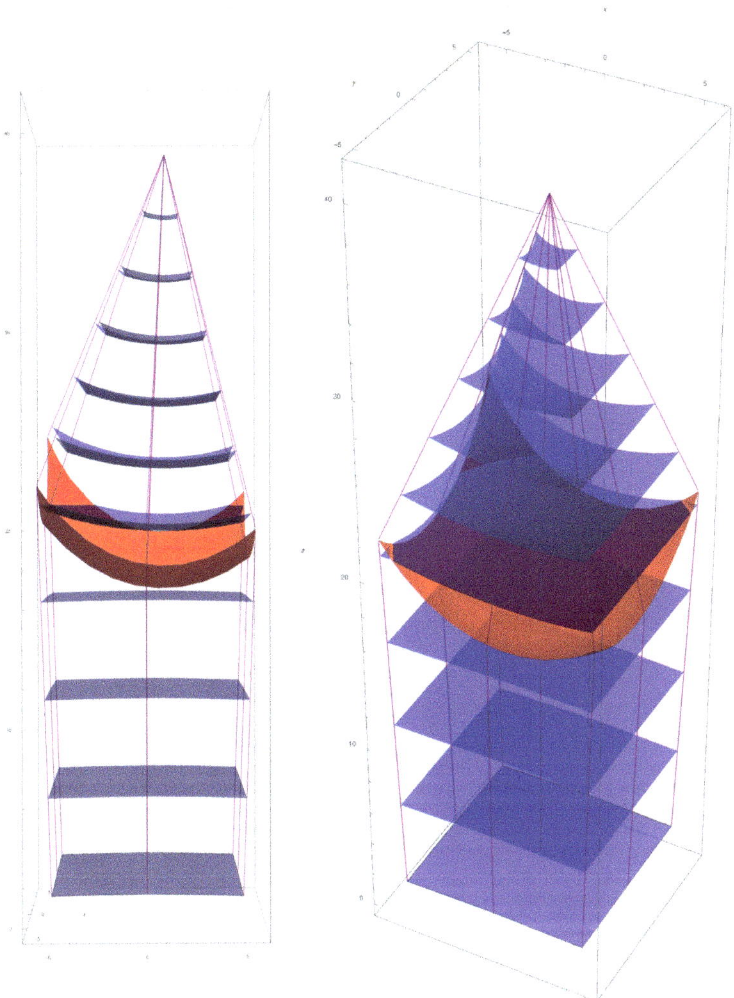

(a) A frontal view point of the adaptative mirror system. (b) A frontal perspective view point of the adaptative mirror system.

Figure 7.6. Surface specifications for an arbitrary input wavefront; the system produces a flat output wavefront: (a) the input wavefront is $(x/5)^2 - (y/3)$, refractive index $n = 1.5$, object distance $z_o = 40$ mm, $r_{max} = 10$ mm, parameter $s = 10$; parameter $p = 0$; $(r_a, z_a) =$ equations (7.15), (7.16), and (7.17), $H(t, r_a) =$ equation (7.29); (b) a frontal perspective of the same configuration.

$$\begin{aligned}
&\lim_{x_i \to sz_i}\Big[n_i\sqrt{(x_a - x_i)^2 + (y_a - y_i)^2 + (z_a - z_i - z_o)^2} \\
&\quad + n_o\sqrt{(z_a - w)^2 + (x_a - x)^2 + (y_a - y)^2} - n_o z_o - n_i\sqrt{z_i^2 + x_i^2 + y_i^2}\Big] \\
&= -n_i\sqrt{x_i^2 + (s^2 + 1)z_i^2} + n_i\sqrt{(x_i - x_a)^2 + (y_a - sz_i)^2 + (-z_a + z_i + z_o)^2} \\
&\quad + n_o\sqrt{(w - z_a)^2 + (x - x_a)^2 + (y - y_a)^2} - n_o z_o,
\end{aligned} \tag{7.31}$$

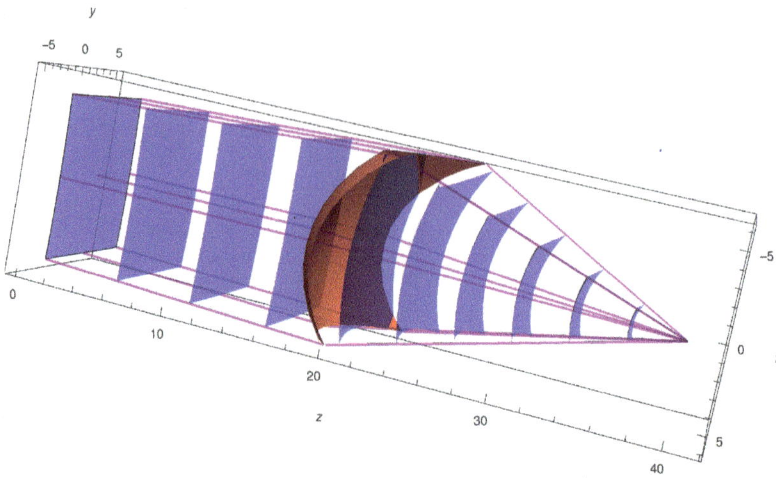

(a) A frontal left view point of the adaptative mirror system.

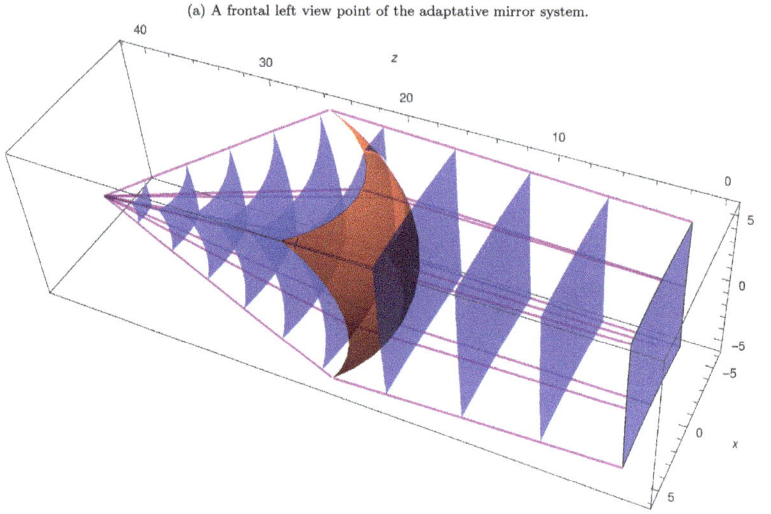

(b) A frontal right view point of the adaptative mirror system.

Figure 7.7. Surface specifications for a parabolic input wavefront; the system produces a flat output wavefront: (a) a frontal left perspective of the same configuration as that shown in figure 7.6; (b) a frontal right perspective of the same configuration as that shown in figure 7.6.

and then apply $y_i \to pz_i$ over equation (7.31):

$$\begin{aligned}&\lim_{y_i \to pz_i}\Big[-n_i\sqrt{x_i^2 + (s^2+1)z_i^2}\\ &\quad + n_i\sqrt{(x_i - x_a)^2 + (y_a - sz_i)^2 + (-z_a + z_i + z_o)^2}\\ &\quad + n_o\sqrt{(w - z_a)^2 + (x - x_a)^2 + (y - y_a)^2} - n_o z_o\Big]\\ &= -n_i\sqrt{z_i^2(p^2 + s^2 + 1)} + n_i\sqrt{(x_a - pz_i)^2 + (y_a - sz_i)^2 + (-z_a + z_i + z_o)^2}\\ &+ n_o\sqrt{(w - z_a)^2 + (x - x_a)^2 + (y - y_a)^2} - n_o z_o.\end{aligned} \tag{7.32}$$

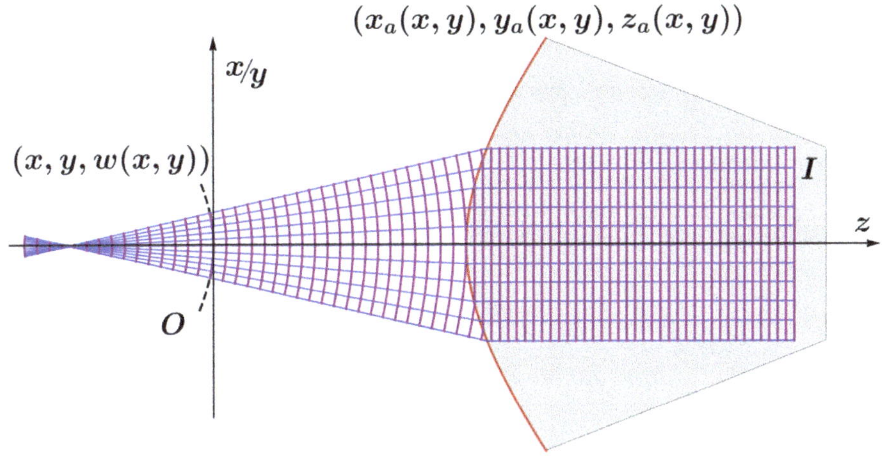

Figure 7.8. Schematic diagram of the configuration of a freeform surface with the image at infinity.

We also recall the cosine directors given by equation (7.4),

$$w_x = -\frac{\partial_x w}{\sqrt{\partial_x w^2 + \partial_y w^2 + 1}}, \quad w_y = -\frac{\partial_y w}{\sqrt{\partial_x w^2 + \partial_y w^2 + 1}}, \quad w_z = \frac{1}{\sqrt{\partial_x w^2 + \partial_y w^2 + 1}}, \tag{7.33}$$

the distance needed to reach the refractive surface, equation (7.5),

$$\vartheta \equiv \sqrt{(x_a - x_i)^2 + (y_a - y_i)^2 + (z_a - z_i - z_o)^2}, \tag{7.34}$$

and the first surface in terms of ϑ and the cosine directors:

$$x_a = x + \vartheta w_x, \quad y_a = y + \vartheta w_y, \quad z_a = w + \vartheta w_z. \tag{7.35}$$

We now replace equations (7.5) and (7.6) over the optical path, equation (7.32),

$$\begin{aligned} &- n_i\sqrt{z_i^2(p^2 + s^2 + 1)} + n_o\vartheta - n_o z_o \\ &+ n_i\sqrt{\left[x + \vartheta w_x - p z_i\right]^2 + \left[y + \vartheta w_y - s z_i\right]^2 + [-(w + \vartheta w_z) + z_i + z_o]^2} = 0. \end{aligned} \tag{7.36}$$

At this point it is appropriate to set the image at infinity, so we compute the limit $z_i \to \infty$ over equation (7.36):

$$\begin{aligned} &\lim_{z_i\to\infty}\Big[-n_i\sqrt{z_i^2\left(p^2 + s^2 + 1\right)} + n_o\vartheta - n_o z_o \\ &+ n_i\sqrt{\left[x + \vartheta w_x - p z_i\right]^2 + \left[y + \vartheta w_y - s z_i\right]^2 + [-(w + \vartheta w_z) + z_i + z_o]^2}\Big] \\ &= -\frac{n_i\left\{p\left[w_x \lim\limits_{z_i\to\infty}(\vartheta) + x\right] + s w_y \lim\limits_{z_i\to\infty}(\vartheta) + s y + w + w_z \lim\limits_{z_i\to\infty}(\vartheta) - z_o\right\}}{\sqrt{p^2 + s^2 + 1}} - n_o z_o + n_o \lim_{z_i\to\infty}(\vartheta). \end{aligned} \tag{7.37}$$

Note that the left-hand side of equation (7.37) is zero, since we are computing the limit over the optical path difference (OPD):

$$-\frac{n_i\left\{p\left[w_x \lim\limits_{z_i\to\infty}(\vartheta) + x\right] + sw_y \lim\limits_{z_i\to\infty}(\vartheta) + sy + w + w_z \lim\limits_{z_i\to\infty}(\vartheta) - z_o\right\}}{\sqrt{p^2+s^2+1}} - n_o z_o \tag{7.38}$$
$$+ n_o \lim_{z_i\to\infty}(\vartheta) = 0.$$

We now solve for $\lim_{z_i\to\infty}(\vartheta)$ in the previous expression:

$$\lim_{z_i\to\infty}(\vartheta) = \frac{n_i\left(px + sy + w - z_o\right) + n_o z_o\sqrt{p^2+s^2+1}}{n_o\sqrt{p^2+s^2+1} - n_i\left(pw_x + sw_y + w_z\right)}. \tag{7.39}$$

Using $\lim_{z_i\to\infty}(\vartheta)$, equation (7.39), and equation (7.6), we can properly describe the refractive surface, as follows: in x

$$\boxed{\lim_{z_i\to\infty}(x_a) = x - \frac{\partial_x w\left(n_i\left(w + px + sy\right) - n_i z_o + n_o z_o\sqrt{p^2+s^2+1}\right)}{n_i\left(p\partial_x w + s\partial_y w - 1\right) + n_o\sqrt{p^2+s^2+1}\sqrt{\partial_x w^2 + \partial_y w^2 + 1}}}, \tag{7.40}$$

in y

$$\boxed{\lim_{z_i\to\infty}(y_a) = y - \frac{\partial_y w\left(n_i\left(w + px + sy\right) - n_i z_o + n_o z_o\sqrt{p^2+s^2+1}\right)}{n_i\left(p\partial_x w + s\partial_y w - 1\right) + n_o\sqrt{p^2+s^2+1}\sqrt{\partial_x w^2 + \partial_y w^2 + 1}}}, \tag{7.41}$$

and in z

$$\boxed{\lim_{z_i\to\infty}(z_a) = w + \frac{n_i\left(w + px + sy\right) - n_i z_o + n_o z_o\sqrt{p^2+s^2+1}}{n_i\left(p\partial_x w + s\partial_y w - 1\right) + n_o\sqrt{p^2+s^2+1}\sqrt{\partial_x w^2 + \partial_y w^2 + 1}}}. \tag{7.42}$$

Equations (7.36), (7.37), and (7.38) are the solution to the problem proposed in this section. Equations (7.36), (7.37), and (7.38) are the same as equations (7.15), (7.16), and (7.17) except that some limits have been applied to them. The properties used in equations (7.36), (7.37), and (7.38) are also shared by equations (7.15), (7.16), and (7.17).

7.7 The wavefront and the collimated output rays

To compute the collimated outputs we need to compute the normal components of the normal vector of the refractive surface. Therefore, we compute equation (7.20), but instead of using equations (7.15), (7.16), and (7.17) as inputs, we use equations (7.36), (7.37), and (7.38) as inputs. The described computation gives the x component of the normal vector of the refractive surface as:

$$\lim_{z_i\to\infty}(\Phi_x) = \left[\partial_x \lim_{z_i\to\infty}(y_a)\partial_y \lim_{z_i\to\infty}(z_a) - \partial_x \lim_{z_i\to\infty}(z_a)\partial_y \lim_{z_i\to\infty}(y_a)\right] \times\left\{\left[\partial_x \lim_{z_i\to\infty}(x_a)\partial_y \lim_{z_i\to\infty}(y_a) - \partial_x \lim_{z_i\to\infty}(y_a)\partial_y \lim_{z_i\to\infty}(x_a)\right]^2 + \left[\partial_x \lim_{z_i\to\infty}(z_a)\partial_y \lim_{z_i\to\infty}(x_a) - \partial_x \lim_{z_i\to\infty}(x_a)\partial_y \lim_{z_i\to\infty}(z_a)\right]^2 + \left[\partial_x \lim_{z_i\to\infty}(y_a)\partial_y \lim_{z_i\to\infty}(z_a) - \partial_x \lim_{z_i\to\infty}(z_a)\partial_y \lim_{z_i\to\infty}(y_a)\right]^2\right\}^{\frac{1}{2}}, \tag{7.43}$$

and the y component of the normal vector of the refractive surface is

$$\lim_{z_i\to\infty}\left(\Phi_y\right) = \left[\partial_x \lim_{z_i\to\infty}(z_a)\partial_y \lim_{z_i\to\infty}(x_a) - \partial_x \lim_{z_i\to\infty}(x_a)\partial_y \lim_{z_i\to\infty}(z_a)\right] \times\left\{\left[\partial_x \lim_{z_i\to\infty}(x_a)\partial_y \lim_{z_i\to\infty}(y_a) - \partial_x \lim_{z_i\to\infty}(y_a)\partial_y \lim_{z_i\to\infty}(x_a)\right]^2 + \left[\partial_x \lim_{z_i\to\infty}(z_a)\partial_y \lim_{z_i\to\infty}(x_a) - \partial_x \lim_{z_i\to\infty}(x_a)\partial_y \lim_{z_i\to\infty}(z_a)\right]^2 + \left[\partial_x \lim_{z_i\to\infty}(y_a)\partial_y \lim_{z_i\to\infty}(z_a) - \partial_x \lim_{z_i\to\infty}(z_a)\partial_y \lim_{z_i\to\infty}(y_a)\right]^2\right\}^{\frac{1}{2}}, \tag{7.44}$$

and the z component of the normal vector of the refractive surface is given by

$$\lim_{z_i\to\infty}(\Phi_z) = \left[\partial_x \lim_{z_i\to\infty}(x_a)\partial_y \lim_{z_i\to\infty}(y_a) - \partial_x \lim_{z_i\to\infty}(y_a)\partial_y \lim_{z_i\to\infty}(x_a)\right] \times\left\{\left[\partial_x \lim_{z_i\to\infty}(x_a)\partial_y \lim_{z_i\to\infty}(y_a) - \partial_x \lim_{z_i\to\infty}(y_a)\partial_y \lim_{z_i\to\infty}(x_a)\right]^2 + \left[\partial_x \lim_{z_i\to\infty}(z_a)\partial_y \lim_{z_i\to\infty}(x_a) - \partial_x \lim_{z_i\to\infty}(x_a)\partial_y \lim_{z_i\to\infty}(z_a)\right]^2 + \left[\partial_x \lim_{z_i\to\infty}(y_a)\partial_y \lim_{z_i\to\infty}(z_a) - \partial_x \lim_{z_i\to\infty}(z_a)\partial_y \lim_{z_i\to\infty}(y_a)\right]^2\right\}^{\frac{1}{2}}. \tag{7.45}$$

So, to compute the cosine directors at the output, we just need to use equations (7.39), (7.40), and (7.41) instead of equations (7.21), (7.22), and (7.23) in equations (7.26), (7.27), and (7.28). Using this procedure we can obtain the direction cosine in x

$$
\begin{aligned}
\lim_{z_i\to\infty}(\wp_x) = -\frac{n_o}{n_i}\Bigg\{ & -w_x\left[\lim_{z_i\to\infty}\left(\Phi_y^2\right) + \lim_{z_i\to\infty}(\Phi_z^2)\right] \\
& + w_y \lim_{z_i\to\infty}(\Phi_x)\lim_{z_i\to\infty}\left(\Phi_y\right) + w_z \lim_{z_i\to\infty}(\Phi_x)\lim_{z_i\to\infty}(\Phi_z)\Bigg\} \\
& + \lim_{z_i\to\infty}(\Phi_x)\left(1 - \frac{n_o^2}{n_i^2}\left\{\left[w_y \lim_{z_i\to\infty}(\Phi_x) - w_x \lim_{z_i\to\infty}\left(\Phi_y\right)\right]^2\right.\right. \\
& + \left[w_z \lim_{z_i\to\infty}(\Phi_x) - w_x \lim_{z_i\to\infty}(\Phi_z)\right]^2 \\
& \left.\left. + \left[w_z \lim_{z_i\to\infty}\left(\Phi_y\right) - w_y \lim_{z_i\to\infty}(\Phi_z)\right]^2\right\}\right)^{\frac{1}{2}},
\end{aligned} \tag{7.46}
$$

the cosine director in y

$$
\begin{aligned}
\lim_{z_i\to\infty}(\wp_y) = -\frac{n_o}{n_i}\Bigg\{\Bigg[& w_y\left[\lim_{z_i\to\infty}(\Phi_x^2) + \lim_{z_i\to\infty}(\Phi_z^2)\right] \\
& - \lim_{z_i\to\infty}\left(\Phi_y\right)\left[w_x \lim_{z_i\to\infty}(\Phi_x) + w_z \lim_{z_i\to\infty}(\Phi_z)\right]\Bigg]\Bigg\} \\
& + \lim_{z_i\to\infty}\left(\Phi_y\right)\left(1 - \frac{n_o^2}{n_i^2}\left\{\left[w_y \lim_{z_i\to\infty}(\Phi_x) - w_x \lim_{z_i\to\infty}\left(\Phi_y\right)\right]^2\right.\right. \\
& + \left[w_z \lim_{z_i\to\infty}(\Phi_x) - w_x \lim_{z_i\to\infty}(\Phi_z)\right]^2 \\
& \left.\left. + \left[w_z \lim_{z_i\to\infty}\left(\Phi_y\right) - w_y \lim_{z_i\to\infty}(\Phi_z)\right]^2\right\}\right)^{\frac{1}{2}},
\end{aligned} \tag{7.47}
$$

and the cosine director in z

$$
\begin{aligned}
\lim_{z_i\to\infty}(\wp_z) = -\frac{n_o}{n_i}\Bigg\{\Bigg[& w_z\left[\lim_{z_i\to\infty}(\Phi_x^2) + \lim_{z_i\to\infty}\left(\Phi_y^2\right)\right] \\
& - \lim_{z_i\to\infty}(\Phi_z)\left[w_x \lim_{z_i\to\infty}(\Phi_x) + w_y \lim_{z_i\to\infty}\left(\Phi_y\right)\right]\Bigg]\Bigg\} \\
& + \lim_{z_i\to\infty}\left(\Phi_y\right)\left(1 - \frac{n_o^2}{n_i^2}\left\{\left[w_y \lim_{z_i\to\infty}(\Phi_x) - w_x \lim_{z_i\to\infty}\left(\Phi_y\right)\right]^2\right.\right. \\
& + \left[w_z \lim_{z_i\to\infty}(\Phi_x) - w_x \lim_{z_i\to\infty}(\Phi_z)\right]^2 \\
& \left.\left. + \left[w_z \lim_{z_i\to\infty}\left(\Phi_y\right) - w_y \lim_{z_i\to\infty}(\Phi_z)\right]^2\right\}\right)^{\frac{1}{2}}.
\end{aligned} \tag{7.48}
$$

Equations (7.36), (7.37), and (7.38) are such that the rays at the output are collimated. Therefore, equations (7.42), (7.43), and (7.44) are equal to equations (7.45), (7.46), and (7.47), so in x

$$\lim_{z_i \to \infty} (\wp_x) = \frac{p}{\sqrt{p^2 + s^2 + 1}}, \tag{7.49}$$

in y

$$\lim_{z_i \to \infty} (\wp_y) = \frac{s}{\sqrt{p^2 + s^2 + 1}}, \tag{7.50}$$

and in z

$$\lim_{z_i \to \infty} (\wp_z) = \frac{1}{\sqrt{p^2 + s^2 + 1}}. \tag{7.51}$$

Using equation (7.4) we can find the cosine directors of the input rays and using equations (7.49), (7.50), and (7.51) we can obtain the cosine directors of the output; therefore, we can now define the eikonal $H(t, x, y)$,

$$\boxed{\lim_{z_i \to \infty} H(t, x, y) = \begin{cases} [x + tw_x, y + tw_y, w + tw_z], & t < \lim\limits_{z_i \to \infty} (\vartheta), \\ \begin{bmatrix} \lim\limits_{z_i \to \infty} (x_a) + \lim\limits_{z_i \to \infty} (\wp_x)\left\{t - \lim\limits_{z_i \to \infty} (\vartheta)\right\}, \\ \lim\limits_{z_i \to \infty} (y_a) + \lim\limits_{z_i \to \infty} (\wp_y)\left\{t - \lim\limits_{z_i \to \infty} (\vartheta)\right\}, \\ \lim\limits_{z_i \to \infty} (z_a) + \lim\limits_{z_i \to \infty} (\wp_z)\left\{t - \lim\limits_{z_i \to \infty} (\vartheta)\right\} \end{bmatrix}, & t > \lim\limits_{z_i \to \infty} (\vartheta), \end{cases}} \tag{7.52}$$

where ϑ is the velocity of the light inside the medium of the refractive surface. The wavefront information is given by $H(t, x, y)$ for all points in time. $\lim_{z_i \to \infty} \vartheta$ is the natural condition that separates the distinct zones of the input and output rays.

7.8 Mathematica code

```
In[11]:=  w[x_,y_]:=4 (BesselJ[0,3Sqrt[0.5(x)^2+0.5(y )^2]/5])/10
e=10^(-5);

wdx[x_,y_]:=(w[x+e,y]-w[x,y])/e

wdy[x_,y_]:=(w[x,y+e]-w[x,y])/e

wx[x_,y_]:=(-wdx[x,y])/√(wdx[x,y]^2+wdy[x,y]^2+1)

wy[x_,y_]:=(-wdy[x,y])/√(wdx[x,y]^2+wdy[x,y]^2+1)

wz[x_,y_]:=1/√(wdx[x,y]^2+wdy[x,y]^2+1)

τz[x_,y_]:=w[x,y]-zi-zo

τx[x_,y_]:=x-hi

τy[x_,y_]:=y-ki


ϑ[x_,y_]:=1/(1-no^2/ni^2) (-( wx[x,y] τx[x,y]+ wy[x,y] τy[x,y]+ wz[x,y] τz[x,y]+(k no)/ni^2)
  -√(( wx[x,y] τx[x,y]+ wy[x,y] τy[x,y]+ wz[x,y] τz[x,y]+(k no)/ni^2)^2
   -(τx[x,y]^2+τy[x,y]^2+τz[x,y]^2-k^2/ni^2)(1-no^2/ni^2)))


xa[x_,y_]:=x+ϑ[x,y]wx[x,y]

ya[x_,y_]:=y+ϑ[x,y]wy[x,y]

za[x_,y_]:=w[x,y]+ϑ[x,y]wz[x,y]
```

```
dxxa[x_,y_]:=(xa[x+e,y]-xa[x,y])/e

dyxa[x_,y_]:=(xa[x,y+e]-xa[x,y])/e

dxya[x_,y_]:=(ya[x+e,y]-ya[x,y])/e

dyya[x_,y_]:=(ya[x,y+e]-ya[x,y])/e

dxza[x_,y_]:=(za[x+e,y]-za[x,y])/e

dyza[x_,y_]:=(za[x,y+e]-za[x,y])/e


Φx[x_,y_]:=(-dxza[x,y] dyya[x,y]+dxya[x,y] dyza[x,y])/
 (√
   ((-dxya[x,y] dyxa[x,y]+dxxa[x,y] dyya[x,y])^2
   +(dxza[x,y] dyxa[x,y]-dxxa[x,y] dyza[x,y])^2
   +(-dxza[x,y] dyya[x,y]+dxya[x,y] dyza[x,y])^2))


Φy[x_,y_]:=(dxza[x,y] dyxa[x,y]-dxxa[x,y] dyza[x,y])
 /(√
  ((-dxya[x,y] dyxa[x,y]+dxxa[x,y] dyya[x,y])^2
  +(dxza[x,y] dyxa[x,y]-dxxa[x,y] dyza[x,y])^2
  +(-dxza[x,y] dyya[x,y]+dxya[x,y] dyza[x,y])^2))


Φz[x_,y_]:=(-dxya[x,y] dyxa[x,y]+dxxa[x,y] dyya[x,y])
/(√
 ((-dxya[x,y] dyxa[x,y]+dxxa[x,y] dyya[x,y])^2
 +(dxza[x,y] dyxa[x,y]-dxxa[x,y] dyza[x,y])^2
 +(-dxza[x,y] dyya[x,y]+dxya[x,y] dyza[x,y])^2))
```

$$\varrho x[x\_,y\_]:=-\frac{1}{ni}no\ (wy[x,y]\ \Phi x[x,y]\ \Phi y[x,y]+wz[x,y]\ \Phi x[x,y]\ \Phi z[x,y]-wx[x,y]\ (\Phi y[x,y]^2+\Phi z[x,y]^2))+\Phi x[x,y]\ \sqrt{}(1-\frac{1}{ni^2}no^2\ ((wy[x,y]\ \Phi x[x,y]-wx[x,y]\ \Phi y[x,y])^2+(wz[x,y]\ \Phi x[x,y]-wx[x,y]\ \Phi z[x,y])^2+(wz[x,y]\ \Phi y[x,y]-wy[x,y]\ \Phi z[x,y])^2))$$

$$\varrho y[x\_,y\_]:=\frac{1}{ni}no\ (-\Phi y[x,y]\ (wx[x,y]\ \Phi x[x,y]+wz[x,y]\ \Phi z[x,y])+wy[x,y]\ (\Phi x[x,y]^2+\Phi z[x,y]^2))+\Phi y[x,y]\ \sqrt{}(1-\frac{1}{ni^2}no^2\ ((wy[x,y]\ \Phi x[x,y]-wx[x,y]\ \Phi y[x,y])^2+(wz[x,y]\ \Phi x[x,y]-wx[x,y]\ \Phi z[x,y])^2+(wz[x,y]\ \Phi y[x,y]-wy[x,y]\ \Phi z[x,y])^2))$$

$$\varrho z[x\_,y\_]:=\frac{1}{ni}no\ (wz[x,y]\ (\Phi x[x,y]^2+\Phi y[x,y]^2)-(wx[x,y]\ \Phi x[x,y]+wy[x,y]\ \Phi y[x,y])\ \Phi z[x,y])+\Phi z[x,y]\ \sqrt{}(1-\frac{1}{ni^2}no^2\ ((wy[x,y]\ \Phi x[x,y]-wx[x,y]\ \Phi y[x,y])^2+(wz[x,y]\ \Phi x[x,y]-wx[x,y]\ \Phi z[x,y])^2+(wz[x,y]\ \Phi y[x,y]-wy[x,y]\ \Phi z[x,y])^2))$$

$$wz2[x\_,y\_]:=\frac{-(za[x,y]-zi-zo)}{\sqrt{(za[x,y]-zi-zo)^2+(xa[x,y]-hi)^2+(ya[x,y]-ki)^2}}$$

$$wx2[x\_,y\_]:=\frac{-(xa[x,y]-hi)}{\sqrt{(za[x,y]-zi-zo)^2+(xa[x,y]-hi)^2+(ya[x,y]-ki)^2}}$$

$$wy2[x\_,y\_]:=\frac{-(ya[x,y]-ki)}{\sqrt{(za[x,y]-zi-zo)^2+(xa[x,y]-hi)^2+(ya[x,y]-ki)^2}};$$

c=1;
v=c/ni;
no=1;
ni=1.5;
rmax=5;
zo=4;
hi=1;
ki=-1;
zi=30;
$k=no\ zo+ni\sqrt{zi^2+hi^2+ki^2};$

```
Homotopy={{
 Piecewise[{
  {wx[x,y]tt+x,tt<ϑ[x,y]},{ v wx2[x,y](tt-ϑ[x,y])+xa[x,y],tt>ϑ[x,y]}}],
 Piecewise[{
  {wy[x,y]tt+y,tt<ϑ[x,y]},{ v wy2[x,y](tt-ϑ[x,y])+ya[x,y],tt>ϑ[x,y]}}],
 Piecewise[{
  {wz[x,y]tt+w[x,y],tt<ϑ[x,y]},{v wz2[x,y](tt-ϑ[x,y])+za[x,y],tt>ϑ[x,y]}}]}};


Waves:=Table[{ParametricPlot3D[Homotopy,{x,-rmax,rmax},{y,-rmax,rmax},
PlotRange→All,PlotStyle→{Blue,Opacity[.5]},PlotPoints→10,
Mesh→False]},{tt,5,k,5}]


Rays:=Table[{ParametricPlot3D[Homotopy,{tt,0,k},
PlotRange→All,PlotStyle→{Purple,Opacity[.5]},PlotPoints→10,
Mesh→False]},{x,-rmax,rmax,rmax},{y,-rmax,rmax,rmax}]


g0=Plot3D[{w[x,y]},{x,-rmax,rmax},{y,-rmax,rmax},Axes→True,
PlotStyle→{Blue,Opacity[.5]},AxesLabel→{x,y,z},AspectRatio→Automatic,
Boxed→True,Mesh→False];


g1=ParametricPlot3D[{xa[x,y],ya[x,y],za[x,y]},{x,-rmax,rmax},{y,-rmax,rmax},
Axes→True,PlotStyle→{Red,Opacity[.75]},Boxed→True,AxesLabel→{x,y,z},
AspectRatio→Automatic,Mesh→False,PlotPoints→10];


Show[g1,g0, Waves,Rays,PlotRange→All]
```

7.9 Examples

We present here the results of the coding example shown in the last section as a gallery composed of images (figures 7.9–7.14) generated by `Mathematica`.

7.10 End notes

In this chapter, we presented general freeform dioptric stigmatic surfaces such that an arbitrary freeform wavefront w is refracted by these surfaces as a spherical wavefront or a flat wavefront. The general freeform dioptric stigmatic surfaces were obtained step by step. As they were derived using Fermat's principle, the surfaces are wholly analytic, lacking any approximations, either numerical or paraxial. The equations describing the surface provided the predicted performance.

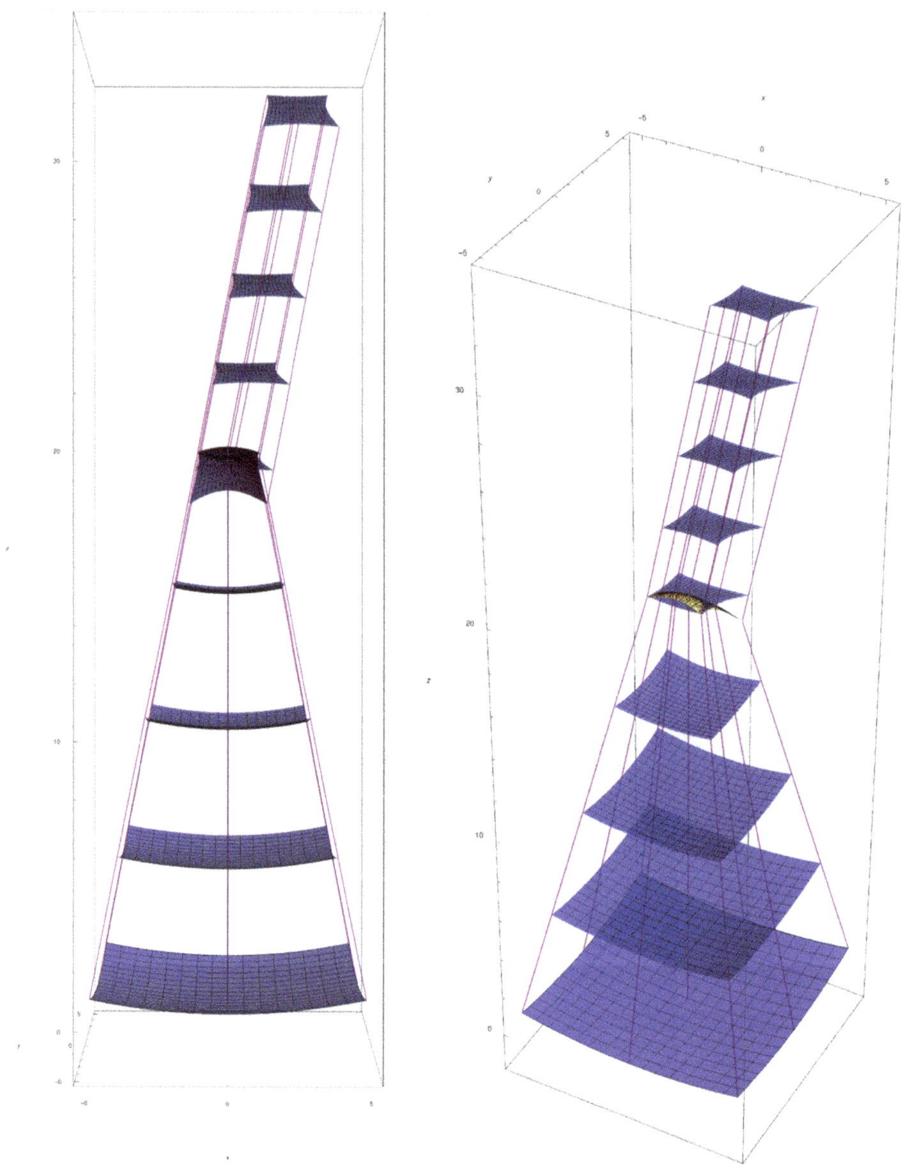

(a) A frontal view point of the adaptative mirror system.

(b) A frontal perspective view point of the adaptative mirror system.

Figure 7.9. Surface specifications for a parabolic input wavefront; the system produces a flat output wavefront; (a) the input wavefront is $-\cos(x/5)^2 + \cos(y/8)^2$, refractive index $n = 1.5$, object distance $z_o = 50$ mm, $r_{max} = 10$ mm, parameter $s = 0.5$; parameter $p = 0$; (r_a, z_a) = equations (7.40), (7.41), and (7.42), $H(t, r_a)$ = equation (7.52); (b) a frontal perspective of the same configuration.

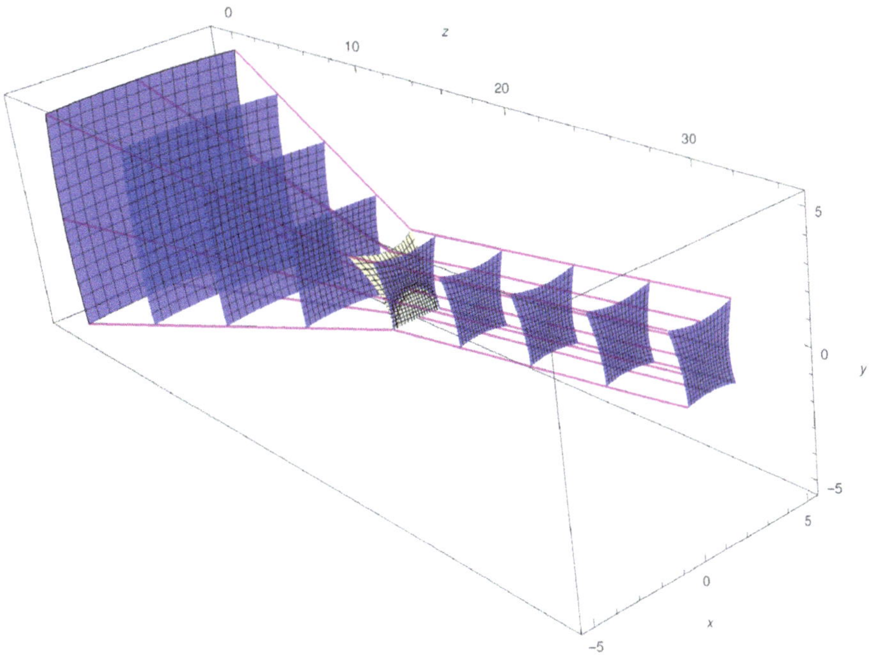

(a) A frontal left view point of the adaptative mirror system.

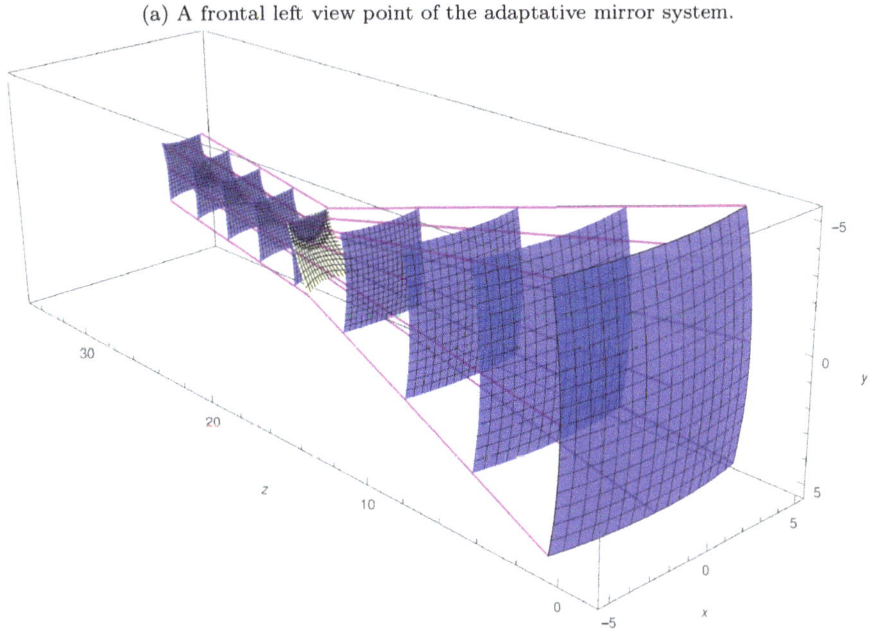

(b) A frontal right view point of the adaptative mirror system.

Figure 7.10. Surface specifications for a parabolic input wavefront; the system produces a flat output wavefront: (a) the input wavefront is $-\cos(x/5)^2 + \cos(y/8)^2$, refractive index $n = 1.5$, object distance $z_o = 50$ mm, $\mathrm{r}_{\max} = 10$ mm, parameter $s = 0.5$; parameter $p = 0$; $(r_a, z_a) =$ equations (7.40), (7.41), and (7.42), $H(t, r_a) =$ equation (7.52); (b) a frontal perspective of the same configuration.

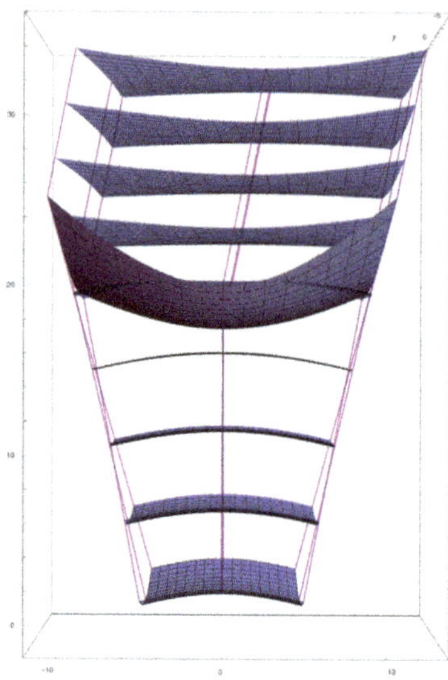

(a) A frontal view point of the adaptative mirror system.

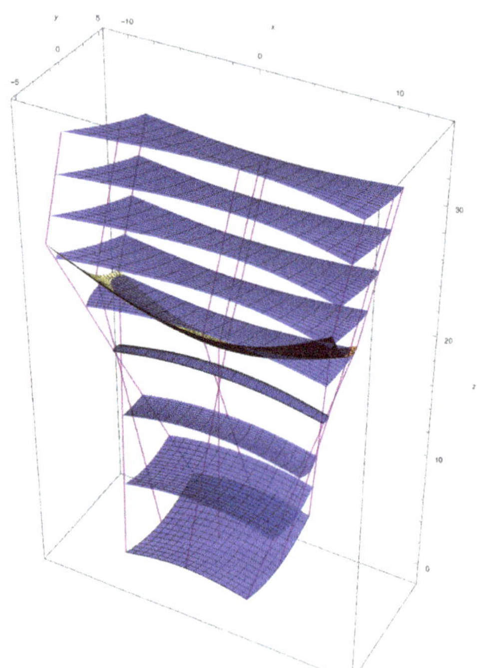

(b) A frontal perspective view point of the adaptative mirror system.

Figure 7.11. Surface specifications for a parabolic input wavefront; the system produces a flat output wavefront: (a) the input wavefront is $-\cos(x/5)^2 + \cos(y/8)^2$, refractive index $n = 1.5$, object distance $z_o = 50$ mm, $r_{\max} = 10$ mm, parameter $s = 0.5$; parameter $p = 0$; $(r_a, z_a) =$ equations (7.40), (7.41), and (7.42), $H(t, r_a) =$ equation (7.52); (b) a frontal perspective of the same configuration.

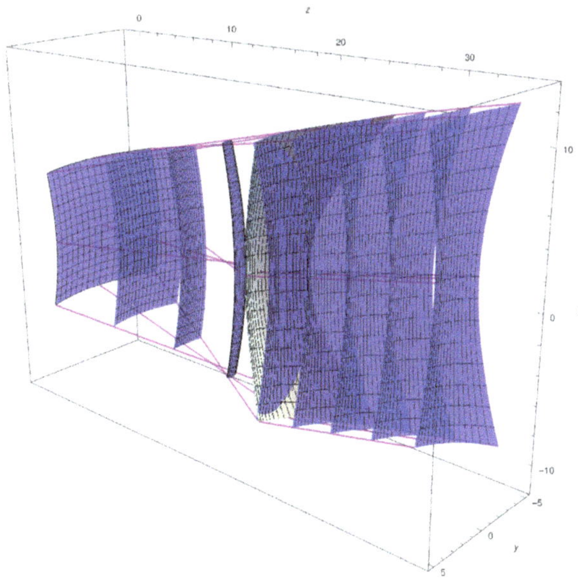

(a) A frontal left view point of the adaptative mirror system.

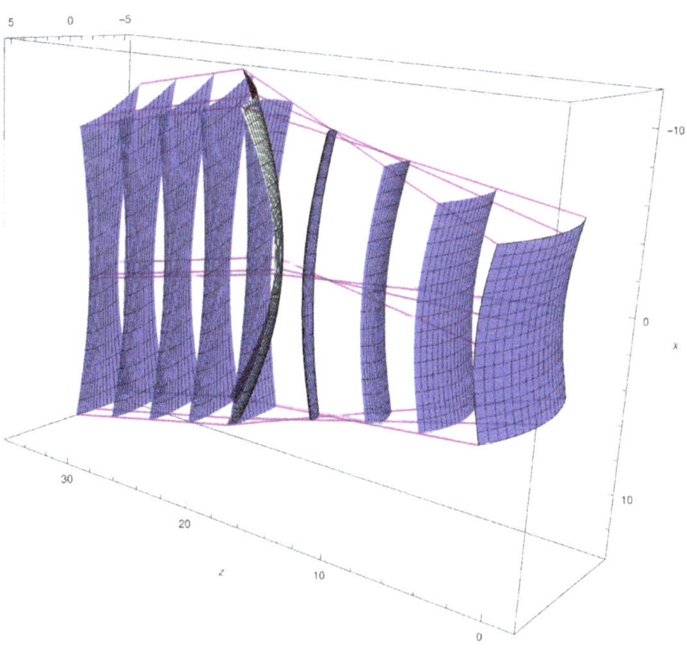

(b) A frontal right view point of the adaptative mirror system.

Figure 7.12. Surface specifications for a parabolic input wavefront; the system produces a flat output wavefront; (a) the input wavefront is $-\cos(x/5)^2 + \cos(y/8)^2$, refractive index $n = 1.5$, object distance $z_o = 50\,\text{mm}$, $\text{r}_{\text{max}} = 10\,\text{mm}$, parameter $s = 0.5$; parameter $p = 0$; $(r_a, z_a) =$ equations (7.40), (7.41), and (7.42), $H(t, r_a) =$ equation (7.52); (b) a frontal perspective of the same configuration.

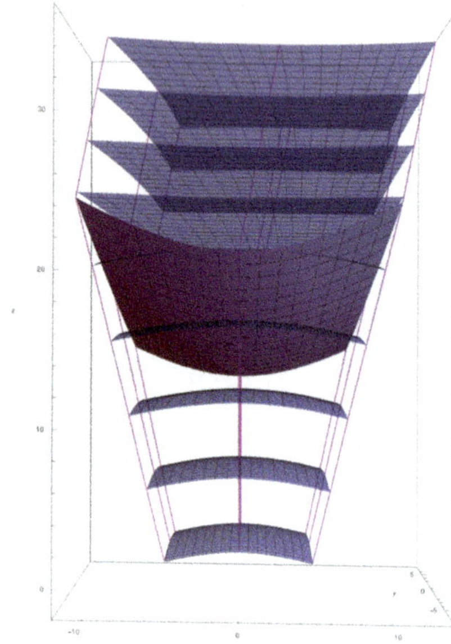

(a) A frontal view point of the adaptative mirror system.

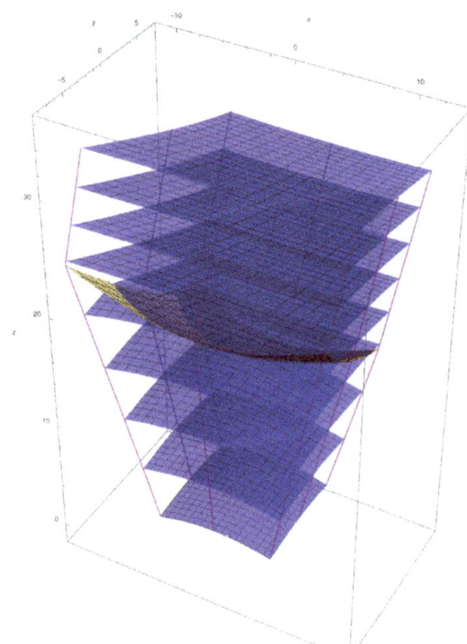

(b) A frontal perspective view point of the adaptative mirror system.

Figure 7.13. Surface specifications for a parabolic input wavefront; the system produces a flat output wavefront: (a) the input wavefront is $-\cos(x/5)^2 + \cos(y/8)^2$, refractive index $n = 1.5$, object distance $z_o = 50$ mm, $r_{max} = 10$ mm, parameter $s = 0.5$; parameter $p = 0$; $(r_a, z_a) =$ equations (7.40), (7.41), and (7.42), $H(t, r_a) =$ equation (7.52); (b) a frontal perspective of the same configuration.

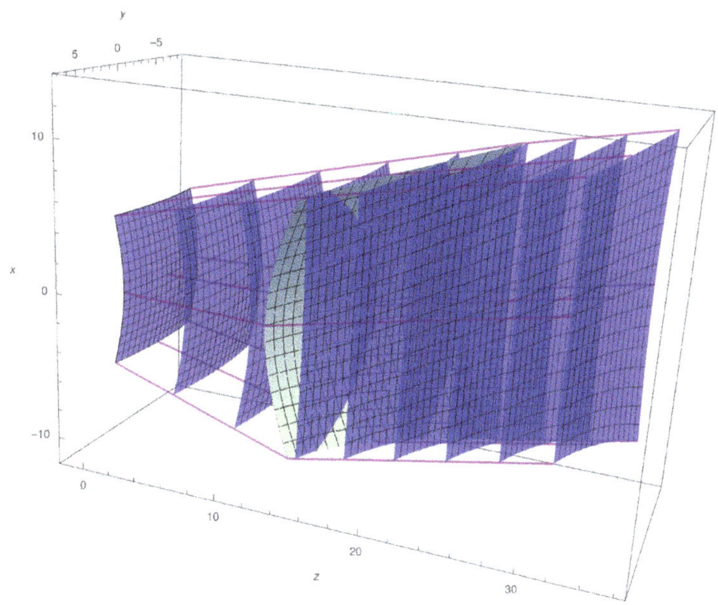

(a) A frontal left view point of the adaptative mirror system.

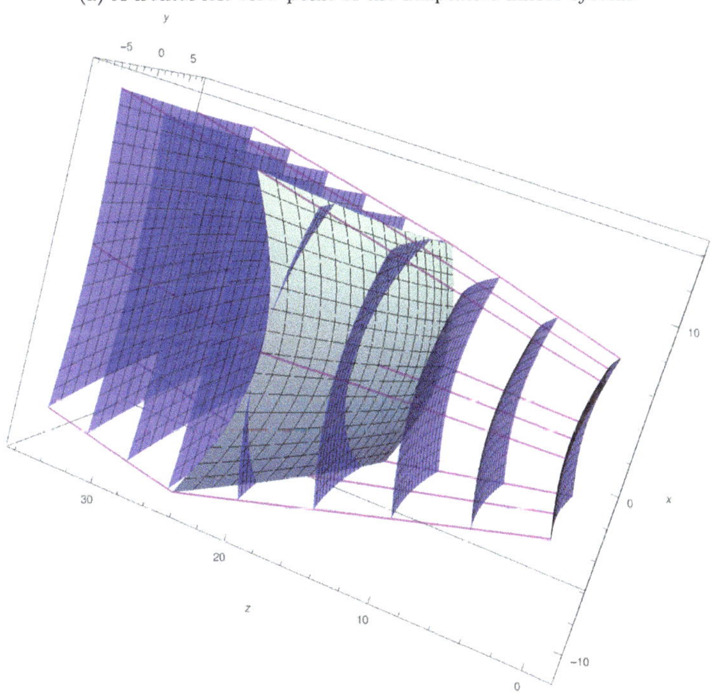

(b) A frontal right view point of the adaptative mirror system.

Figure 7.14. Surface specifications for a parabolic input wavefront; the system produces a flat output wavefront: (a) the input wavefront is $-\cos(x/5)^2 + \cos(y/8)^2$, refractive index $n = 1.5$, object distance $z_o = 50$ mm, $r_{\max} = 10$ mm, parameter $s = 0.5$; parameter $p = 0$; $(r_a, z_a) =$ equations (7.40), (7.41), and (7.42), $H(t, r_a) =$ equation (7.52); (b) a frontal perspective of the same configuration.

Chapter 8

The freeform transfer function lens

In this chapter, we present a mathematical model that allows a lens to be designed that transforms a given freeform wavefront into another given freeform wavefront. This mathematical model is based solely on Fermat's principle; no other assumptions were implemented.

8.1 Introduction

One of the most natural, challenging, and fundamental problems is the following: what system is required in order to obtain a given output from a given input? This problem is general and it is presented in many theories and branches of natural and social sciences.

Optical science is no exception, and the same problem can be formulated for the applications of optical science, for example, diffraction theory and physical optics, photonics and quantum optics, optical communications, optical signal progressing, optical fibres, etc. are similar versions of the same problem.

Geometrical optics and optical path theory are not exceptions. This treatise is only based on geometrical optics and optical path theory. This chapter is dedicated to solving the problem proposed in the first paragraph from the perspective of geometrical optics and optical path theory.

The problem can be stated as follows: given an input wavefront and output wavefront, define a lens that refracts the input wavefront and transforms it into the output wavefront. The input and output wavefronts can have no symmetry, thus they can be considered to be freeform.

8.2 Mathematical model

The diagram of the lens in which are interested is displayed in figure 8.1. The Cartesian coordinate system's origin is placed at the vertex of the input wavefront

doi:10.1088/978-0-7503-4705-1ch8

w_a. w_a is a continuous smooth function that depends on the independent variables x and y, so $w_a(x, y)$ (see figure 8.1).

The distance from the vertex of w_a to the vertex of the first surface is denoted by z_o. The lens has a central thickness of τ, and the distance from the second surface to the output wavefront is z_i. Let n_o, n, and n_i be the refractive indexes of the object, lens, and image space, respectively. The output wavefront is denoted by w_a and only depends on the independent variables x and y. Therefore, x and y are the independent variables of the system. The problem to find shapes of the first and second surfaces such that w_a changes into w_b. First, we are going to find the first surface such that it collimates the rays from w_a and then we are going to obtain the second surface such that it collimates the rays from w_b. The first surface is denoted by (x_a, y_a, z_a), where x_a is the x component, y_a is the y component, and z_a is the z component. The second surface is denoted by (x_b, y_b, z_b), where x_a, y_a, and z_a are the components in x, y, and z, respectively. Both (x_a, y_a, z_a) and (x_b, y_b, z_b) must depend only on the independent variables x and y, the input constants z_o, τ, and z_i, and the refractive indexes n_o, n, and n_i.

8.2.1 The first surface

We begin with the first surface; we have mentioned that we want it to collimate the light along the optical axis, so for the moment, we are not considering the second surface and its related parameters. We have learned from the previous chapters that we can obtain a refractive surface that collimates light using Fermat's principle for a finite image and then compute the limit when the image is very far away. The distance from the first surface to this finite image is z_{in}, where the subscript *'in'* denotes the inside, i.e. inside the lens.

From Fermat's principle we know that the optical path of each ray must be the same and constant. Let k be this constant, so the optical path length (OPL) of a non-axial ray is given by

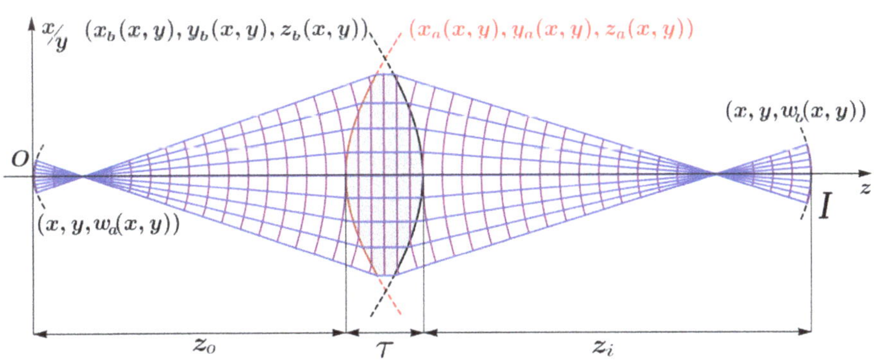

Figure 8.1. Schematic diagram of the configuration of a lens that performs a transfer function between the given wavefronts.

$$n_i\sqrt{x_a^2 + y_a^2 + (z_a - z_{in} - z_o)^2} + n_o\sqrt{(z_a - w_a)^2 + (x_a - x)^2 + (y_a - y)^2} = k. \quad (8.1)$$

The OPL of the last expression starts where the ray leaves w_a and finishes where the ray meets the image at $z_o + z_{in}$. Using the same strategy, we can obtain the OPL of the axial ray, as follows:

$$n_i z_{in} + n_o z_o = k. \quad (8.2)$$

In fact, equations (8.1) and (8.2) are equal, so we can write

$$n_i\sqrt{x_a^2 + y_a^2 + (z_a - z_{in} - z_o)^2} + n_o\sqrt{(z_a - w_a)^2 + (x_a - x)^2 + (y_a - y)^2} = n_i z_{in} + n_o z_o. \quad (8.3)$$

We can reduce the length of the previous expression by letting ϑ_a be the second square root of the left-hand side of equation (8.3):

$$\vartheta_a \equiv \sqrt{(z_a - w_a)^2 + (x_a - x)^2 + (y_a - y)^2}, \quad (8.4)$$

and by substituting equation (8.4) into equation (8.3), we obtain

$$n_i\sqrt{x_a^2 + y_a^2 + (z_a - z_{in} - z_o)^2} + n_o\vartheta = k. \quad (8.5)$$

The next step is like that of chapter 7, where we expressed the first surface in terms of the input cosine directors given by

$$w_{a_x} = -\frac{\partial_x w_a}{\sqrt{\partial_x w_a^2 + \partial_y w_a^2 + 1}}, \quad w_{a_y} = -\frac{\partial_y w_a}{\sqrt{\partial_x w_a^2 + \partial_y w_a^2 + 1}}, \quad w_{a_z} = \frac{1}{\sqrt{\partial_x w_a^2 + \partial_y w_a^2 + 1}}, \quad (8.6)$$

where w_{a_x}, w_{a_y}, w_{a_z} are the cosine directors of the input rays in the z, x, and y directions, respectively. The partial derivatives of w are given by $\partial_x w_a \equiv \partial w/\partial x$ and $\partial_y w_a \equiv \partial w/\partial y$. Therefore, the refractive surface (x_a, y_a, z_a) can be expressed as

$$x_a = x + \vartheta_a w_{a_x}, \quad y_a = y + \vartheta_a w_{a_y}, \quad z_a = w_a + \vartheta_a w_{a_z}. \quad (8.7)$$

By substituting equation (8.7) into equation (8.5), we obtain

$$-(n_i z_{in} + n_o z_o) + n_i\sqrt{(w_{a_x}\vartheta_a + x)^2 + (w_{a_y}\vartheta_a + y)^2 + (z_a - z_{in} - z_o)^2} + n_o\vartheta_a = 0. \quad (8.8)$$

At this point, it is appropriate to define the image as being far away, so we compute the limit $z_{in} \rightarrow \infty$ over equation (8.8):

$$\lim_{z_{in}\rightarrow\infty}\Big[-(n_i z_{in} + n_o z_o) + n_i\sqrt{(w_{a_x}\vartheta_a + x)^2 + (w_{a_y}\vartheta_a + y)^2 + (z_a - z_{in} - z_o)^2} + n_o\vartheta_a\Big] = n_o(\vartheta_a - z_o) - n_i(w_a + w_{a_z}\vartheta_a - z_o). \tag{8.9}$$

The computation of this limit gives

$$n_o(\vartheta_a - z_o) - n_i(w_a + w_{a_z}\vartheta_a - z_o) = 0. \tag{8.10}$$

Note the zero on the right-hand side of this equation; it appears because we applied the limit $z_{in} \rightarrow \infty$ over equation (8.10), which is the optical path difference, which is zero. Solving for ϑ_a, we obtain

$$\vartheta_a = \frac{n_i w - n_i z_o + n_o z_o}{n_o - n_i w_{a_z}}. \tag{8.11}$$

Thus, the first surface can be obtained by substituting equation (8.11) into equation (8.7). The component x of the first refractive surface is

$$\boxed{x_a = x - \frac{\partial_x w_a(n w_a + z_o(n_o - n))}{n - n_o\sqrt{\partial_x w_a^2 + \partial_y w_a^2 + 1}}}, \tag{8.12}$$

the component y of the first refractive surface is

$$\boxed{y_a = y - \frac{\partial_y w_a\big(n w_a + z_o\big(n_o - n\big)\big)}{n - n_o\sqrt{\partial_x w_a^2 + \partial_y w_a^2 + 1}}}, \tag{8.13}$$

and the component z of the first refractive surface is

$$\boxed{z_a = w_a + \frac{n w_a + z_o(n_o - n)}{n_o\sqrt{\partial_x w_a^2 + \partial_y w_a^2 + 1} - n}}. \tag{8.14}$$

Equations (8.12), (8.13), and (8.14) are closed form and have very interesting properties that are shared with the second refractive surface. We will discuss these properties along with the second refractive surface. First, we focus on the design of the second surface.

8.2.2 The second surface

Using equations (8.12), (8.13), and (8.14) we can also design a closed-form expression for the second surface. We only need to replace some parameters in the structures of equations (8.12), (8.13), and (8.14). We will list these parameters; the first set of them consists of the cosine directors of the rays coming from or going to w_b.

The direction cosine of w_b in the x direction is

$$w_{b_x} = -\frac{\partial_x w_b}{\sqrt{\partial_x w_b^2 + \partial_y w_b^2 + 1}}, \tag{8.15}$$

the direction cosine of w_b in the y direction is given by

$$w_{b_y} = -\frac{\partial_y w_b}{\sqrt{\partial_x w_b^2 + \partial_y w_b^2 + 1}}, \tag{8.16}$$

and finally, the direction cosine in the z direction is

$$w_{b_z} = \frac{1}{\sqrt{\partial_x w_b^2 + \partial_y w_b^2 + 1}}. \tag{8.17}$$

We can now substitute expressions (8.15), (8.16), and (8.17) into (8.11) to obtain ϑ_b for the second surface, and by also substituting $z_o \to z_i$, $n \to n_i$, and $n_o \to n$, we get

$$\vartheta_b = \frac{nw_b + z_i(n_i - n)}{n_i - nw_{b_z}}. \tag{8.18}$$

The second surface has the same structure as equation (8.7), so the x component of the second refractive surface is

$$x_b = x + w_{b_x}\vartheta_b, \tag{8.19}$$

the y component of the second refractive surface is

$$y_b = y + w_{b_y}\vartheta_b, \tag{8.20}$$

and the z component of the second refractive surface is

$$z_b = -w_b - w_{b_z}\vartheta_b + \tau + z_i + z_o. \tag{8.21}$$

Substituting equation (8.18) into equations (8.19), (8.20), and (8.21) yields, in x

$$x_b = x + \frac{\partial_x w_b(nw_b + z_i(n_i - n))}{n - n_i\sqrt{\partial_x w_b^2 + \partial_y w_b^2 + 1}}, \tag{8.22}$$

in y

$$y_b = y + \frac{\partial_y w_b\big(nw_b + z_i(n_i - n)\big)}{n - n_i\sqrt{\partial_x w_b^2 + \partial_y w_b^2 + 1}}, \tag{8.23}$$

and in z

$$z_b = -w_b + \tau + z_i + z_o - \frac{nw_b + z_i(n_i - n)}{n_i\sqrt{\partial_x w_b^2 + \partial_y w_b^2 + 1} - n}. \tag{8.24}$$

Note that the negative signs of $-w_{b_z}\vartheta_b$ and $-w_b$ in equation (8.21) invert the second surface with respect to the first surface. Thus, equation (8.21) expresses the second surface as required, i.e. a surface that receives collimated rays and generates the wavefront w_b at the output plane. Since the coordinate system is positioned at the vertex of the first wavefront in equation (8.21), τ is a necessary offset for the output wavefront and the second surface.

A wavefront inside a medium with refractive index n is collimated by these surfaces, similarly to the collimation performed by Cartesian ovals for spherical waterfronts. In the same way that Cartesian ovals are not physically limited, neither are these surfaces. In practice, however, both surfaces are measured where they touch, and this measurement is proportional to τ.

As a result, knowing the boundary conditions, we affirm that these equations have a unique solution. It can be concluded that it is determined by the limit $z_i \to \infty$, which results in an algebraic equation when the square roots are removed.

8.3 Ray tracing of light passing through the transfer function lens

The next step is to perform ray tracing for the system based on the lens designed using equations (8.12), (8.13), (8.14), (8.22), (8.23), and (8.24).

8.3.1 Ray tracing at the first surface

We begin with the rays that leave w_a; these just follow the cosine directors of equation (8.6) until they reach (x_a, y_a, z_a). When they reach (x_a, y_a, z_a), they are refracted according to Snell's law:

$$\boxed{\vec{\mathbf{v}}_2 = \frac{n_o}{n}[\vec{\mathbf{v}}_1 - (\vec{\mathbf{n}}_a \cdot \vec{\mathbf{v}}_1)\vec{\mathbf{n}}_a] - \vec{\mathbf{n}}_a\sqrt{1 - \frac{n_o^2}{n^2}(\vec{\mathbf{n}}_a \times \vec{\mathbf{v}}_1)^2} \quad \text{for} \quad \vec{\mathbf{v}}_2, \vec{\mathbf{v}}_1, \vec{\mathbf{n}}_a \in \mathbb{R}^3,} \tag{8.25}$$

where $\vec{\mathbf{v}}_1$ is the unit vector of the incident ray, $\vec{\mathbf{v}}_2$ is unit vector of the refracted ray, and $\vec{\mathbf{n}}_a$ is the normal vector of the first surface:

$$\begin{aligned}
\vec{\mathbf{v}}_1 &= w_{a_x}\vec{\mathbf{e}}_1 + w_{a_y}\vec{\mathbf{e}}_2 + w_{a_z}\vec{\mathbf{e}}_3, \\
\vec{\mathbf{v}}_2 &= \wp_x\vec{\mathbf{e}}_1 + \wp_y\vec{\mathbf{e}}_2 + \wp_z\vec{\mathbf{e}}_3, \\
\vec{\mathbf{n}}_a &= \Phi_{a_x}\vec{\mathbf{e}}_1 + \Phi_{a_y}\vec{\mathbf{e}}_2 + \Phi_{a_z}\vec{\mathbf{e}}_3,
\end{aligned} \tag{8.26}$$

where $\vec{\mathbf{e}}_1$ denotes the x direction, $\vec{\mathbf{e}}_2$ denotes the y direction and $\vec{\mathbf{e}}_3$ denotes the z direction. The normal vector of (x_a, y_a, z_a) is computed using the following expression:

$$\begin{aligned}
\vec{\mathbf{n}}_a &= \Phi_{a_x}\vec{\mathbf{e}}_1 + \Phi_{a_y}\vec{\mathbf{e}}_2 + \Phi_{a_z}\vec{\mathbf{e}}_3 \\
&= \pm\frac{\dfrac{\partial(x_a\vec{\mathbf{e}}_1 + y_a\vec{\mathbf{e}}_2 + z_a\vec{\mathbf{e}}_3)}{\partial x} \times \dfrac{\partial(x_a\vec{\mathbf{e}}_1 + y_a\vec{\mathbf{e}}_2 + z_a\vec{\mathbf{e}}_3)}{\partial y}}{\left|\dfrac{\partial(x_a\vec{\mathbf{e}}_1 + y_a\vec{\mathbf{e}}_2 + z_a\vec{\mathbf{e}}_3)}{\partial x} \times \dfrac{\partial(x_a\vec{\mathbf{e}}_1 + y_a\vec{\mathbf{e}}_2 + z_a\vec{\mathbf{e}}_3)}{\partial y}\right|}.
\end{aligned} \tag{8.27}$$

If we compute equation (8.27) and separate the components of the normal vector of (x_a, y_a, z_a), then for x, we obtain

$$\Phi_{a_x} = \frac{\partial_x y_a \partial_y z_a - \partial_x z_a \partial_y y_a}{\sqrt{\left(\partial_x x_a \partial_y y_a - \partial_x y_a \partial_y x_a\right)^2 + \left(\partial_x z_a \partial_y x_a - \partial_x x_a \partial_y z_a\right)^2 + \left(\partial_x y_a \partial_y z_a - \partial_x z_a \partial_y y_a\right)^2}}, \tag{8.28}$$

for y we get

$$\Phi_{a_y} = \frac{\partial_x z_a \partial_y x_a - \partial_x x_a \partial_y z_a}{\sqrt{\left(\partial_x x_a \partial_y y_a - \partial_x y_a \partial_y x_a\right)^2 + \left(\partial_x z_a \partial_y x_a - \partial_x x_a \partial_y z_a\right)^2 + \left(\partial_x y_a \partial_y z_a - \partial_x z_a \partial_y y_a\right)^2}}, \tag{8.29}$$

and for z we obtain

$$\Phi_{a_z} = \frac{\partial_x x_a \partial_y y_a - \partial_x y_a \partial_y x_a}{\sqrt{\left(\partial_x x_a \partial_y y_a - \partial_x y_a \partial_y x_a\right)^2 + \left(\partial_x z_a \partial_y x_a - \partial_x x_a \partial_y z_a\right)^2 + \left(\partial_x y_a \partial_y z_a - \partial_x z_a \partial_y y_a\right)^2}}. \tag{8.30}$$

Equations (8.28), (8.29), and (8.30) define the normal vector of (x_a, y_a, z_a). Note that equations (8.28), (8.29), and (8.30) depend on the partial derivatives of (x_a, y_a, z_a). We have computed these partial derivatives of (x_a, y_a, z_a); for the derivative of x_a with respect to x, we obtain

$$\begin{aligned}
\partial_x x_a = 1 &+ \frac{n \partial_x w_a^2}{n - n_o\sqrt{\partial_x w_a^2 + \partial_y w_a^2 + 1}} + \frac{\partial_{xx} w_a[nw + z_o(n_o - n)]}{n - n_o\sqrt{\partial_x w_a^2 + \partial_y w_a^2 + 1}} \\
&+ \frac{\partial_x w_a\left(\partial_x w_a \partial_{xx} w_a + \partial_y w_a \partial_{yx} w_a\right)\left[z_o\left(n - n_o\right) - nw\right]}{\left(\partial_x w_a^2 + \partial_y w_a^2 + 1\right)\left(n - n_o\sqrt{\partial_x w_a^2 + \partial_y w_a^2 + 1}\right)} \\
&+ \frac{n \partial_x w_a\left(\partial_x w_a \partial_{xx} w_a + \partial_y w_a \partial_{yx} w_a\right)\left[nw + z_o\left(n_o - n\right)\right]}{\left[n\sqrt{\partial_x w_a^2 + \partial_y w_a^2 + 1} - n_o\left(\partial_x w_a^2 + \partial_y w_a^2 + 1\right)\right]^2},
\end{aligned} \tag{8.31}$$

for the derivative of x_a with respect to y, we obtain

$$\begin{aligned}
\partial_y x_a = \Big\{ &\partial_x w_a\Big[n \partial_y w_a\left(\partial_y w_a^2 + 1\right)\left(n - n_o\sqrt{\partial_x w_a^2 + \partial_y w_a^2 + 1}\right) \\
&+ n_o \partial_y w_a \partial_{yy} w_a \sqrt{\partial_x w_a^2 + \partial_y w_a^2 + 1}\left(nw - nz_o + n_o z_o\right)\Big] \\
&+ n \partial_y w_a \partial_x w_a^3\left(n - n_o\sqrt{\partial_x w_a^2 + \partial_y w_a^2 + 1}\right) \\
&- \partial_{xy} w_a\left(\partial_y w_a^2 + 1\right)\left[z_o(n - n_o) - nw\right]\left(n - n_o\sqrt{\partial_x w_a^2 + \partial_y w_a^2 + 1}\right) \\
&+ n \partial_{xy} w_a \partial_x w_a^2\left[nw + z_o(n_o - n)\right]\Big\} \\
\times &\left\{n\sqrt{\partial_x w_a^2 + \partial_y w_a^2 + 1} - n_o\left(\partial_x w_a^2 + \partial_y w_a^2 + 1\right)\right\}^{-2},
\end{aligned} \tag{8.32}$$

for the derivative of y_a with respect to x, we obtain

$$\begin{aligned}\partial_x y_a = &\Big\{\partial_y w_a\Big[n\partial_x w_a(\partial_x w_a^2 + 1)\Big(n - n_o\sqrt{\partial_x w_a^2 + \partial_y w_a^2 + 1}\Big)\\ &+ n_o\partial_x w_a\partial_{xx}w_a\sqrt{\partial_x w_a^2 + \partial_y w_a^2 + 1}\Big(nw - nz_o + n_o z_o\Big)\Big]\\ &+ n\partial_x w_a\partial_y w_a^3\Big(n - n_o\sqrt{\partial_x w_a^2 + \partial_y w_a^2 + 1}\Big)\\ &- \Big(\partial_x w_a^2 + 1\Big)\partial_{yx}w_a\Big[z_o(n - n_o) - nw\Big]\Big(n - n_o\sqrt{\partial_x w_a^2 + \partial_y w_a^2 + 1}\Big)\\ &+ n\partial_{yx}w_a\partial_y w_a^2\Big[nw + z_o(n_o - n)\Big]\Big\}\\ &\times\Big\{n\sqrt{\partial_x w_a^2 + \partial_y w_a^2 + 1} - n_o\Big(\partial_x w_a^2 + \partial_y w_a^2 + 1\Big)\Big\}^{-2},\end{aligned} \tag{8.33}$$

for the derivative of y_a with respect to y, we obtain

$$\begin{aligned}\partial_y y_a = 1 &+ \frac{n\partial_y w_a\partial_y w_a}{n - n_o\sqrt{\partial_x w_a^2 + \partial_y w_a^2 + 1}} + \frac{\partial_{yy}w_a\Big[nw + z_o\Big(n_o - n\Big)\Big]}{n - n_o\sqrt{\partial_x w_a^2 + \partial_y w_a^2 + 1}}\\ &+ \frac{\partial_y w_a\Big(\partial_x w_a\partial_{xy}w_a + \partial_y w_a\partial_{yy}w_a\Big)\Big[z_o\Big(n - n_o\Big) - nw\Big]}{\Big(\partial_x w_a^2 + \partial_y w_a^2 + 1\Big)\Big(n - n_o\sqrt{\partial_x w_a^2 + \partial_y w_a^2 + 1}\Big)}\\ &+ \frac{n\partial_y w_a\Big(\partial_x w_a\partial_{xy}w_a + \partial_y w_a\partial_{yy}w_a\Big)\Big[nw + z_o\Big(n_o - n\Big)\Big]}{\Big(n\sqrt{\partial_x w_a^2 + \partial_y w_a^2 + 1} - n_o\Big[\partial_x w_a^2 + \partial_y w_a^2 + 1\Big]\Big)^2},\end{aligned} \tag{8.34}$$

for the derivative of z_a with respect to x, we obtain

$$\begin{aligned}\partial_x z_a = &\Big\{n_o\Big[\Big(\partial_x w_a\partial_{xx}w_a + \partial_y w_a\partial_{yx}w_a\Big)\Big[z_o\Big(n - n_o\Big) - nw\Big]\\ &- \partial_x w_a\Big(\partial_x w_a^2 + \partial_y w_a^2 + 1\Big)\Big(n - n_o\sqrt{\partial_x w_a^2 + \partial_y w_a^2 + 1}\Big)\Big]\Big\}\\ &\times\Big\{\sqrt{\partial_x w_a^2 + \partial_y w_a^2 + 1}\Big(n - n_o\sqrt{\partial_x w_a^2 + \partial_y w_a^2 + 1}\Big)^2\Big\}^{-1},\end{aligned} \tag{8.35}$$

and for the derivative of z_a with respect to y, we obtain

$$\begin{aligned}\partial_y z_a = &\Big\{n_o\Big[\Big(\partial_x w_a\partial_{xy}w_a + \partial_y w_a\partial_{yy}w_a\Big)\Big[z_o\Big(n - n_o\Big) - nw\Big]\\ &- \partial_y w_a\Big(\partial_x w_a^2 + \partial_y w_a^2 + 1\Big)\Big(n - n_o\sqrt{\partial_x w_a^2 + \partial_y w_a^2 + 1}\Big)\Big]\Big\}\\ &\times\Big\{\sqrt{\partial_x w_a^2 + \partial_y w_a^2 + 1}\Big(n - n_o\sqrt{\partial_x w_a^2 + \partial_y w_a^2 + 1}\Big)^2\Big\}^{-1}.\end{aligned} \tag{8.36}$$

The gigantic expression for the normal vector of the refractive surface (x_a, y_a, z_a) is a composite function of equations (8.31), (8.32), (8.33), (8.34), (8.35), and (8.36) with equations (8.28), (8.29), and (8.30) and finally with equation (8.27). We do not show this final gigantic expression for the normal vector for size reasons, but shall call it 'the vector of equation (8.27)', while considering that the final substitution yields a very large expression.

We now return to Snell's last at the first surface, equation (8.25). We compute $[\vec{\mathbf{v}}_1 - (\vec{\mathbf{n}}_a \cdot \vec{\mathbf{v}}_1)\vec{\mathbf{n}}_a]$, (only considering equation (8.27) for the normal vector); from the last paragraph, we have already an idea of how gigantic this expression is, accordingly:

$$\begin{aligned}\left[\vec{\mathbf{v}}_1 - (\vec{\mathbf{n}}_a \cdot \vec{\mathbf{v}}_1)\vec{\mathbf{n}}_a\right] &= \left[-w_{a_x}\left(\Phi_{a_y}^2 + \Phi_{a_z}^2\right) + w_{a_y}\Phi_{a_x}\Phi_{a_y} + w_{a_z}\Phi_{a_x}\Phi_{a_z}\right]\vec{\mathbf{e}}_1 \\ &+ \left[w_{a_y}\left(\Phi_{a_x}^2 + \Phi_{a_z}^2\right) - \Phi_{a_y}\left(w_{a_x}\Phi_{a_x} + w_{a_z}\Phi_{a_z}\right)\right]\vec{\mathbf{e}}_2 \\ &+ \left[w_{a_z}\left(\Phi_{a_x}^2 + \Phi_{a_y}^2\right) - \Phi_{a_z}\left(w_{a_x}\Phi_{a_x} + w_{a_y}\Phi_{a_y}\right)\right]\vec{\mathbf{e}}_3.\end{aligned} \tag{8.37}$$

The next step is the square root term of equation (8.25); the computation gives

$$\begin{aligned}-\vec{\mathbf{n}}_a\sqrt{1 - \frac{n_o^2}{n^2}(\vec{\mathbf{n}}_a \times \vec{\mathbf{v}}_1)^2} &= (\Phi_{a_x}\vec{\mathbf{e}}_1 + \Phi_{a_y}\vec{\mathbf{e}}_2 + \Phi_{a_z}\vec{\mathbf{e}}_3) \\ \times \sqrt{1 - \frac{n_o^2}{n^2}\left[\left(w_{a_y}\Phi_{a_x} - w_{a_x}\Phi_{a_y}\right)^2 + (w_{a_z}\Phi_{a_x} - w_{a_x}\Phi_{a_z})^2 + \left(w_{a_z}\Phi_{a_y} - w_{a_y}\Phi_{a_z}\right)^2\right]}&.\end{aligned} \tag{8.38}$$

By summing equations (8.37) and (8.38) and separating components, we obtain the cosine directors of the ray inside the lens, in x,

$$\begin{aligned}\wp_x &= \frac{n_o}{n}\left[-w_{a_x}\left(\Phi_{a_y}^2 + \Phi_{a_z}^2\right) + w_{a_y}\Phi_{a_x}\Phi_{a_y} + w_{a_z}\Phi_{a_x}\Phi_{a_z}\right] \\ &+ \Phi_{a_x}\sqrt{1 - \frac{n_o^2}{n^2}\left[\left(w_{a_y}\Phi_{a_x} - w_{a_x}\Phi_{a_y}\right)^2 + (w_{a_z}\Phi_{a_x} - w_{a_x}\Phi_{a_z})^2 + \left(w_{a_z}\Phi_{a_y} - w_{a_y}\Phi_{a_z}\right)^2\right]},\end{aligned} \tag{8.39}$$

in y,

$$\begin{aligned}\wp_y &= \frac{n_o}{n}\left[w_{a_y}\left(\Phi_{a_x}^2 + \Phi_{a_z}^2\right) - \Phi_{a_y}\left(w_{a_x}\Phi_{a_x} + w_{a_z}\Phi_{a_z}\right)\right] \\ &+ \Phi_{a_y}\sqrt{1 - \frac{n_o^2}{n^2}\left[\left(w_{a_y}\Phi_{a_x} - w_{a_x}\Phi_{a_y}\right)^2 + (w_{a_z}\Phi_{a_x} - w_{a_x}\Phi_{a_z})^2 + \left(w_{a_z}\Phi_{a_y} - w_{a_y}\Phi_{a_z}\right)^2\right]},\end{aligned} \tag{8.40}$$

and in z,

$$\begin{aligned}\wp_z &= \frac{n_o}{n}\left[w_{a_z}\left(\Phi_{a_x}^2 + \Phi_{a_y}^2\right) - \Phi_{a_z}\left(w_{a_x}\Phi_{a_x} + w_{a_y}\Phi_{a_y}\right)\right] \\ &+ \Phi_{a_z}\sqrt{1 - \frac{n_o^2}{n^2}\left[\left(w_{a_y}\Phi_{a_x} - w_{a_x}\Phi_{a_y}\right)^2 + (w_{a_z}\Phi_{a_x} - w_{a_x}\Phi_{a_z})^2 + \left(w_{a_z}\Phi_{a_y} - w_{a_y}\Phi_{a_z}\right)^2\right]}.\end{aligned} \tag{8.41}$$

Equations (8.39), (8.40), and (8.41) can actually be reduced to

$$\wp_x = 0, \quad \wp_x = 0, \quad \wp_z = 1, \tag{8.42}$$

since (x_a, y_a, z_a) is such that it collimates the rays coming from w_a. If this normal vector, (8.27), is substituted into equations (8.39), (8.40) and (8.41), the result is equation (8.42). The procedure is extremely long, but it can be computed using symbolic software such as `Mathematica`.

8.3.2 Ray tracing at the second surface

The rays refracted by the second surface can be computed as follows: the collimated vectors inside the lens strike the second surface and are refracted as rays that form the output wavefront w_b. Second, we can compute the rays arriving from wavefront w_b that strike (x_b, y_b, z_b), which are collimated inside the lens. We are going to use this perspective because it is similar to that implemented for the first refractive surface. Using the experience gained in the last section, this second part of the ray tracing becomes the simplest.

We start by recalling Snell's law at (x_b, y_b, z_b),

$$\boxed{\overrightarrow{\mathbf{v}_3} = \frac{n}{n_i}[\overrightarrow{\mathbf{v}_3} - (\overrightarrow{\mathbf{n}_b} \cdot \overrightarrow{\mathbf{v}_2})\overrightarrow{\mathbf{n}_b}] - \overrightarrow{\mathbf{n}_b}\sqrt{1 - \frac{n^2}{n_i^2}(\overrightarrow{\mathbf{n}_b} \times \overrightarrow{\mathbf{v}_3})^2} \quad \text{for} \quad \overrightarrow{\mathbf{v}_2}, \overrightarrow{\mathbf{v}_3}, \overrightarrow{\mathbf{n}_b} \in \mathbb{R}^3,} \tag{8.43}$$

where $\overrightarrow{\mathbf{v}_2}$ is the unit vector of the incident ray, $\overrightarrow{\mathbf{v}_2}$ is the unit vector of the refracted ray, and $\overrightarrow{\mathbf{n}_b}$ is the normal vector of the first surface:

$$\begin{aligned}
\overrightarrow{\mathbf{v}_3} &= w_{b_x}\overrightarrow{\mathbf{e}_1} + w_{b_y}\overrightarrow{\mathbf{e}_2} + w_{b_z}\overrightarrow{\mathbf{e}_3},\\
\overrightarrow{\mathbf{v}_2} &= \wp_x\overrightarrow{\mathbf{e}_1} + \wp_y\overrightarrow{\mathbf{e}_2} + \wp_z\overrightarrow{\mathbf{e}_3},\\
\overrightarrow{\mathbf{n}_b} &= \Phi_{b_x}\overrightarrow{\mathbf{e}_1} + \Phi_{b_y}\overrightarrow{\mathbf{e}_2} + \Phi_{b_z}\overrightarrow{\mathbf{e}_3},
\end{aligned} \tag{8.44}$$

where $\overrightarrow{\mathbf{e}_1}$ denotes the x direction, $\overrightarrow{\mathbf{e}_2}$ denotes the y direction and $\overrightarrow{\mathbf{e}_3}$ denotes the z direction. The normal vector of (x_b, y_b, z_b) can be computed in a similar way to the normal vector of (x_a, y_a, z_a), so we can write

$$\begin{aligned}
\overrightarrow{\mathbf{n}_b} &= \Phi_{b_x}\overrightarrow{\mathbf{e}_1} + \Phi_{b_y}\overrightarrow{\mathbf{e}_2} + \Phi_{b_z}\overrightarrow{\mathbf{e}_3}\\
&= \pm\frac{\dfrac{\partial(x_b\overrightarrow{\mathbf{e}_1} + y_b\overrightarrow{\mathbf{e}_2} + z_b\overrightarrow{\mathbf{e}_3})}{\partial x} \times \dfrac{\partial(x_b\overrightarrow{\mathbf{e}_1} + y_b\overrightarrow{\mathbf{e}_2} + z_b\overrightarrow{\mathbf{e}_3})}{\partial y}}{\left|\dfrac{\partial(x_b\overrightarrow{\mathbf{e}_1} + y_b\overrightarrow{\mathbf{e}_2} + z_b\overrightarrow{\mathbf{e}_3})}{\partial x} \times \dfrac{\partial(x_b\overrightarrow{\mathbf{e}_1} + y_b\overrightarrow{\mathbf{e}_2} + z_b\overrightarrow{\mathbf{e}_3})}{\partial y}\right|},
\end{aligned} \tag{8.45}$$

where $\overrightarrow{\mathbf{n}_b}$, the normal vector of (x_b, y_b, z_b), is as follows in x:

$$\Phi_{b_x} = \frac{\partial_x y_b \partial_y z_b - \partial_x z_b \partial_y y_b}{\sqrt{\left(\partial_x x_b \partial_y y_b - \partial_x y_b \partial_y x_b\right)^2 + \left(\partial_x z_b \partial_y x_b - \partial_x x_b \partial_y z_b\right)^2 + \left(\partial_x y_b \partial_y z_b - \partial_x z_b \partial_y y_b\right)^2}}, \tag{8.46}$$

in y, it is

$$\Phi_{b_y} = \frac{\partial_x z_b \partial_y x_b - \partial_x x_b \partial_y z_b}{\sqrt{\left(\partial_x x_b \partial_y y_b - \partial_x y_b \partial_y x_b\right)^2 + \left(\partial_x z_b \partial_y x_b - \partial_x x_b \partial_y z_b\right)^2 + \left(\partial_x y_b \partial_y z_b - \partial_x z_b \partial_y y_b\right)^2}}, \tag{8.47}$$

and in z, it is

$$\Phi_{b_z} = \frac{\partial_x x_b \partial_y y_b - \partial_x y_b \partial_y x_b}{\sqrt{\left(\partial_x x_b \partial_y y_b - \partial_x y_b \partial_y x_b\right)^2 + \left(\partial_x z_b \partial_y x_b - \partial_x x_b \partial_y z_b\right)^2 + \left(\partial_x y_b \partial_y z_b - \partial_x z_b \partial_y y_b\right)^2}}. \quad (8.48)$$

The components of the normal vector of (x_b, y_b, z_b), equations (8.46), (8.47), and (8.48), depend on the partial derivatives of the normal vector of (x_b, y_b, z_b), which are listed below. The derivative of x_b with respect to x,

$$\begin{aligned}
\partial_x x_b = {} & \frac{1}{\left[n - n_i\sqrt{\partial_x w_b^2 + \partial_y w_b^2 + 1}\right]^2}\left\{\frac{n_i \partial_x w_b^2 \partial_{xx} w_b\left[n w_b + z_i\left(n_i - n\right)\right]}{\sqrt{\partial_x w_b^2 + \partial_y w_b^2 + 1}}\right. \\
& \left(n - n_i\sqrt{\partial_x w_b^2 + \partial_y w_b^2 + 1}\right)\left(\partial_{xx} w_b[n w_b + z_i(n_i - n)] - n_i\sqrt{\partial_x w_b^2 + \partial_y w_b^2 + 1} + n\right) \\
& \left.\partial_x w_b n \partial_x w_b\left(n - n_i\sqrt{\partial_x w_b^2 + \partial_y w_b^2 + 1}\right) + \partial_x w_b \frac{n_i \partial_y w_b \partial_{yx} w_b\left[n w_b + z_i\left(n_i - n\right)\right]}{\sqrt{\partial_x w_b^2 + \partial_y w_b^2 + 1}}\right\},
\end{aligned} \quad (8.49)$$

the derivative of x_b with respect to y,

$$\begin{aligned}
\partial_y x_b = {} & \left\{\partial_x w_b\left[n \partial_y w_b\left(\partial_y w_b^2 + 1\right)\left(n - n_i\sqrt{\partial_x w_b^2 + \partial_y w_b^2 + 1}\right)\right.\right. \\
& \left. + n_i \partial_y w_b \partial_{yy} w_b \sqrt{\partial_x w_b^2 + \partial_y w_b^2 + 1}\left(n w_b - n z_i + n_i z_i\right)\right] \\
& + n \partial_y w_b \partial_x w_b^3\left(n - n_i\sqrt{\partial_x w_b^2 + \partial_y w_b^2 + 1}\right) \\
& - \partial_{xy} w_b\left(\partial_y w_b^2 + 1\right)[z_i(n - n_i) - n w_b]\left(n - n_i\sqrt{\partial_x w_b^2 + \partial_y w_b^2 + 1}\right) \\
& \left. + n \partial_{xy} w_b \partial_x w_b^2[n w_b + z_i(n_i - n)]\right\} \\
& \left\{n\sqrt{\partial_x w_b^2 + \partial_y w_b^2 + 1} - n_i\left(\partial_x w_b^2 + \partial_y w_b^2 + 1\right)\right\}^{-2},
\end{aligned} \quad (8.50)$$

the derivative of y_b with respect to x,

$$\begin{aligned}
\partial_x y_b = {} & \left\{\partial_y w_b\left[n \partial_x w_b\left(\partial_x w_b^2 + 1\right)\left(n - n_i\sqrt{\partial_x w_b^2 + \partial_y w_b^2 + 1}\right)\right.\right. \\
& \left. + n_i \partial_x w_b \partial_{xx} w_b \sqrt{\partial_x w_b^2 + \partial_y w_b^2 + 1}\,(n w_b - n z_i + n_i z_i)\right] \\
& + n \partial_x w_b \partial_y w_b^3\left(n - n_i\sqrt{\partial_x w_b^2 + \partial_y w_b^2 + 1}\right) \\
& - \left(\partial_x w_b^2 + 1\right)\partial_{yx} w_b[z_i(n - n_i) - n w_b]\left(n - n_i\sqrt{\partial_x w_b^2 + \partial_y w_b^2 + 1}\right) \\
& \left. + n \partial_{yx} w_b \partial_y w_b^2[n w_b + z_i(n_i - n)]\right\} \\
& \left\{n\sqrt{\partial_x w_b^2 + \partial_y w_b^2 + 1} - n_i\left(\partial_x w_b^2 + \partial_y w_b^2 + 1\right)\right\}^{-2},
\end{aligned} \quad (8.51)$$

the derivative of y_b with respect to y,

$$\partial_y y_b = \frac{1}{\left(n - n_i\sqrt{\partial_x w_b^2 + \partial_y w_b^2 + 1}\right)^2}\Bigg\{\frac{n_i\partial_y w_b^2\partial_{yy}w_b\left[nw_b + z_i\left(n_i - n\right)\right]}{\sqrt{\partial_x w_b^2 + \partial_y w_b^2 + 1}}$$
$$+ \partial_y w_b\left[n\partial_y w_b\left(n - n_i\sqrt{\partial_x w_b^2 + \partial_y w_b^2 + 1}\right) + \frac{n_i\partial_x w_b\partial_{xy}w_b\left[nw_b + z_i(n_i - n)\right]}{\sqrt{\partial_x w_b^2 + \partial_y w_b^2 + 1}}\right] \quad (8.52)$$
$$+\left(n - n_i\sqrt{\partial_x w_b^2 + \partial_y w_b^2 + 1}\right)\Big(\partial_{yy}w_b[nw_b + z_i(n_i - n)]$$
$$- n_i\sqrt{\partial_x w_b^2 + \partial_y w_b^2 + 1} + n\Big)\Bigg\},$$

the derivative of z_b with respect to x,

$$\partial_x z_b = \Big\{n_i\partial_x w_b\left(\partial_x w_b^2 + \partial_y w_b^2 + 1\right)\left(n - n_i\sqrt{\partial_x w_b^2 + \partial_y w_b^2 + 1}\right)$$
$$+ n_i\left(\partial_x w_b\partial_{xx}w_b + \partial_y w_b\partial_{yx}w_b\right)[nw_b + z_i(n_i - n)]\Big\} \quad (8.53)$$
$$\times\left\{\sqrt{\partial_x w_b^2 + \partial_y w_b^2 + 1}\left(n - n_i\sqrt{\partial_x w_b^2 + \partial_y w_b^2 + 1}\right)^2\right\}^{-1},$$

and the derivative of z_b with respect to y,

$$\partial_y z_b = \Big\{n_i\partial_y w_b\left(\partial_x w_b^2 + \partial_y w_b^2 + 1\right)\left(n - n_i\sqrt{\partial_x w_b^2 + \partial_y w_b^2 + 1}\right)$$
$$+ n_i\left(\partial_x w_b\partial_{xy}w_b + \partial_y w_b\partial_{yy}w_b\right)[nw_b + z_i(n_i - n)]\Big\} \quad (8.54)$$
$$\times\left\{\sqrt{\partial_x w_b^2 + \partial_y w_b^2 + 1}\left(n - n_i\sqrt{\partial_x w_b^2 + \partial_y w_b^2 + 1}\right)^2\right\}^{-1}.$$

So, if we substitute equations (8.49), (8.50), (8.51), (8.52), (8.53), and (8.54), into equations (8.46), (8.47), and (8.48), we get the components of the normal vector at (x_b, y_b, z_b). With these components we can return to Snell's law for (x_b, y_b, z_b), equation (8.43), and compute $[\vec{\mathbf{v}}_3 - (\vec{\mathbf{n}}_b \cdot \vec{\mathbf{v}}_3)\vec{\mathbf{n}}_b]$,

$$[\vec{\mathbf{v}}_3 - (\vec{\mathbf{n}}_b \cdot \vec{\mathbf{v}}_3)\vec{\mathbf{n}}_b] = \left\{\begin{array}{l}\left[-w_{b_x}\left(\Phi_{b_y}^2 + \Phi_{b_z}^2\right) + w_{b_y}\Phi_{b_x}\Phi_{b_y} + w_{b_z}\Phi_{b_x}\Phi_{b_z}\right]\vec{\mathbf{e}}_1\\ +\left[w_{b_y}\left(\Phi_{b_x}^2 + \Phi_{b_z}^2\right) - \Phi_{b_y}\left(w_{b_x}\Phi_{b_x} + w_{b_z}\Phi_{b_z}\right)\right]\vec{\mathbf{e}}_2\\ +\left[w_{b_z}\left(\Phi_{b_x}^2 + \Phi_{b_y}^2\right) - \Phi_{b_z}\left(w_{b_x}\Phi_{b_x} + w_{b_y}\Phi_{b_y}\right)\right]\vec{\mathbf{e}}_3\end{array}\right\}, \quad (8.55)$$

and we then compute $-\vec{\mathbf{n}}_b\sqrt{1 - \frac{n^2}{n_i^2}(\vec{\mathbf{n}}_b \times \vec{\mathbf{v}}_3)^2}$ of equation (8.43):

$$-\overline{\mathbf{n}}_b\sqrt{1-\frac{n^2}{n_i^2}(\overline{\mathbf{n}}_b\times\overline{\mathbf{v}}_3)^2}=\left(\Phi_{b_x}\overline{\mathbf{e}}_1+\Phi_{b_y}\overline{\mathbf{e}}_2+\Phi_{b_z}\overline{\mathbf{e}}_3\right)$$
$$\times\sqrt{1-\frac{n^2}{n_i^2}\left[\left(w_{b_y}\Phi_{b_x}-w_{b_x}\Phi_{b_y}\right)^2+\left(w_{b_z}\Phi_{b_x}-w_{b_x}\Phi_{b_z}\right)^2+\left(w_{b_z}\Phi_{b_y}-w_{b_y}\Phi_{b_z}\right)^2\right]}. \tag{8.56}$$

Equations (8.55) and (8.56) can be concatenated and their components separated to obtain the cosine directors of the refracted rays, as follows: in x,

$$-\wp_x=\frac{n}{n_i}\left[-w_{b_x}\left(\Phi_{b_y}^2+\Phi_{b_z}^2\right)+w_{b_y}\Phi_{b_x}\Phi_{b_y}+w_{b_z}\Phi_{b_x}\Phi_{b_z}\right]$$
$$+\Phi_{b_x}\sqrt{1-\frac{n^2}{n_i^2}\left[\left(w_{b_y}\Phi_{b_x}-w_{b_x}\Phi_{b_y}\right)^2+\left(w_{b_z}\Phi_{b_x}-w_{b_x}\Phi_{b_z}\right)^2+\left(w_{b_z}\Phi_{b_y}-w_{b_y}\Phi_{b_z}\right)^2\right]}, \tag{8.57}$$

in y,

$$-\wp_y=\frac{n}{n_i}\left[w_{b_y}\left(\Phi_{b_x}^2+\Phi_{b_z}^2\right)-\Phi_{b_y}\left(w_{b_x}\Phi_{b_x}+w_{b_z}\Phi_{b_z}\right)\right]$$
$$+\Phi_{b_y}\sqrt{1-\frac{n^2}{n_i^2}\left[\left(w_{b_y}\Phi_{b_x}-w_{b_x}\Phi_{b_y}\right)^2+\left(w_{b_z}\Phi_{b_x}-w_{b_x}\Phi_{b_z}\right)^2+\left(w_{b_z}\Phi_{b_y}-w_{b_y}\Phi_{b_z}\right)^2\right]}, \tag{8.58}$$

and in z,

$$-\wp_z=\frac{n}{n_i}\left[w_{b_z}\left(\Phi_{b_x}^2+\Phi_{b_y}^2\right)-\Phi_{b_z}\left(w_{b_x}\Phi_{b_x}+w_{b_y}\Phi_{b_y}\right)\right]$$
$$+\Phi_{b_z}\sqrt{1-\frac{n^2}{n_i^2}\left[\left(w_{b_y}\Phi_{b_x}-w_{b_x}\Phi_{b_y}\right)^2+\left(w_{b_z}\Phi_{b_x}-w_{b_x}\Phi_{b_z}\right)^2+\left(w_{b_z}\Phi_{b_y}-w_{b_y}\Phi_{b_z}\right)^2\right]}. \tag{8.59}$$

Note that the negative signs on the left-hand sides of equations (8.57), (8.58), and (8.59) are caused by the inverted order of the ray tracing. We started at w_b instead of starting with the rays inside the lens. As was the case for equations (8.39), (8.40), and (8.41), equations (8.57), (8.58), and (8.59) can be reduced to

$$\wp_x=0,\quad \wp_x=0,\quad \wp_z=1. \tag{8.60}$$

Equations (8.6), (8.57), (8.58), (8.59), (8.15), (8.16), and (8.17) provide the details of the cosine directors in the object, lens, and image spaces, respectively. Using these, we can obtain an expression for the wavefront at all points and all times. However, something very important needs to be discussed: both surfaces, i.e. (x_a, y_a, z_a) and (x_b, y_b, z_b), are expressed in terms of the independent variables x and y with the main condition that the rays inside the lens are collimated; therefore, $x_a = x_b$. This is something that should be borne in mind (together with $y_a = y_b$) in order to understand the code provided in the next section.

8.4 Mathematica code

In this section we present the `Mathematica` code to compute a lens (x_a, y_a, z_a) and (x_b, y_b, z_b) as explained in the introduction using equations (8.12), (8.13), (8.14), (8.22), (8.23), and (8.24).

```
In[12]:= w[x_,y_]:=((x)^2+(y)^2)/50

e=10^(-5);

wdx[x_,y_]:=(w[x+e,y]-w[x,y])/e

wdy[x_,y_]:=(w[x,y+e]-w[x,y])/e

wx[x_,y_]:=-wdx[x,y]/Sqrt[wdx[x,y]^2+wdy[x,y]^2+1]

wy[x_,y_]:=-wdy[x,y]/Sqrt[wdx[x,y]^2+wdy[x,y]^2+1]

wz[x_,y_]:=1/Sqrt[wdx[x,y]^2+wdy[x,y]^2+1]

ϑ[x_,y_]:=(n w[x,y]-n zo+no zo)/(no-n wz[x,y])

xa[x_,y_]:=x+ϑ[x,y]wx[x,y]

ya[x_,y_]:=y+ϑ[x,y]wy[x,y]

za[x_,y_]:=w[x,y]+ϑ[x,y]wz[x,y]

wb[x_,y_]:=((x)^2+(y)^2)/50

wbdx[x_,y_]:=(wb[x+e,y]-wb[x,y])/e

wbdy[x_,y_]:=(wb[x,y+e]-wb[x,y])/e
```

$$\text{wxb[x\_,y\_]:=}\frac{\text{-wbdx[x,y]}}{\sqrt{\text{wbdx[x,y]}^2+\text{wbdy[x,y]}^2+1}}$$

$$\text{wyb[x\_,y\_]:=}\frac{\text{-wbdy[x,y]}}{\sqrt{\text{wbdx[x,y]}^2+\text{wbdy[x,y]}^2+1}}$$

$$\text{wzb[x\_,y\_]:=}\frac{1}{\sqrt{\text{wbdx[x,y]}^2+\text{wbdy[x,y]}^2+1}}$$

$$\vartheta\text{b[x\_,y\_]:=}\frac{\text{n wb[x,y]+zi(ni-n)}}{\text{ni-n wzb[x,y]}}$$

```
xb[x_,y_]:=x+ϑb[x,y]wxb[x,y]

yb[x_,y_]:=y+ϑb[x,y]wyb[x,y]

zb[x_,y_]:=zo+τ+zi-wb[x,y]-ϑb[x,y]wzb[x,y]

wz2[x_,y_]:=1

wx2[x_,y_]:=0

wy2[x_,y_]:=0

c=1;
v=c/n;
no=1;
ni=1;
n=1.5;
τ=10;
rmax=6;
zo=40;
```

```
zi=40;

Homotopy={
{Piecewise[{
{wx[x,y]tt+x,tt<ϑ[x,y]},{ v wx2[x,y](tt-ϑ[x,y])+xa[x,y],tt>ϑ[x,y]}}],

Piecewise[{
{wy[x,y]tt+y,tt<ϑ[x,y]},{ v wy2[x,y](tt-ϑ[x,y])+ya[x,y],tt>ϑ[x,y]}}],

Piecewise[{
{wz[x,y]tt+w[x,y],tt<ϑ[x,y]},{v wz2[x,y](tt-ϑ[x,y])+za[x,y],tt>ϑ[x,y]}}]}};

Waves:=Table[{ParametricPlot3D[Homotopy,{x,-rmax,rmax},{y,-rmax,rmax},
PlotRange→All,PlotStyle→{Blue,Opacity[.5]}, Mesh→True]},{tt,5,zo+n τ-5,5}]

HomotopyB={{
Piecewise[{
{wxb[x,y]tt+x,tt<ϑb[x,y]},{ v wx2b[x,y](tt-ϑb[x,y])+xb[x,y],tt>ϑb[x,y]}}],

Piecewise[{
{wyb[x,y]tt+y,tt<ϑb[x,y]},{ v wy2b[x,y](tt-ϑb[x,y])+yb[x,y],tt>ϑb[x,y]}}],

Piecewise[{
{-wzb[x,y]tt+zo+τ+zi-wb[x,y],tt<ϑb[x,y]},
{v wz2b[x,y](tt-ϑb[x,y])+zb[x,y],tt>ϑb[x,y]}}]}};

WavesB:=Table[{ParametricPlot3D[HomotopyB,{x,-rmax,rmax},{y,-rmax,rmax},
PlotRange→All,PlotStyle→{Gray,Opacity[.5]}, Mesh→True]},{tt,5,zo+n τ-5,5}]

l[x_,y_]:=Graphics3D[{Purple,Line[{{x,y,w[x,y]},{xa[x,y],ya[x,y],za[x,y]}}]}];

l2[x_,y_]:=Graphics3D[{Purple,Line[{{xa[x,y],ya[x,y],za[x,y]},
        {xa[x,y],ya[x,y],zb[x,y]}}]}];
```

```
l3[x_,y_]:=Graphics3D[{Purple,Line[{{xa[x,y],ya[x,y],zb[x,y]},
        {x,y,zo+τ+zi-wb[x,y]}}]}];
```

```
g0=Plot3D[{w[x,y]},{x,-rmax,rmax},{y,-rmax,rmax},Axes→True,
PlotStyle→{Blue,Opacity[.5]},AxesLabel→{x,y,z},AspectRatio→Automatic,
Boxed→True];
```

```
g1=ParametricPlot3D[{xa[x,y],ya[x,y],za[x,y]},{x,-rmax,rmax},{y,-rmax,rmax},
Axes→True, PlotStyle→LightYellow,Boxed→True,AxesLabel→{x,y,z},
AspectRatio→Automatic];
```

```
g2=Plot3D[{zo+τ+zi-wb[x,y]},{x,-rmax,rmax},{y,-rmax,rmax},Axes→True,
PlotStyle→{Blue,Opacity[.5]},AxesLabel→{x,y,z}, AspectRatio→Automatic,
Boxed→True];
```

```
g3=ParametricPlot3D[{xb[x,y],yb[x,y],zb[x,y]},{x,-rmax,rmax},{y,-rmax,rmax},
Axes→True, PlotStyle→LightYellow,  Boxed→True, AxesLabel→{x,y,z},
AspectRatio→Automatic];
```

```
Show[g1,g0,g2,g3, Waves,WavesB,
      Table[l[x,y],{x,-rmax,rmax,3},{y,-rmax,rmax,3}],
      Table[l2[x,y],{x,-rmax,rmax,3},{y,-rmax,rmax,3}],
      Table[l3[x,y],{x,-rmax,rmax,3},{y,-rmax,rmax,3}],
      PlotRange→All, ImageSize->{640,Automatic}]
```

8.5 Examples

Examples of refractive lenses designed using equations (8.12). (8.13), (8.14), (8.22), (8.23), and (8.24) are presented in figures 8.2, 8.3, 8.4, 8.5, 8.6, and 8.7.

8.6 End notes

In order to couple two arbitrary wavefronts, it is possible to analytically determine the shape of the required lens's refractive surface. From the examples, it is clear that

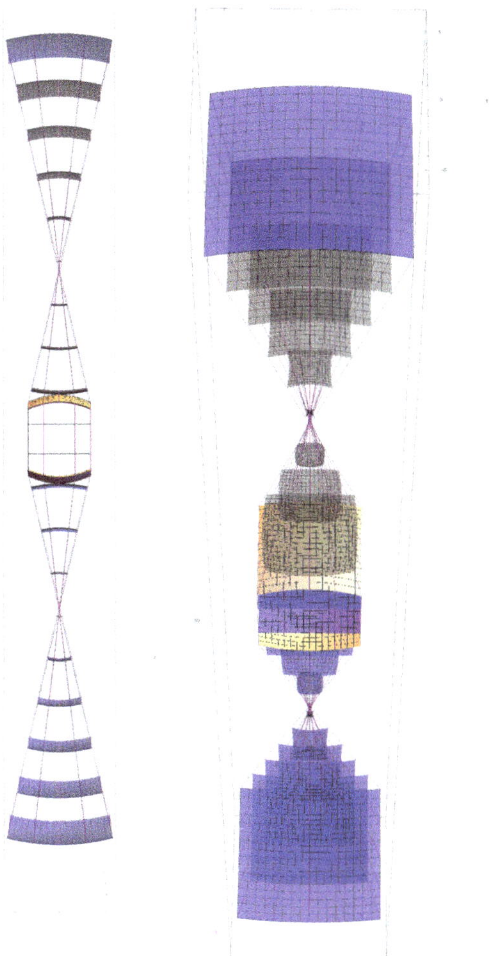

(a) A frontal view point of the adaptative mirror system.

(b) A frontal perspective view point of the adaptative mirror system.

Figure 8.2. Lens design specifications for a parabolic input wavefront; the system produces a flat output wavefront: (a) the input wavefront is $\frac{x^2+y^2}{50}$, the output wavefront is $\frac{x^2+y^2}{50}$, refractive index $n = 1.5$, object distance $z_o = 40$ mm, image distance $z_i = 40$ mm, thickness τ: 10 mm, $r_{max} = 5$ mm, (x_a, y_a, z_a) = equations (8.12), (8.13), and (8.14), (x_a, y_a, z_a) = equations (8.22), (8.23), and (8.24); (b) a frontal perspective of the same configuration.

both surfaces obey Snell's law from beginning to end. Furthermore, the formal approach used does not rely on numerical methods that assume paraxiality. Inputs and outputs can be used arbitrarily, except that they need to be sufficiently smooth that rays do not cross over one another. Our test of equations (8.13), (8.14), (8.22),

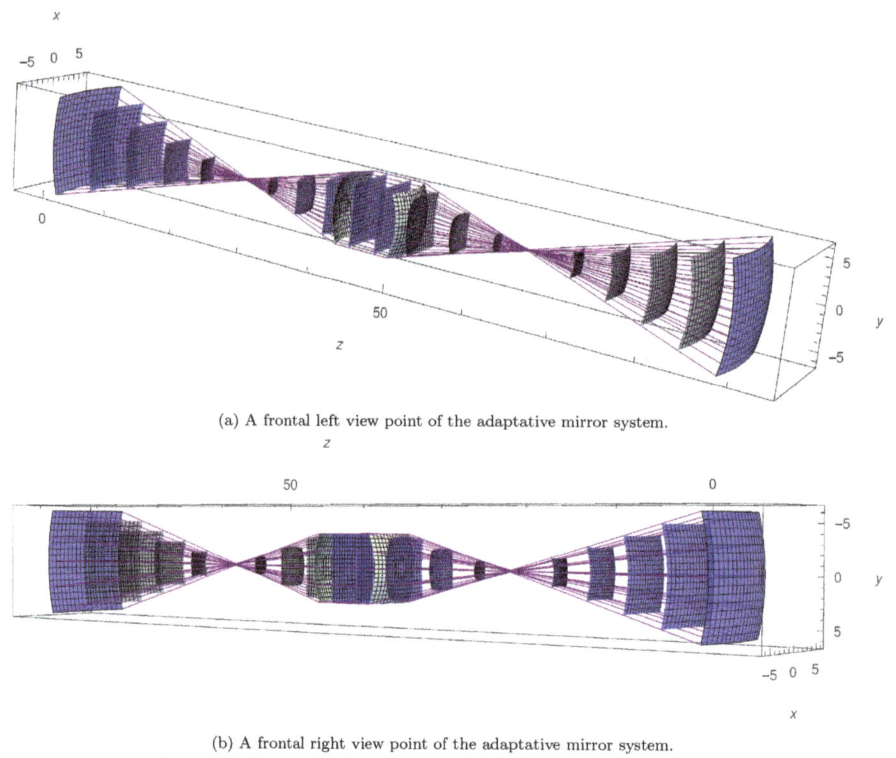

Figure 8.3. Lens design specifications for a parabolic input wavefront; the system produces a flat output wavefront: (a) a frontal left perspective of the same configuration as that used for figure 8.4; (b) a frontal right perspective of the same configuration as that used for figure 8.4.

(8.23), and (8.24) was consistently successful for a wide variety of wavefronts and experimental conditions. From an analytical perspective, the results of this work provide a paradigm shift in how optical systems are analysed.

When mathematical approximations are used, a loss of mathematical and physical meaning can occur. As a result, when mathematical approximations are not used, the nature of the problem becomes clearer. Therefore, the proposed method may be more useful than other numerical approaches in comprehending the problem. We are able to efficiently produce our analytical closed-form solution, which makes equations (8.13), (8.14), (8.22), (8.23), and (8.24) unique. Thus, the result of this analysis is the same as the results of previous chapters. As the steps in the deduction of the equations demonstrate, this fact can easily be verified. Equations (8.13), (8.14), (8.22), (8.23), and (8.24) are just natural implications of the optical path theory.

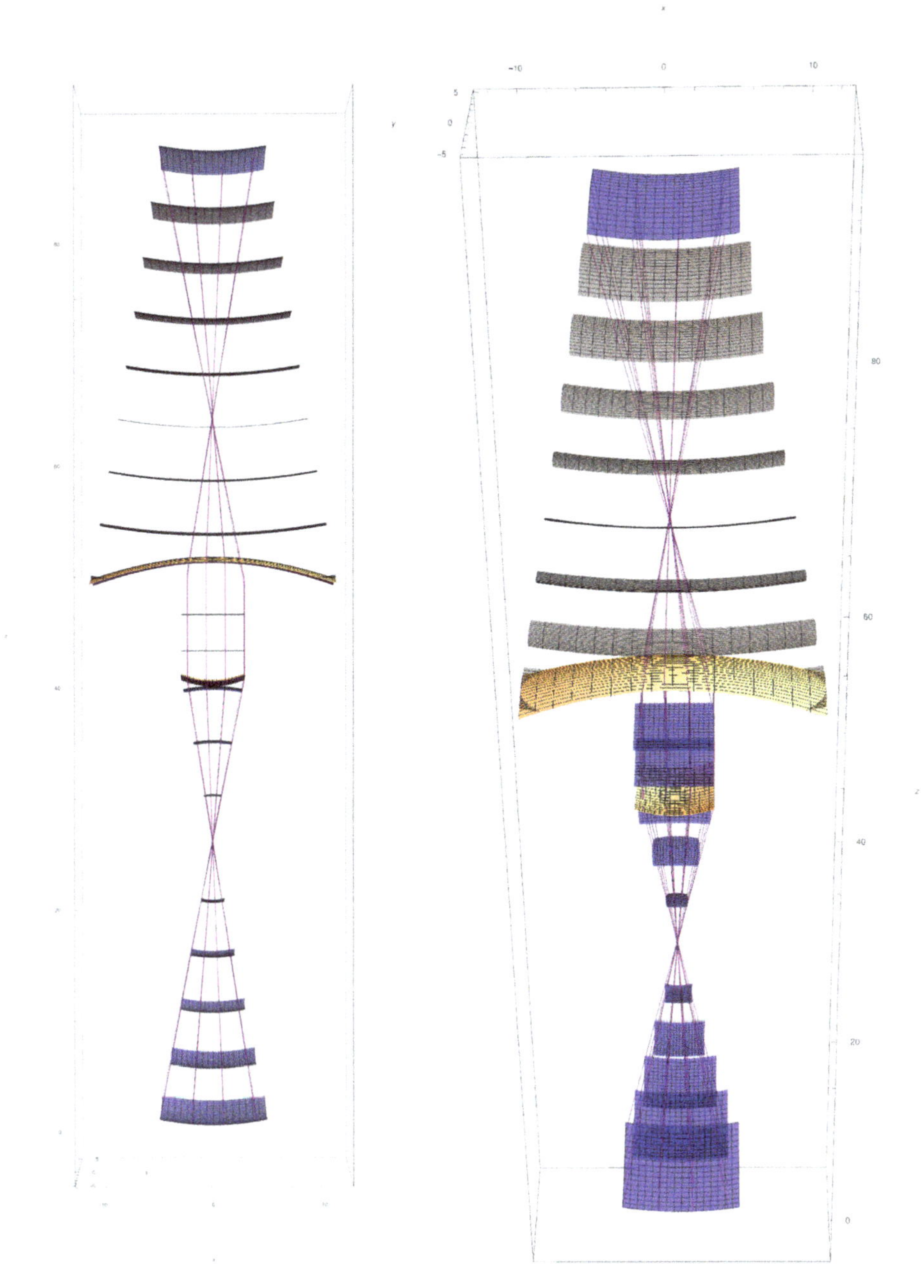

(a) A frontal view point of the adaptative mirror system.

(b) A frontal perspective view point of the mirror system.

Figure 8.4. Lens design specifications for a parabolic input wavefront; the system produces a flat output wavefront: (a) the input wavefront is $\frac{x^2+y^2}{50}$, the output wavefront is $\cos(x/5) + \frac{y^2}{50}$, refractive index $n = 1.5$, object distance $z_o = 40$ mm, image distance $z_i = 40$ mm, thickness τ: 10 mm, $r_{\max} = 5$ mm, $(x_a, y_a, z_a) =$ equations (8.12), (8.13), and (8.14), $(x_a, y_a, z_a) =$ equations(8.22), (8.23), and (8.24); (b) a frontal perspective of the same configuration.

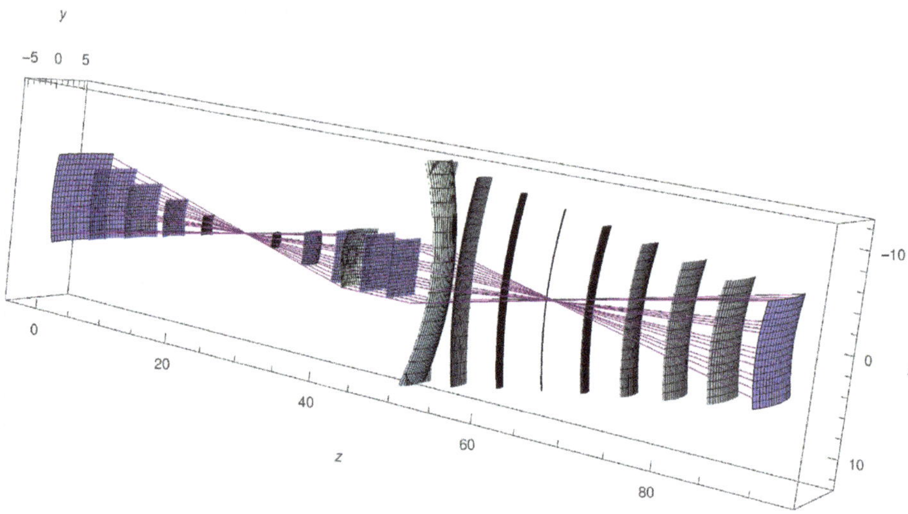

(a) A frontal left view point of the adaptative mirror system.

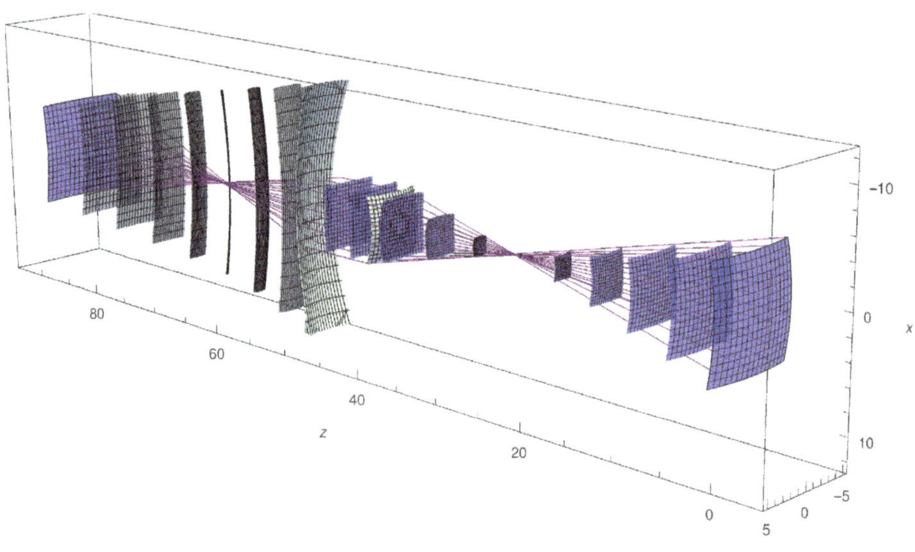

(b) A frontal right view point of the adaptative mirror system.

Figure 8.5. Lens design specifications for a parabolic input wavefront; the system produces a flat output wavefront: (a) a frontal left perspective of the same configuration as that used for figure 8.4; (b) a frontal right perspective of the same configuration as that used for figure 8.4.

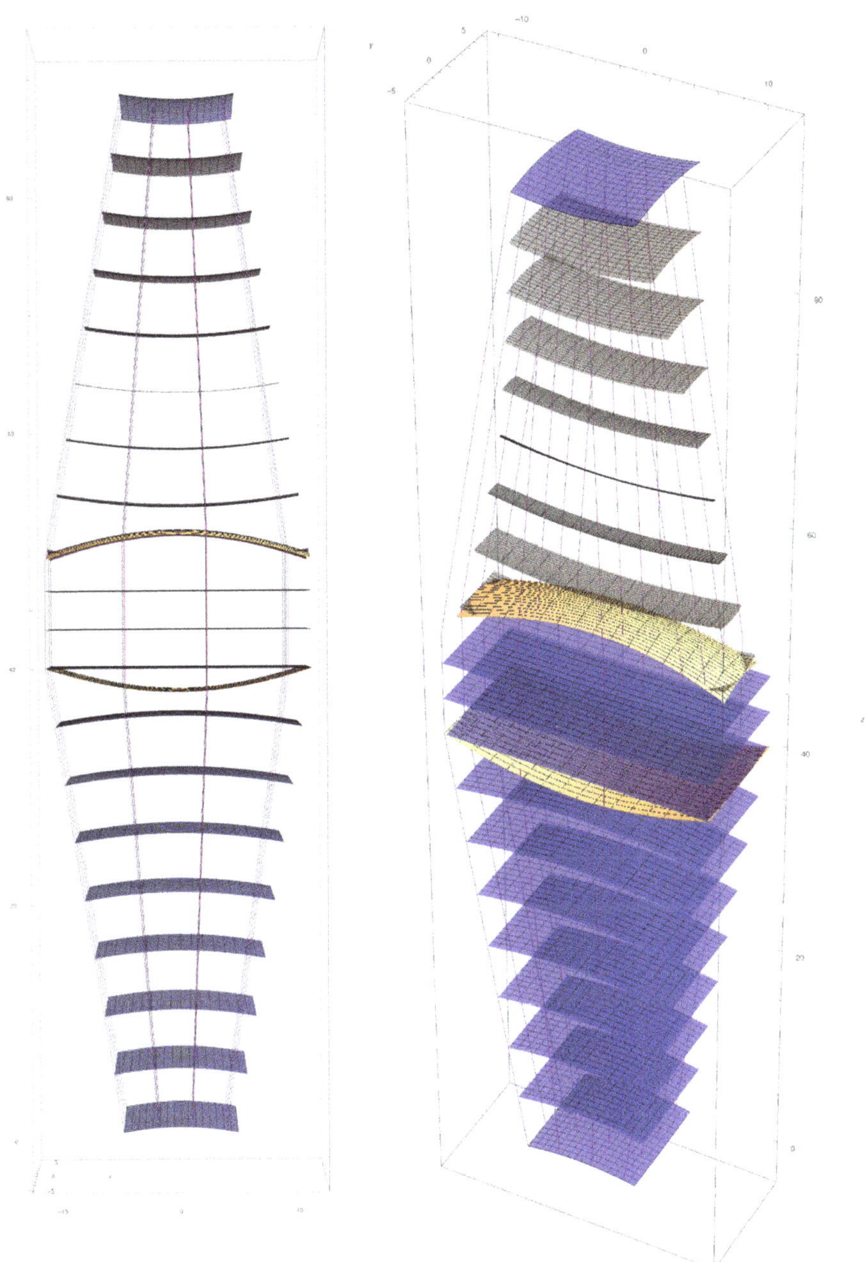

(a) A frontal view point of the adaptative mirror system.

(b) A frontal perspective view point of the adaptative mirror system.

Figure 8.6. Lens design specifications for a parabolic input wavefront; the system produces a flat output wavefront: (a) the input wavefront is $\cos(x/5) + \frac{\cos(y/3)}{50}$, the output wavefront is $\cos(x/5) + \frac{y^2}{50}$, refractive index $n = 1.5$, object distance $z_o = 40$ mm, image distance $z_i = 40$ mm, thickness τ: 10 mm, $r_{\max} = 5$ mm, (x_a, y_a, z_a) = equations (8.12), (8.13), and (8.14), (x_a, y_a, z_a) = equations (8.22), (8.23), and (8.24); (b) a frontal perspective of the same configuration.

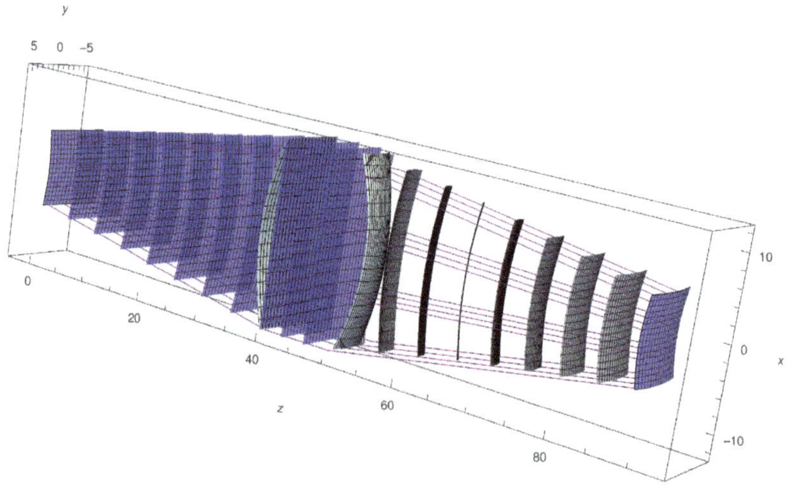

(a) A frontal left view point of the adaptative mirror system.

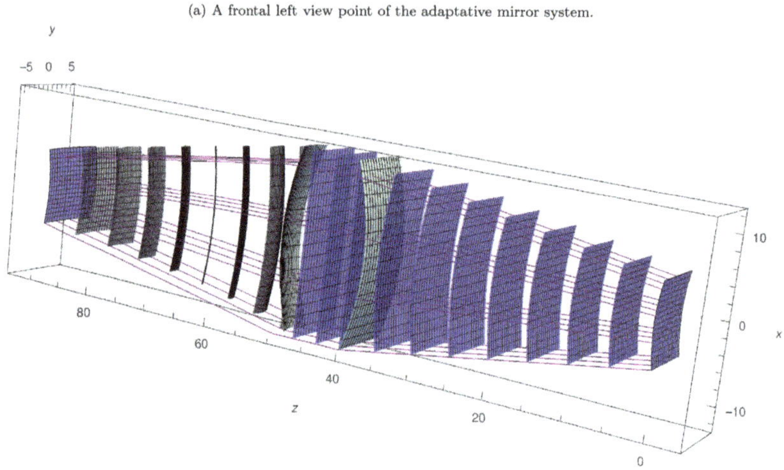

(b) A frontal right view point of the adaptative mirror system.

Figure 8.7. Lens design specifications for a parabolic input wavefront; the system produces a flat output wavefront: (a) a frontal left perspective of the same configuration as that used for figure 8.4; (b) a frontal right perspective of the same configuration as that used for figure 8.4.

IOP Publishing

Optical Path Theory
Fundamentals to freeform adaptive optics
Rafael G González-Acuña and Héctor A Chaparro-Romo

Chapter 9

General equation of the freeform wavefront generator lens

We present a wavefront generator lens design that refracts a given arbitrary input wavefront to a single image point. The arbitrary input wavefront can be considered to be one produced by a test mirror, and its shape can be freeform. A closed-form analytical equation is used in the design of the aforementioned lens. Several scenarios have been tested, and this equation performs as expected.

9.1 Introduction

In this treatise, we have used only Fermat's principle to obtain equations that define lenses with potential applications in adaptive optics and optical metrology. The treatise has been divided into three parts: the introduction, aspherical designs, and freeform designs. This is the last chapter of this treatise, and we are going to focus on the general equation for the freeform wavefront generator lens, which can potentially be used to design null lenses.

As mentioned in the previous chapters, a null lens is a lens that diverges rays in such a way that when they strike a test mirror, they are in parallel to the normal vectors of the mirror. The rays then follow the same path as the path they arrived on, but in the opposite direction. As a result, the display of the interferometer shows a uniform interferometric pattern. If the lens or mirror has defects in its surface, then the interferometer will display a nonuniform interferometric pattern proportional to the defects. See figure 1.17.

We called this chapter the 'general equation of the freeform wavefront generator lens' because we will find that the set of lenses designed using the equation derived in this chapter contains all possible null freeform lenses. We call it 'freeform' because the wavefront generated can be freeform as well. We will derive the general equation for the freeform wavefront generator lens without using paraxial approximation or iterations, in line with the paradigm that we have followed throughout this treatise.

9.2 Mathematical model for freeform wavefront generator lenses

An illustration of the problem is shown in figure 9.1. In designing a lens, the aim is for all the light coming from the input wavefront to converge into a single point image. Furthermore, all rays should have the same optical path length.

9.2.1 The first surface in terms of the input rays

The wavefront is denoted by w, where x and y are the only independent variables of the model, so $w(x, y)$. The first refractive surface of the lens is given by (x_a, y_a, z_a), where x_a, y_a, and z_a are the x, y, and z components, respectively; x_a, y_a, z_a only depend on x and y. The second surface is denoted by (x_b, y_b, z_b), where x_b, y_b, and z_a are the x, y, and z, components respectively; x_b, y_b, and z_b also only depend on x and y. The vertex of this first surface (x_a, y_a, z_a) is located at the origin of the coordinate system. The distance from the vertex of the wavefront w to the first refractive surface's vertex is z_o. Let τ be the central thickness of the freeform lens. The distance from the second surface's vertex to the image point is z_i. The refractive indexes n_o, n, and n_i are constant and belong to the object, lens, and image spaces, respectively.

The cosine directors of the rays of w are given by the cosine directors of its normal vector, as follows: in x,

$$W_x = -\frac{\partial_x w}{\sqrt{\partial_x w^2 + \partial_y w^2 + 1}}, \tag{9.1}$$

in y,

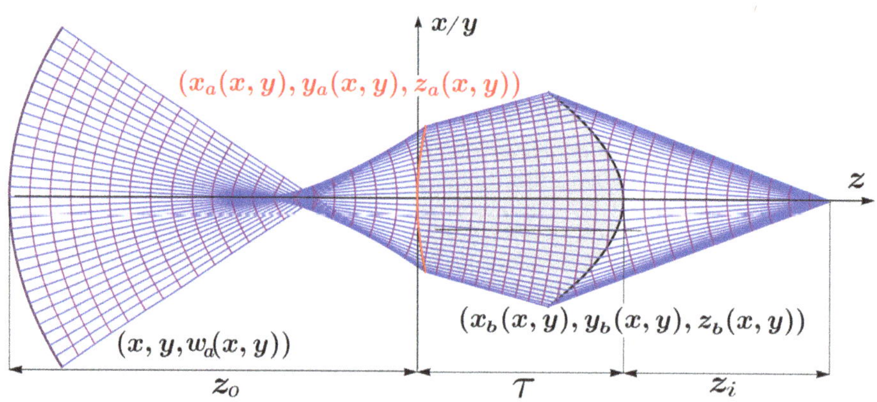

Figure 9.1. Schematic diagram of the configuration of a freeform-wavefront-generating lens.

$$w_y = -\frac{\partial_y w}{\sqrt{\partial_x w^2 + \partial_y w^2 + 1}}, \tag{9.2}$$

and in z,

$$w_z = \frac{1}{\sqrt{\partial_x w^2 + \partial_y w^2 + 1}}. \tag{9.3}$$

The first surface, which is given by the user, can then be defined in x,

$$x_a = x + w_x \vartheta_a, \tag{9.4}$$

in y,

$$y_a = y + w_y \vartheta_a, \tag{9.5}$$

and in z,

$$z_a = w + w_z \vartheta_a, \tag{9.6}$$

where the term ϑ_a is as it appears in equations (9.4), (9.5), and (9.6); it is the distance each ray from w has to travel in order to reach (x_a, y_a, z_a). So, equations (9.4), (9.5), and (9.6) express the first surface (x_a, y_a, z_a) in terms of the input rays.

9.2.2 About the parameter ϑ_a

We begin with the concept of ϑ_a. As a reminder of its definition, ϑ_a is the distance each input ray has to travel from the wavefront to the first surface. Also, remember that the first surface is given. Therefore, for a given first surface, we can always solve ϑ_a for a given w. The sagitta of the first surface (x_a, y_a, z_a) is given by

$$z_a(x_a, y_a) = w + w_z \vartheta_a, . \tag{9.7}$$

By substituting equations (9.4) and (9.5) into the previous expression, we obtain

$$z_a(x + w_x \vartheta_a, y + w_y \vartheta_a) = w + w_z \vartheta_a, \tag{9.8}$$

and we can finally solve this for ϑ_a. As an example, let be the first surface be a parabola with a focus of f_a; we can use equation (9.8) to get ϑ_a:

$$\left[(w_x \vartheta_a + x)^2 + (w_y \vartheta_a + y)^2\right]\frac{1}{f_a} == w + w_z \vartheta_a; \tag{9.9}$$

Solving for ϑ_a yields

$$\begin{aligned} \vartheta_a = \Big\{ & - \sqrt{[w_z - 2\psi(x w_x + y w_y)]^2 - 4\psi[\psi(x^2 + y^2) - w]\left(w_x^2 + w_y^2\right)} \\ & - 2\psi(x w_x + y w_y) + w_z \Big\} \times [2\psi\left(w_x^2 + w_y^2\right)]^{-1}, \end{aligned} \tag{9.10}$$

where $\psi = 1/f_a$. We can apply a similar approach for other curves that model the first surface.

9.2.3 The refracted rays inside the lens

In order to get (x_b, y_b, z_b), we first compute Snell's law at (x_a, y_a, z_a), as follows:

$$\boxed{\vec{\mathbf{v}}_2 = \frac{n_o}{n}[\vec{\mathbf{v}}_1 - (\vec{\mathbf{n}}_a \cdot \vec{\mathbf{v}}_1)\vec{\mathbf{n}}_a] - \vec{\mathbf{n}}_a\sqrt{1 - \frac{n_o^2}{n^2}(\vec{\mathbf{n}}_a \times \vec{\mathbf{v}}_1)^2} \quad \text{for} \quad \vec{\mathbf{v}}_2, \vec{\mathbf{v}}_1, \vec{\mathbf{n}}_a \in \mathbb{R}^3,} \tag{9.11}$$

where $\vec{\mathbf{v}}_1$ is the unit vector of the incident ray, $\vec{\mathbf{v}}_2$ is the unit vector of the refracted ray, and $\vec{\mathbf{n}}_a$ is the normal vector of the first surface:

$$\begin{aligned} \vec{\mathbf{v}}_1 &= w_x\vec{\mathbf{e}}_1 + w_y\vec{\mathbf{e}}_2 + w_z\vec{\mathbf{e}}_3, \\ \vec{\mathbf{v}}_2 &= \wp_x\vec{\mathbf{e}}_1 + \wp_y\vec{\mathbf{e}}_2 + \wp_z\vec{\mathbf{e}}_3, \\ \vec{\mathbf{n}}_a &= \Phi_x\vec{\mathbf{e}}_1 + \Phi_y\vec{\mathbf{e}}_2 + \Phi_z\vec{\mathbf{e}}_3, \end{aligned} \tag{9.12}$$

where $\vec{\mathbf{e}}_1$ denotes the x direction, $\vec{\mathbf{e}}_2$ denotes the y direction and $\vec{\mathbf{e}}_3$ denotes the z direction. We need to compute the normal vector of the first surface (x_a, y_a, z_a), i.e. $\vec{\mathbf{n}}_a$. To compute it, we use the following mathematical expression:

$$\begin{aligned} \vec{\mathbf{n}}_a &= \Phi_x\vec{\mathbf{e}}_1 + \Phi_y\vec{\mathbf{e}}_2 + \Phi_z\vec{\mathbf{e}}_3 \\ &= \pm\frac{\dfrac{\partial(x_a\vec{\mathbf{e}}_1 + y_a\vec{\mathbf{e}}_2 + z_a\vec{\mathbf{e}}_3)}{\partial x} \times \dfrac{\partial(x_a\vec{\mathbf{e}}_1 + y_a\vec{\mathbf{e}}_2 + z_a\vec{\mathbf{e}}_3)}{\partial y}}{\left|\dfrac{\partial(x_a\vec{\mathbf{e}}_1 + y_a\vec{\mathbf{e}}_2 + z_a\vec{\mathbf{e}}_3)}{\partial x} \times \dfrac{\partial(x_a\vec{\mathbf{e}}_1 + y_a\vec{\mathbf{e}}_2 + z_a\vec{\mathbf{e}}_3)}{\partial y}\right|}. \end{aligned} \tag{9.13}$$

If we compute the last expression, we can get the components of $\vec{\mathbf{n}}_a$; for x we get

$$\Phi_x = \frac{\partial_x y_a \partial_y z_a - \partial_x z_a \partial_y y_a}{\sqrt{(\partial_x x_a \partial_y y_a - \partial_x y_a \partial_y x_a)^2 + (\partial_x z_a \partial_y x_a - \partial_x x_a \partial_y z_a)^2 + (\partial_x y_a \partial_y z_a - \partial_x z_a \partial_y y_a)^2}}, \tag{9.14}$$

for y we have

$$\Phi_y = \frac{\partial_x z_a \partial_y x_a - \partial_x x_a \partial_y z_a}{\sqrt{(\partial_x x_a \partial_y y_a - \partial_x y_a \partial_y x_a)^2 + (\partial_x z_a \partial_y x_a - \partial_x x_a \partial_y z_a)^2 + (\partial_x y_a \partial_y z_a - \partial_x z_a \partial_y y_a)^2}}, \tag{9.15}$$

and for z,

$$\Phi_z = \frac{\partial_x x_a \partial_y y_a - \partial_x y_a \partial_y x_a}{\sqrt{(\partial_x x_a \partial_y y_a - \partial_x y_a \partial_y x_a)^2 + (\partial_x z_a \partial_y x_a - \partial_x x_a \partial_y z_a)^2 + (\partial_x y_a \partial_y z_a - \partial_x z_a \partial_y y_a)^2}}. \tag{9.16}$$

Once we have the components of $\vec{\mathbf{n}}_a$, we can substitute them along with equations (9.14), (9.15), and (9.16) into equation (9.13). The procedure is quite long, we first compute term $[\vec{\mathbf{v}}_1 - (\vec{\mathbf{n}}_a \cdot \vec{\mathbf{v}}_1)\vec{\mathbf{n}}_a]$ of equation (9.11):

$$
\begin{aligned}
[\vec{\mathbf{v}}_1 - (\vec{\mathbf{n}}_a \cdot \vec{\mathbf{v}}_1)\vec{\mathbf{n}}_a] = & \left[-w_x\left(\Phi_y^2 + \Phi_z^2\right) + w_y\Phi_x\Phi_y + w_z\Phi_x\Phi_z\right]\vec{\mathbf{e}}_1 \\
& + \left[w_y(\Phi_x^2 + \Phi_z^2) - \Phi_y(w_x\Phi_x + w_z\Phi_z)\right]\vec{\mathbf{e}}_2 \\
& + \left[w_z\left(\Phi_x^2 + \Phi_y^2\right) - \Phi_z(w_x\Phi_x + w_y\Phi_y)\right]\vec{\mathbf{e}}_3
\end{aligned}
\tag{9.17}
$$

and then compute the square root of equation (9.11):

$$
\begin{aligned}
-\vec{\mathbf{n}}_a\sqrt{1 - \frac{n_o^2}{n^2}(\vec{\mathbf{n}}_a \times \vec{\mathbf{v}}_1)^2} = & (\Phi_x\vec{\mathbf{e}}_1 + \Phi_y\vec{\mathbf{e}}_2 + \Phi_z\vec{\mathbf{e}}_3) \\
& \times \sqrt{1 - \frac{n_o^2}{n^2}\left[(w_y\Phi_x - w_x\Phi_y)^2 + (w_z\Phi_x - w_x\Phi_z)^2 + (w_z\Phi_y - w_y\Phi_z)^2\right]}.
\end{aligned}
\tag{9.18}
$$

If we concatenate equations (9.17) and (9.18) and separate their components, we obtain the cosine directors of the refracted rays; for x, we obtain

$$
\begin{aligned}
\wp_x = & \frac{n_o}{n}\left[-w_x\left(\Phi_y^2 + \Phi_z^2\right) + w_y\Phi_x\Phi_y + w_z\Phi_x\Phi_z\right] \\
& + \Phi_x\sqrt{1 - \frac{n_o^2}{n^2}\left[(w_y\Phi_x - w_x\Phi_y)^2 + (w_z\Phi_x - w_x\Phi_z)^2 + (w_z\Phi_y - w_y\Phi_z)^2\right]},
\end{aligned}
\tag{9.19}
$$

for y, we obtain

$$
\begin{aligned}
\wp_y = & \frac{n_o}{n}\left[w_y\left(\Phi_x^2 + \Phi_z^2\right) - \Phi_y(w_x\Phi_x + w_z\Phi_z)\right] \\
& + \Phi_y\sqrt{1 - \frac{n_o^2}{n^2}\left[(w_y\Phi_x - w_x\Phi_y)^2 + (w_z\Phi_x - w_x\Phi_z)^2 + (w_z\Phi_y - w_y\Phi_z)^2\right]},
\end{aligned}
\tag{9.20}
$$

and for z, we get

$$
\begin{aligned}
\wp_z - & \frac{n_o}{n}\left[w_z\left(\Phi_x^2 + \Phi_y^2\right) - \Phi_z(w_x\Phi_x + w_y\Phi_y)\right] \\
& + \Phi_z\sqrt{1 - \frac{n_o^2}{n^2}\left[(w_y\Phi_x - w_x\Phi_y)^2 + (w_z\Phi_x - w_x\Phi_z)^2 + (w_z\Phi_y - w_y\Phi_z)^2\right]}.
\end{aligned}
\tag{9.21}
$$

Note that the cosine directors have the same direction for the different points of the first surface and the points of the second surface, thus, in x

$$
\frac{x_b - x_a}{\sqrt{(x_b - x_a)^2 + (y_b - y_a)^2 + (z_b - z_a)^2}} = \wp_x, \tag{9.22}
$$

in the y direction,

$$
\frac{y_b - y_a}{\sqrt{(x_b - x_a)^2 + (y_b - y_a)^2 + (z_b - z_a)^2}} = \wp_y, \tag{9.23}
$$

and, on the optical axis,

$$\frac{z_b - z_a}{\sqrt{(x_b - x_a)^2 + (y_b - y_a)^2 + (z_b - z_a)^2}} = \wp_z. \tag{9.24}$$

We can now simplify the left-hand sides of equations (9.22), (9.23), and (9.24) using the following parameter,

$$\vartheta \equiv \sqrt{(x_b - x_a)^2 + (y_b - y_a)^2 + (z_b - z_a)^2}. \tag{9.25}$$

Let ϑ be the distance travelled by the ray inside the lens. By substituting ϑ into equations (9.22), (9.23), and (9.24), we get the following: in x,

$$\frac{x_b - x_a}{\vartheta} = \wp_x, \tag{9.26}$$

in the y direction,

$$\frac{y_b - y_a}{\vartheta} = \wp_y, \tag{9.27}$$

finally, on the optical axis,

$$\frac{z_b - z_a}{\vartheta} = \wp_z, \tag{9.28}$$

which leads to the following equations, (9.29), (9.30), and (9.31):

$$x_b = x_a + \vartheta\wp_x, \tag{9.29}$$

$$y_b = y_a + \vartheta\wp_y, \tag{9.30}$$

and,

$$z_b = z_a + \vartheta\wp_z. \tag{9.31}$$

The above expressions are the solution we are looking for. They explain, in detail, how the second surface of the lens must be developed in such a way as to obtain the wavefront generator for the given w. The only problem with these equations we that we do not know the length of ϑ. ϑ can be combined with Fermat's principle to find the solution. We will address this in the next subsection.

9.2.4 The optical path and the solution

Observe that ϑ is expressed in terms of the unknowns (x_b, y_b, z_b). By comparing the optical path lengths (OPLs) of axial and non-axial rays, we can express it in terms of known parameters. We are able to do this because the input and refracted rays of the input and second surfaces are respectively expressed in terms of the first and second surfaces. Therefore, we compare the OPL of the axial ray,

$$\text{OPL} = -n_o z_o + n\tau + n_i z_i, \tag{9.32}$$

and the OPL of the non-axial ray,

$$\begin{aligned}\text{OPL} = & n\sqrt{(x_b - x_a)^2 + (y_b - y_a)^2 + (z_b - z_a)^2} + n_o\sqrt{(x_a - x)^2 + (y_a - y)^2 + (z_a - w)^2} \\ & + n_i\sqrt{x_b^2 + r_b^2 + (z_b - z_i - \tau)^2},\end{aligned} \tag{9.33}$$

thus

$$\begin{aligned}-n_o z_o + n\tau + n_i z_i = & n\sqrt{(x_b - x_a)^2 + (y_b - y_a)^2 + (z_b - z_a)^2} \\ & + n_o\sqrt{(x_a - x)^2 + (y_a - y)^2 + (z_a - w)^2} + n_i\sqrt{x_b^2 + r_b^2 + (z_b - z_i - \tau)^2}.\end{aligned} \tag{9.34}$$

By substituting equations (9.25), (9.29), (9.30), and (9.31) into the last expression, we obtain

$$\begin{aligned}-n_o z_o + n\tau + n_i z_i = & n_o\sqrt{(x_a - x)^2 + (y_a - y)^2 + (z_a - w)^2} + n\vartheta \\ & + n_i\sqrt{(x_a + \vartheta\wp_x)^2 + (y_a + \vartheta\wp_y)^2 + (z_a + \vartheta\wp_z - z_i - \tau)^2}.\end{aligned} \tag{9.35}$$

To decrease the size of the above equation, we define f as follows:

$$f \equiv -n_o z_o + n\tau + n_i z_i - n_o\sqrt{(x_a - x)^2 + (y_a - y)^2 + (z_a - w)^2}, \tag{9.36}$$

and substituting f into equation (9.35) yields

$$f = n\vartheta + n_i\sqrt{(x_a + \vartheta\wp_x)^2 + (y_a + \vartheta\wp_y)^2 + (z_a + \vartheta\wp_z - z_i - \tau)^2}. \tag{9.37}$$

We also define ζ as follows:

$$\zeta = z_a - z_i - \tau; \tag{9.38}$$

by replacing ζ in equation (9.37), we obtain

$$f = n\vartheta + n_i\sqrt{(x_a + \vartheta\wp_x)^2 + (y_a + \vartheta\wp_y)^2 + (\zeta + \vartheta\wp_z)^2}. \tag{9.39}$$

We then square the previous equation:

$$(f - n\vartheta)^2 = n_i^2[(x_a + \vartheta\wp_x)^2 + (y_a + \vartheta\wp_y)^2 + (\zeta + \vartheta\wp_z)^2], \tag{9.40}$$

expand it:

$$\begin{aligned}f^2 - 2nf\vartheta + n^2\vartheta^2 = & n_i^2[\vartheta^2\wp_z^2 + 2\vartheta\wp_z\zeta + \zeta^2 + x_a^2 + 2x_a\vartheta\wp_x + \vartheta^2\wp_x^2 \\ & + y_a^2 + 2y_a\vartheta\wp_y + \vartheta^2\wp_y^2],\end{aligned} \tag{9.41}$$

reorder the terms:

$$\begin{aligned}& (n^2 - n_i^2)\vartheta^2 - 2\vartheta[\wp_z\zeta n_i^2 + \wp_x x_a n_i^2 + \wp_y y_a n_i^2 + nf] + f^2 - n_i^2\zeta^2 - n_i^2 x_a^2 \\ & - n_i^2 y_a^2 = 0,\end{aligned} \tag{9.42}$$

and finally solve for ϑ:

$$\vartheta = \frac{nf + n_i^2\zeta\wp_z + n_i^2 x_a\wp_x + n_i^2 y_a\wp_y}{\left(n^2 + n_i^2\right)} - \frac{\sqrt{\left(-nf - n_i^2\zeta\wp_z - n_i^2 x_a\wp_x - n_i^2 y_a\wp_y\right)^2 - \left(n^2 + n_i^2\right)\left(f^2 - n_i^2\zeta^2 - n_i^2 x_a^2 - n_i^2 y_a^2\right)}}{\left(n^2 + n_i^2\right)}. \tag{9.43}$$

If we use an equation such as (9.43) to obtain ϑ, we can express the second surface in terms of the given parameters, as follows: in x,

$$\boxed{x_b = x_a - \frac{\partial_x w}{\sqrt{\partial_x w^2 + \partial_y w^2 + 1}}\left[\frac{nf + n_i^2\zeta\wp_z + n_i^2 x_a\wp_x + n_i^2 y_a\wp_y}{\left(n^2 + n_i^2\right)} - \frac{\sqrt{\left(-nf - n_i^2\zeta\wp_z - n_i^2 x_a\wp_x - n_i^2 y_a\wp_y\right)^2 - \left(n^2 + n_i^2\right)\left(f^2 - n_i^2\zeta^2 - n_i^2 x_a^2 - n_i^2 y_a^2\right)}}{\left(n^2 + n_i^2\right)}\right]}, \tag{9.44}$$

in y,

$$\boxed{y_b = y_a - \frac{\partial_y w}{\sqrt{\partial_x w^2 + \partial_y w^2 + 1}}\left[\frac{nf + n_i^2\zeta\wp_z + n_i^2 x_a\wp_x + n_i^2 y_a\wp_y}{\left(n^2 + n_i^2\right)} - \frac{\sqrt{\left(-nf - n_i^2\zeta\wp_z - n_i^2 x_a\wp_x - n_i^2 y_a\wp_y\right)^2 - \left(n^2 + n_i^2\right)\left(f^2 - n_i^2\zeta^2 - n_i^2 x_a^2 - n_i^2 y_a^2\right)}}{\left(n^2 + n_i^2\right)}\right]}, \tag{9.45}$$

and,

$$\boxed{z_b = z_a + \frac{1}{\sqrt{\partial_x w^2 + \partial_y w^2 + 1}}\left[\frac{nf + n_i^2\zeta\wp_z + n_i^2 x_a\wp_x + n_i^2 y_a\wp_y}{\left(n^2 + n_i^2\right)} - \frac{\sqrt{\left(-nf - n_i^2\zeta\wp_z - n_i^2 x_a\wp_x - n_i^2 y_a\wp_y\right)^2 - \left(n^2 + n_i^2\right)\left(f^2 - n_i^2\zeta^2 - n_i^2 x_a^2 - n_i^2 y_a^2\right)}}{\left(n^2 + n_i^2\right)}\right]}, \tag{9.46}$$

Equations (9.44), (9.45), and (9.46) are the most important equations of this chapter. Given the input wavefront and the first surface, the second surface is such that the lens behaves as a wavefront generator freeform lens for the wavefront w.

The reader will have noticed that we did not use an approximation to obtain equations (9.44), (9.45), and (9.46). These equations are analytical solutions to the proposed problem. In other words, equations (9.44), (9.45), and (9.46) are a natural extension of Fermat's principle. No paraxial approximation, iteration, or optimization was used. Since the derivation of equations (9.44), (9.45), and (9.46) was based on Fermat's principle, equations (9.44), (9.45), and (9.46) were as predicted by the optical path theory. As long as Fermat's principle remains the fundamental axiom, equations of the type described above are theorems that have been derived from it. Moreover, we should note that equations (9.44), (9.45) and (9.46) have only one solution given a set of input values. In other words, they are also uniqueness theorems. There is only one deterministic path of light in optical path theory.

9.3 The wavefront produced by the wavefront generator lens for finite images

At this point, we know the cosine directors of the input rays, as given by equations (9.1), (9.2), and (9.3) and the cosine directors inside the lens, given by equations (9.19), (9.20), and (9.21). To complete the analytic ray tracing, we need to obtain equations for the output rays. First, we compute Snell's law at the second surface:

$$\boxed{\vec{\mathbf{v}}_3 = \frac{n}{n_i}[\vec{\mathbf{v}}_2 - (\vec{\mathbf{n}}_b \cdot \vec{\mathbf{v}}_2)\vec{\mathbf{n}}_b] - \vec{\mathbf{n}}_b\sqrt{1 - \frac{n^2}{n_i^2}(\vec{\mathbf{n}}_b \times \vec{\mathbf{v}}_2)^2} \quad \text{for} \quad \vec{\mathbf{v}}_3, \vec{\mathbf{v}}_2, \vec{\mathbf{n}}_b \in \mathbb{R}^3,} \tag{9.47}$$

where $\vec{\mathbf{v}}_2$ is the unit vector of the incident ray, $\vec{\mathbf{v}}_3$ is the unit vector of the refracted ray, and $\vec{\mathbf{n}}_b$ is the normal vector of the second surface:

$$\begin{aligned} \vec{\mathbf{v}}_2 &= \wp_x\vec{\mathbf{e}}_1 + \wp_y\vec{\mathbf{e}}_2 + \wp_z\vec{\mathbf{e}}_3, \\ \vec{\mathbf{v}}_3 &= \xi_x\vec{\mathbf{e}}_1 + \xi_y\vec{\mathbf{e}}_2 + \xi\vec{\mathbf{e}}_3, \\ \vec{\mathbf{n}}_b &= \Xi_x\vec{\mathbf{e}}_1 + \Xi_y\vec{\mathbf{e}}_2 + \Xi_z\vec{\mathbf{e}}_3, \end{aligned} \tag{9.48}$$

where $\vec{\mathbf{e}}_1$ denotes the x direction, $\vec{\mathbf{e}}_2$ denotes the y direction and $\vec{\mathbf{e}}_3$ denotes the z direction. In order to properly use equation (9.47), we first need to compute $\vec{\mathbf{n}}_b$ and its components using the following mathematical expression,

$$\begin{aligned} \vec{\mathbf{n}}_b &= \Xi_x\vec{\mathbf{e}}_1 + \Xi_y\vec{\mathbf{e}}_2 + \Xi_z\vec{\mathbf{e}}_3 \\ &= \pm\frac{\dfrac{\partial(x_b\vec{\mathbf{e}}_1 + y_b\vec{\mathbf{e}}_2 + z_b\vec{\mathbf{e}}_3)}{\partial x} \times \dfrac{\partial(x_b\vec{\mathbf{e}}_1 + y_b\vec{\mathbf{e}}_2 + z_b\vec{\mathbf{e}}_3)}{\partial y}}{\left|\dfrac{\partial(x_b\vec{\mathbf{e}}_1 + y_b\vec{\mathbf{e}}_2 + z_b\vec{\mathbf{e}}_3)}{\partial x} \times \dfrac{\partial(x_b\vec{\mathbf{e}}_1 + y_b\vec{\mathbf{e}}_2 + z_b\vec{\mathbf{e}}_3)}{\partial y}\right|}, \end{aligned} \tag{9.49}$$

after computing the previous expression, we obtain the components of $\vec{\mathbf{n}}_b$, as follows: in x,

$$\Xi_x = \frac{\partial_x y_b \partial_y z_b - \partial_x z_b \partial_y y_b}{\sqrt{(\partial_x x_b \partial_y y_b - \partial_x y_b \partial_y x_b)^2 + (\partial_x z_b \partial_y x_b - \partial_x x_b \partial_y z_b)^2 + (\partial_x y_b \partial_y z_b - \partial_x z_b \partial_y y_b)^2}}, \tag{9.50}$$

in y,

$$\Xi_y = \frac{\partial_x z_b \partial_y x_b - \partial_x x_b \partial_y z_b}{\sqrt{(\partial_x x_b \partial_y y_b - \partial_x y_b \partial_y x_b)^2 + (\partial_x z_b \partial_y x_b - \partial_x x_b \partial_y z_b)^2 + (\partial_x y_b \partial_y z_b - \partial_x z_b \partial_y y_b)^2}}, \tag{9.51}$$

and in z,

$$\Xi_z = \frac{\partial_x x_b \partial_y y_b - \partial_x y_b \partial_y x_b}{\sqrt{(\partial_x x_b \partial_y y_b - \partial_x y_b \partial_y x_b)^2 + (\partial_x z_b \partial_y x_b - \partial_x x_b \partial_y z_b)^2 + (\partial_x y_b \partial_y z_b - \partial_x z_b \partial_y y_b)^2}}, \tag{9.52}$$

where the symbols ∂_x and ∂_y denote the partial derivations with respect to the independent variables x and y. Using equations (9.50), (9.51), and (9.52) we can construct $\vec{\mathbf{n}}_b$, and therefore we can compute Snell's law at the second surface, equation (9.47). We begin with the term $[\vec{\mathbf{v}}_2 - (\vec{\mathbf{n}}_b \cdot \vec{\mathbf{v}}_2)\vec{\mathbf{n}}_b]$ of equation (9.47):

$$\begin{aligned}[\vec{\mathbf{v}}_2 - (\vec{\mathbf{n}}_b \cdot \vec{\mathbf{v}}_2)\vec{\mathbf{n}}_b] &= \left[-\wp_x\left(\Xi_y^2 + \Xi_z^2\right) + \wp_y \Xi_x \Xi_y + \wp_z \Xi_x \Xi_z\right]\vec{\mathbf{e}}_1 \\ &+ \left[\wp_y\left(\Xi_x^2 + \Xi_z^2\right) - \Xi_y\left(\wp_x \Xi_x + \wp_z \Xi_z\right)\right]\vec{\mathbf{e}}_2 + \left[\wp_z\left(\Xi_x^2 + \Xi_y^2\right) - \Xi_z\left(\wp_x \Xi_x + \wp_y \Xi_y\right)\right]\vec{\mathbf{e}}_3 \\ &;\end{aligned} \tag{9.53}$$

we then focus on the square root term of equation (9.47):

$$\begin{aligned}-\vec{\mathbf{n}}_b\sqrt{1 - \frac{n^2}{n_i^2}(\vec{\mathbf{n}}_b \times \vec{\mathbf{v}}_2)^2} &= (\Xi_x \vec{\mathbf{e}}_1 + \Xi_y \vec{\mathbf{e}}_2 + \Xi_z \vec{\mathbf{e}}_3) \\ &\times \sqrt{1 - \frac{n^2}{n_i^2}\left[(\wp_y \Xi_x - \wp_x \Xi_y)^2 + (\wp_z \Xi_x - \wp_x \Xi_z)^2 + (\wp_z \Xi_y - \wp_y \Xi_z)^2\right]}.\end{aligned} \tag{9.54}$$

Adding equation (9.53) and equation (9.54) and then separating the Cartesian components, we obtain the output cosine directors, as follows: in x,

$$\begin{aligned}\xi_x &= \frac{n}{n_i}\left[-\wp_x\left(\Xi_y^2 + \Xi_z^2\right) + \wp_y \Xi_x \Xi_y + \wp_z \Xi_x \Xi_z\right] \\ &+ \Xi_x\sqrt{1 - \frac{n^2}{n_i^2}\left[(\wp_y \Xi_x - \wp_x \Xi_y)^2 + (\wp_z \Xi_x - \wp_x \Xi_z)^2 + (\wp_z \Xi_y - \wp_y \Xi_z)^2\right]},\end{aligned} \tag{9.55}$$

in y,

$$\begin{aligned}\xi_y &= \frac{n}{n_i}\left[\wp_y\left(\Xi_x^2 + \Xi_z^2\right) - \Xi_y(\wp_x \Xi_x + \wp_z \Xi_z)\right] \\ &+ \Xi_y\sqrt{1 - \frac{n^2}{n_i^2}\left[(\wp_y \Xi_x - \wp_x \Xi_y)^2 + (\wp_z \Xi_x - \wp_x \Xi_z)^2 + (\wp_z \Xi_y - \wp_y \Xi_z)^2\right]},\end{aligned} \tag{9.56}$$

and in z,

$$\begin{aligned}\xi_z &= \frac{n}{n_i}\left[\wp_z\left(\Xi_x^2 + \Xi_y^2\right) - \Xi_z(\wp_x \Xi_x + \wp_y \Xi_y)\right] \\ &+ \Xi_z\sqrt{1 - \frac{n^2}{n_i^2}\left[(\wp_y \Xi_x - \wp_x \Xi_y)^2 + (\wp_z \Xi_x - \wp_x \Xi_z)^2 + (\wp_z \Xi_y - \wp_y \Xi_z)^2\right]}.\end{aligned} \tag{9.57}$$

Equations (9.44), (9.45), and (9.46) are such that they convert the output to an image located at $\tau + z_i$. In fact, equations (9.55), (9.56), and (9.57) can be reduced. In x, equation (9.55) changes into

$$\xi_x = \frac{x_b}{\sqrt{x_b^2 + r_b^2 + (z_b - z_i - \tau)^2}}, \tag{9.58}$$

in y, equation (9.56) becomes

$$\xi_y = \frac{y_b}{\sqrt{x_b^2 + r_b^2 + (z_b - z_i - \tau)^2}}, \tag{9.59}$$

and on the optical axis, equation (9.57) is reduced to

$$\xi_z = \frac{z_b - z_i - \tau}{\sqrt{x_b^2 + r_b^2 + (z_b - z_i - \tau)^2}}. \tag{9.60}$$

Using equations (9.1), (9.2), (9.3), (9.19), (9.20), (9.21), (9.58), (9.59), and (9.60), we can now obtain the cosine directors of all the spaces, i.e. the object space, the lens space and the image space. Using a piecewise function we can directly use these cosine directors to get the eikonal of this optical system. This piecewise function is denoted by $H(t, x, y)$:

$$\boxed{H(t, x, y) = \begin{cases} [x + tw_x, y + tw_y, w + tw_z], & t < \vartheta_a, \\ \begin{pmatrix} x_a + v(t - \vartheta_a)\wp_x, \\ y_a + v(t - \vartheta_a)\wp_y, \\ z_a + v(t - \vartheta_a)\wp_z \end{pmatrix}, & \vartheta_a < t < \vartheta_a + n\vartheta, \\ \begin{pmatrix} x_b + (t - \vartheta_a + n\vartheta)\xi_x, \\ y_b + (t - \vartheta_a + n\vartheta)\xi_y, \\ z_b + (t - \vartheta_a + n\vartheta)\xi_z \end{pmatrix}, & t > \vartheta_a + n\vartheta. \end{cases}} \tag{9.61}$$

where $v = c/n$ is the speed of light inside the lens, and c is the speed of light in vacuum (outside the lens, we set $n_o = n_i = 1$). Observe that the separation conditions between the object space, the lens space and the image space are just ϑ_a and $\vartheta_a + n\vartheta$; this is an expected result. Since ϑ_a is the distance covered by each ray that needs to travel from w to the first surface and $n\vartheta$ is the OPL inside the lens, so the conditions in $H(t, x, y)$ make a scene.

9.4 Mathematica code

This section presents `Mathematica` code for the freeform wavefront generator lens based on equations (9.44), (9.45), (9.46), and (9.61).

```
In[13]:= w[x_,y_]:=((x)^2+(y)^2)/40+zo;

         e=10^(-5);

         wdx[x_,y_]:=(w[x+e,y]-w[x,y])/e;

wdy[x_,y_]:=(w[x,y+e]-w[x,y])/e;

wx[x_,y_]:=-wdx[x,y]/Sqrt[wdx[x,y]^2+wdy[x,y]^2+1];

wy[x_,y_]:=-wdy[x,y]/Sqrt[wdx[x,y]^2+wdy[x,y]^2+1];

wz[x_,y_]:=1/Sqrt[wdx[x,y]^2+wdy[x,y]^2+1];

ϑa[x_,y_]:=-w[x,y]/wz[x,y]

xa[x_,y_]:=x+ϑa[x,y]wx[x,y]

ya[x_,y_]:=y+ϑa[x,y]wy[x,y]

za[x_,y_]:=w[x,y]+ϑa[x,y]wz[x,y]

dxxa[x_,y_]:=(xa[x+e,y]-xa[x,y])/e
dyxa[x_,y_]:=(xa[x,y+e]-xa[x,y])/e
dxya[x_,y_]:=(ya[x+e,y]-ya[x,y])/e
dyya[x_,y_]:=(ya[x,y+e]-ya[x,y])/e
dxza[x_,y_]:=(za[x+e,y]-za[x,y])/e
dyza[x_,y_]:=(za[x,y+e]-za[x,y])/e

Φx[x_,y_]:=0

Φy[x_,y_]:=0

Φz[x_,y_]:=1
```

$$\varrho x[x\_,y\_]:=\frac{no\ wx[x,y]}{n}$$

$$\varrho y[x\_,y\_]:=\frac{no\ wy[x,y]}{n}$$

$$\varrho z[x\_,y\_]:=\sqrt{1-\frac{no^2\ (wx[x,y]^2+wy[x,y]^2)}{n^2}}$$

$$\zeta[x\_,y\_]:=za[x,y]-zi-\tau$$

$$f[x\_,y\_]:=-zo\ no+\tau\ n+zi\ ni-no\sqrt{(xa[x,y]-x)^2+(ya[x,y]-y)^2+(za[x,y]-w[x,y])^2}$$

$$\vartheta[x\_,y\_]:=\frac{1}{2\ (n^2-ni^2)}(2\ n\ f[x,y]+2\ ni^2\ xa[x,y]\ \varrho x[x,y]+2\ ni^2\ ya[x,y]\ \varrho y[x,y]+ 2\ ni^2\ \zeta[x,y]\ \varrho z[x,y]-\surd(-4\ (n^2-ni^2)\ (f[x,y]^2-ni^2\ xa[x,y]^2-ni^2\ ya[x,y]^2-ni^2\ \zeta[x,y]^2)+ (-2\ n\ f[x,y]-2\ ni^2\ xa[x,y]\ \varrho x[x,y]-2\ ni^2\ ya[x,y]\ \varrho y[x,y]-2\ ni^2\ \zeta[x,y]\ \varrho z[x,y])^2))$$

$$xb[x\_,y\_]:=xa[x,y]+\vartheta[x,y]\varrho x[x,y]$$

$$yb[x\_,y\_]:=ya[x,y]+\vartheta[x,y]\varrho y[x,y]$$

$$zb[x\_,y\_]:=za[x,y]+\vartheta[x,y]\varrho z[x,y]$$

$$\Xi x[x\_,y\_]:=\frac{-xb[x,y]}{\sqrt{(xb[x,y])^2+(yb[x,y])^2+(zb[x,y]-zi-\tau)^2}}$$

$$\Xi y[x\_,y\_]:=\frac{-yb[x,y]}{\sqrt{(xb[x,y])^2+(yb[x,y])^2+(zb[x,y]-zi-\tau)^2}}$$

$$\Xi z[x\_,y\_]:=\frac{-(zb[x,y]-zi-\tau)}{\sqrt{(xb[x,y])^2+(yb[x,y])^2+(zb[x,y]-zi-\tau)^2}}$$

```
no=1;
n=1.5;
ni=1;
c=1;
```

```
v=c/n;
τ=10;
zo=-40;
zi=40;
rmax=10;

g0=Plot3D[{w[x,y]},{x,-rmax,rmax},{y,-rmax,rmax},Axes→True,
PlotStyle→{Red},AxesLabel→{x,y,z},AspectRatio→Automatic,
Boxed→True,Mesh→False];

g1=ParametricPlot3D[{xa[x,y],ya[x,y],za[x,y]},{x,-rmax,rmax},{y,-rmax,rmax},
Axes→True,PlotStyle→Black,Boxed→True,AxesLabel→{x,y,z},
AspectRatio→Automatic, Mesh→False];

g2=ParametricPlot3D[{xb[x,y],yb[x,y],zb[x,y]},{x,-rmax,rmax},{y,-rmax,rmax},
Axes→True,PlotStyle→Black,Boxed→True,AxesLabel→{x,y,z},
AspectRatio→Automatic,PlotPoints→10,Mesh→False];

Homotopy=

{Piecewise[{
  {wx[x,y]tt+x,tt<ϑa[x,y]},

  {v ϱx[x,y](tt-ϑa[x,y])+xa[x,y],n ϑ[x,y]+ϑa[x,y]>tt>ϑa[x,y]},

  {Ξx[x,y](tt-ϑa[x,y]-n ϑ[x,y])+xb[x,y],tt>n ϑ[x,y]+ϑa[x,y]}}],

Piecewise[{
  {wy[x,y]tt+y,tt<ϑa[x,y]},

  { v ϱy[x,y](tt-ϑa[x,y])+ya[x,y],n ϑ[x,y]+ϑa[x,y]>tt>ϑa[x,y]},

  {Ξy[x,y](tt-ϑa[x,y]-n ϑ[x,y])+yb[x,y],tt>n ϑ[x,y]+ϑa[x,y]}}],

Piecewise[{
  {wz[x,y]tt+w[x,y],tt<ϑa[x,y]},

  {v ϱz[x,y](tt-ϑa[x,y])+za[x,y],n ϑ[x,y]+ϑa[x,y]>tt>ϑa[x,y]},

  {Ξz[x,y](tt-ϑa[x,y]-n ϑ[x,y])+zb[x,y],tt>n ϑ[x,y]+ϑa[x,y]}}]};
```

```
Waves:=Table[{ParametricPlot3D[Homotopy,{x,-rmax,rmax},{y,-rmax,rmax},
PlotRange→All,PlotStyle→{Blue,Opacity[.5]}, Mesh→False,PlotPoints→10]},
{tt,5,-zo+τ n+zi,5}]
```

```
Rays:=Table[{ParametricPlot3D[Homotopy,{tt,0,-zo+τ n+zi},
PlotRange→All,PlotStyle→{Purple,Opacity[.5]},
Mesh→False,PlotPoints→10]},
{x,-rmax,rmax,5},{y,-rmax,rmax,5}]
```

```
Show[g1,g0,g2,Rays,Waves,PlotRange→All,ImageSize→{640, Automatic}]
```

9.5 Examples

Examples produced by `Mathematica` based on equations (9.44), (9.45), and (9.46) are presented in figures 9.2, 9.3, 9.4, 9.5, 9.6, and 9.7.

Equations (9.44), (9.45), and (9.46) give precise results. These equations perform so well because they are fully analytic and closed form. Neither paraxial optics, optimization nor numerical approximation was used in its deduction. Considering the input parameters for these equations, there can only be one solution. Its uniqueness is clear from examining these equations.

These equations have the limitation that two rays cannot touch at any point on the first or second surfaces. These equations are derived by comparing the axial and non-axial OPLs of the rays. A neighbouring ray's OPL does not enter into these equations. Thus, input rays strike the first and second surfaces at points that have a one-to-one relationship with their paths. In other words, if two rays intersect at the same part of the second surface, the second surface overlaps.

The lens sections displayed can be viewed as wavefront generator lenses, where the input w is the wavefront that results from striking a mirror, as mentioned in the introduction of this chapter.

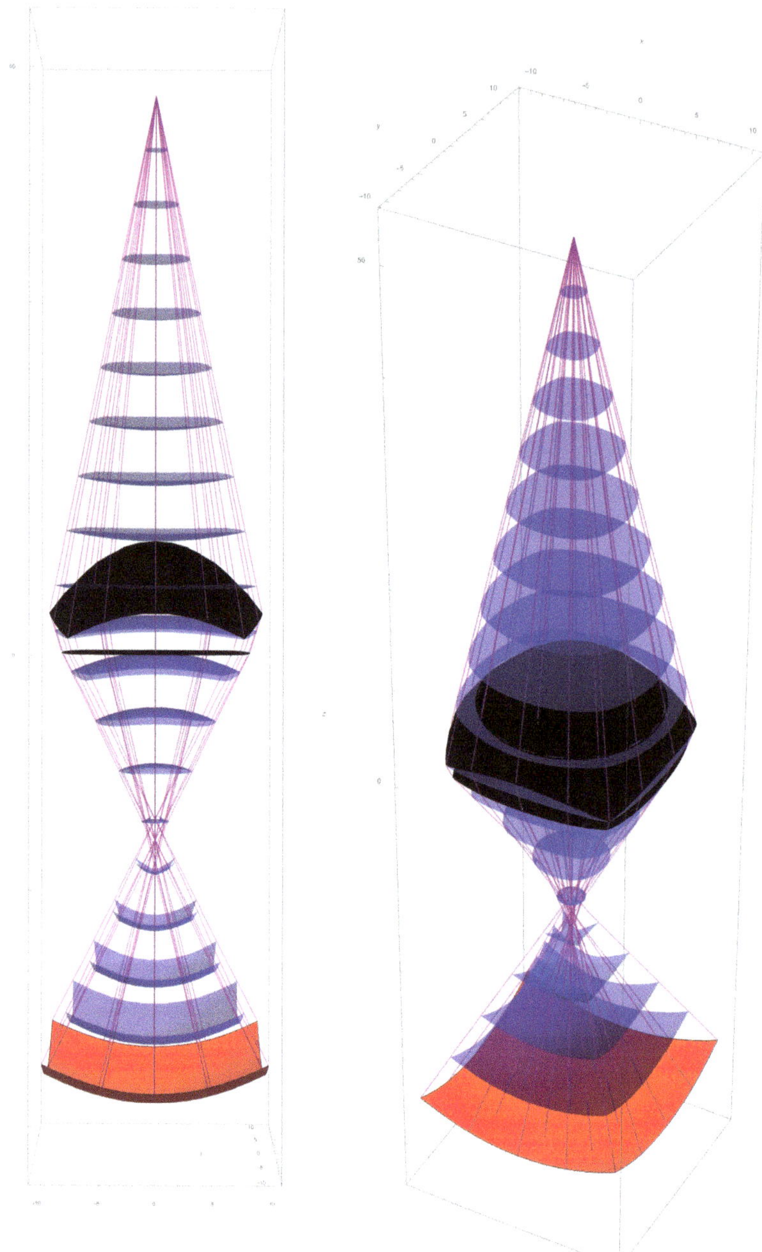

(a) A frontal view point of the adaptative mirror system.

(b) A frontal perspective view point of the adaptative mirror system.

Figure 9.2. Lens design specifications for a parabolic input wavefront; the system produces a flat output wavefront: (a) the input wavefront is $\frac{(x^2+y^2)}{40}$, refractive index $n = 1.5$, object distance $z_o = -40$ mm, image distance $z_i = 40$ mm, thickness $\tau = 10$ mm, $\mathrm{r}_{\max} = 10$ mm, $(x_b, y_b, z_b) =$ equations (9.44), (9.45), and (9.46); (b) a frontal perspective of the same configuration.

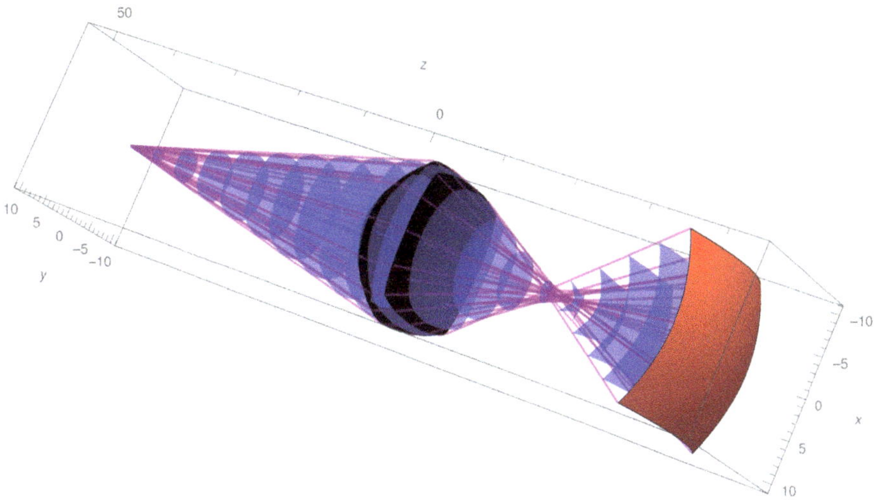

(a) A frontal left view point of the adaptative mirror system.

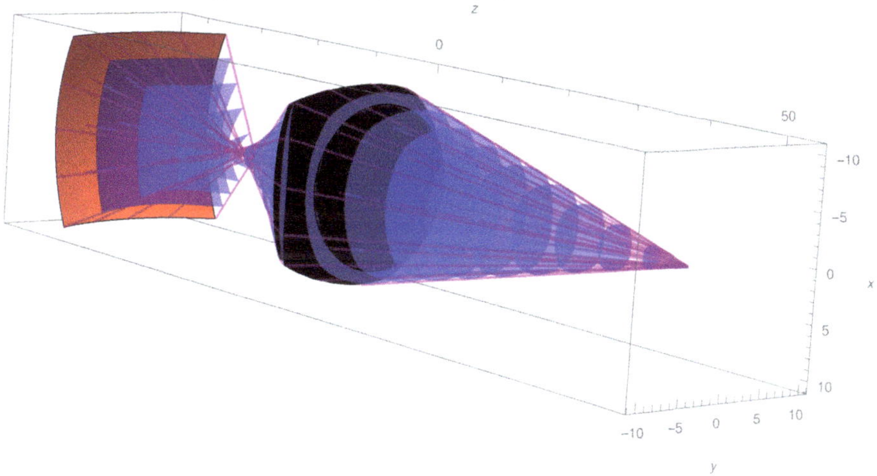

(b) A frontal right view point of the adaptative mirror system.

Figure 9.3. Lens design specifications for a parabolic input wavefront; the system produces a flat output wavefront: (a) a frontal left perspective of the same configuration as that shown in figure 9.6; (b) a frontal right perspective of the same configuration as that shown in figure 9.6.

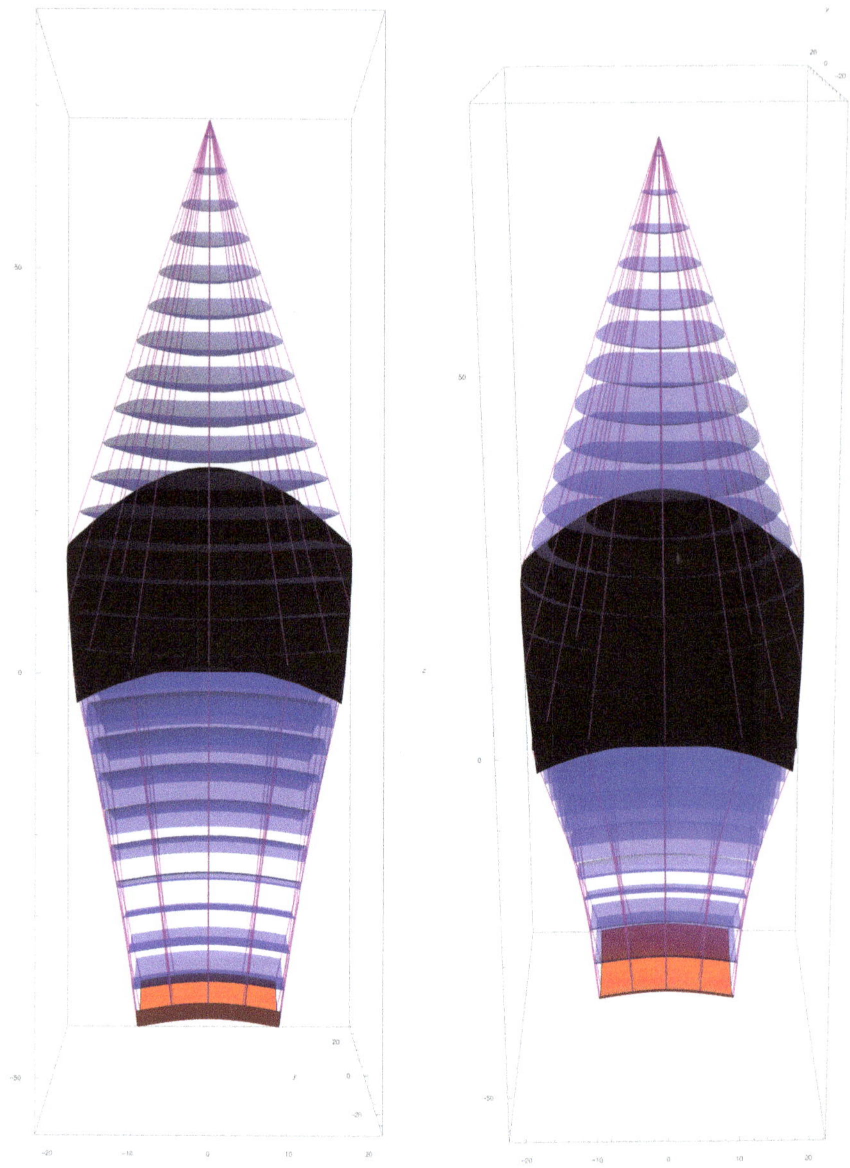

(a) A frontal view point of the adaptative mirror system.

(b) A frontal perspective view point of the adaptative mirror system.

Figure 9.4. Lens design specifications for a parabolic input wavefront; the system produces a flat output wavefront: (a) the input wavefront is $\cos(x/6) + \frac{y^2}{50} + zo$, refractive index $n = 1.5$, object distance $z_o = -50$ mm, image distance $z_i = 50$ mm, thickness $\tau = 25$ mm, $r_{max} = 10$ mm, $(x_b, y_b, z_b) =$ equations (9.44), (9.45), and (9.46); (b) a frontal perspective of the same configuration.

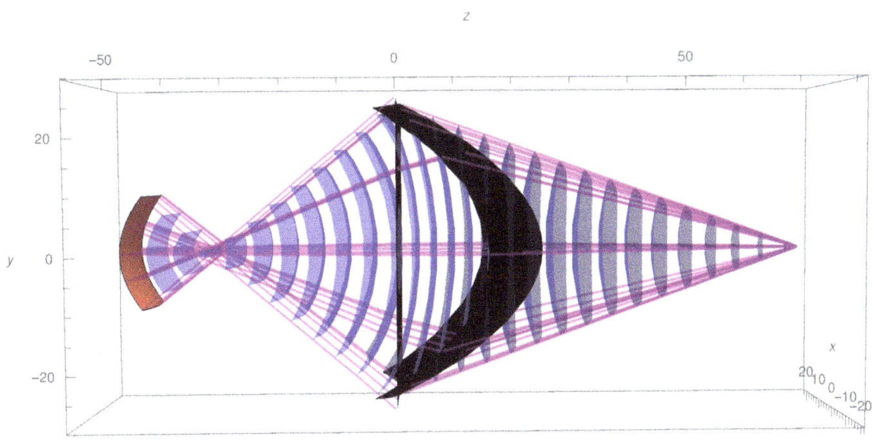

(a) A frontal left view point of the adaptative mirror system.

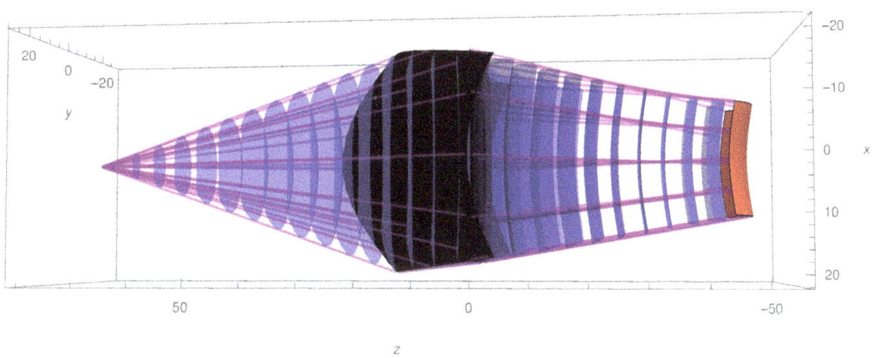

(b) A frontal right view point of the adaptative mirror system.

Figure 9.5. Lens design specifications for a parabolic input wavefront; the system produces a flat output wavefront: (a) a frontal left perspective of the same configuration as that shown in figure 9.6; (b) a frontal right perspective of the same configuration as that shown in figure 9.6.

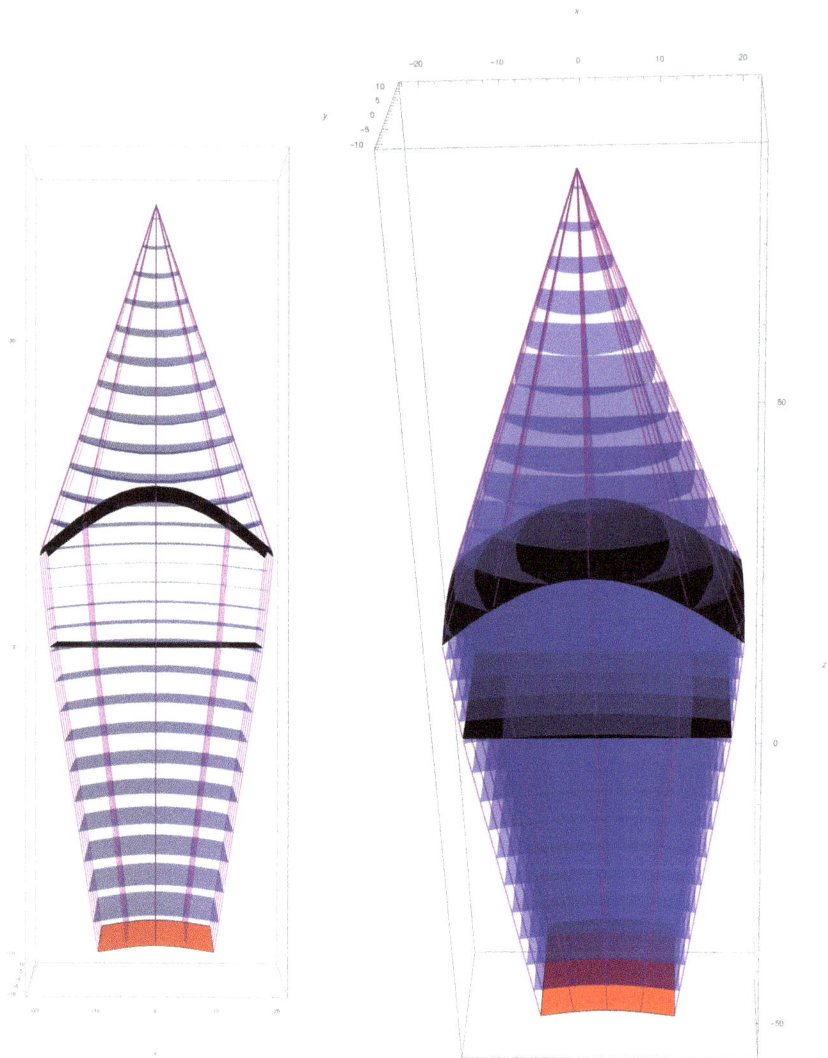

(a) A frontal view point of the adaptative mirror system.

(b) A frontal perspective view point of the adaptative mirror system.

Figure 9.6. Lens design specifications for a parabolic input wavefront; the system produces a flat output wavefront: (a) the input wavefront is $\cos(x/6) + \frac{\cos(y/4)}{25} + zo$, refractive index $n = 1.5$, object distance $z_o = -50$ mm, image distance $z_o = 50$ mm, thickness $\tau = 25$ mm, $r_{max} = 10$ mm, $(x_b, y_b, z_b) =$ equations (9.44), (9.45), and (9.46); (b) a frontal perspective of the same configuration.

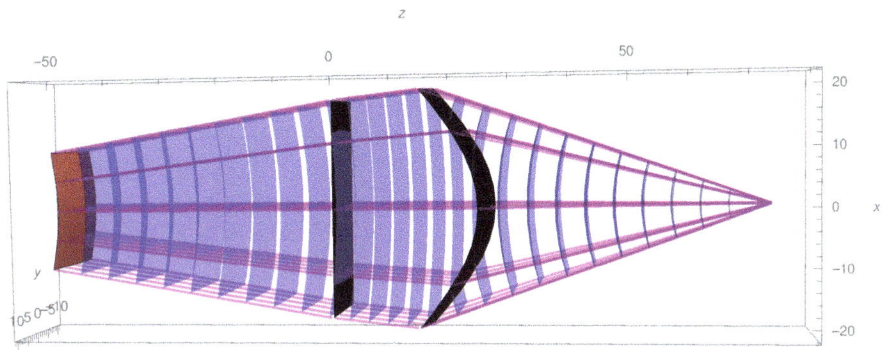

(a) A frontal left view point of the adaptative mirror system.

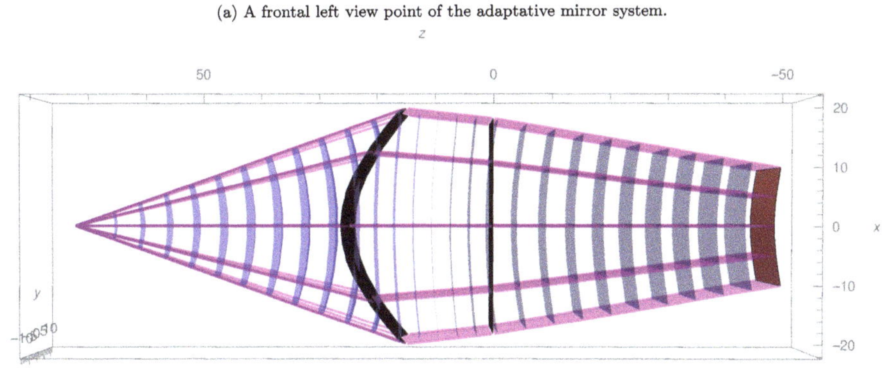

(b) A frontal right view point of the adaptative mirror system.

Figure 9.7. Lens design specifications for a parabolic input wavefront; the system produces a flat output wavefront: (a) a frontal left perspective of the same configuration as that shown in figure 9.6; (b) a frontal right perspective of the same configuration as that shown in figure 9.6.

9.6 End notes

The objective of this chapter was to obtain an equation that allows one to design a lens based on an input wavefront to create a stigmatic image point. A closed-form solution to the general equation was given here. This equation has been tested using ray-tracing techniques, and its performance matched expectations. A discussion of its implications, properties, and limitations was conducted. These equations, i.e. equation (9.44), (9.45), and (9.46), are part of the optical path theory, a theory that uses Fermat's principle as its only axiom.

Further reading

Campbell L and Garnett W 1882 *The Life of James Clerk Maxwell* (London: Macmillan and Co.) https://www.unav.es/gep/JCMaxwell.pdf

Chkhalo N I, Malyshev I V, Pestov A E, Polkovnikov V N, Salashchenko N N, Toropov M N and Soloviev A A 2016 Problems in the application of a null lens for precise measurements of aspheric mirrors *Appl. Opt.* **55** 619–25

Conrady A E 2013 *Applied Optics and Optical Design, Part One* (North Chelmsford, MA: Courier Corporation)

Descartes R 1637a *De la nature des lignes courbes La Géométrie* 2 (Leiden: Jan Maire)

Descartes R 1637b *Discours de la Methode La Géométrie* 1 (Leiden: Jan Maire)

Doskolovich L L, Bykov D A, Andreeva K V and Kazanskiy N L 2018 Design of an axisymmetrical refractive optical element generating required illuminance distribution and wavefront *J. Opt. Soc. Am.* A **35** 1949–53

Doskolovich L L, Bykov D A, Bezus E A and Greisukh G I 2021a Design of a stigmatic lens implementing a required ray mapping *Appl. Opt.* **60** 9138–45

Doskolovich L L, Bykov D A, Greisukh G I, Strelkov Y S and Bezus E A 2021b Designing stigmatic lenses with minimal Fresnel losses *J. Opt. Soc. Am.* A **38** 855–61

Euler L 1770 *Pars Secvnda: De Constrvctione Telescopiorvm Dioptricorvm : Cvm Appendice De Constrvctione Telescopiorvm Catoptrico-Dioptricorvm Dioptrica* 2 (St. Petersburg: Imperial Academy of Sciences)

Gauss C F 1841 *Dioptrische Untersuchungen* (Göttingen: Druck und Verlag der Dieterichschen Buchhandlung)

Geary J M 1999 Wavefront sensors *Adaptive Optics Engineering Handbook* (Boca Raton, FL: CRC Press) pp 129–56

Geary J M, Yoo K M, Davila P S, Wirth A, Jankevics A J, Ruda M C, Zielinski R J and Petrilli L J 1992 Comparison of wavefront-sensor techniques *Proc. SPIE* **1776** 58–72

Geary J M *et al* 1995 *Introduction to Wavefront Sensors* vol 18 (Bellingham, WA: SPIE Optical Engineering Press)

González-Acuña R G 2021a Exact equations to design aplanatic sequential optical systems *Appl. Opt.* **60** 9263–8

González-Acuña R G 2021b Surface solution to correct a freeform wavefront *Appl. Opt.* **60** 9887–91

González-Acuña R G 2021c Two-mirror system for tunable apodization *Appl. Opt.* **60** 10756–60
González-Acuña R G 2022 Stigmatic singlet with a user-defined apodization pupil function *J. Opt. Soc. Am.* A **39** 213–8
González-Acuña R G and Chaparro-Romo H A 2018 General formula for bi-aspheric singlet lens design free of spherical aberration *Appl. Opt.* **57** 9341–5
González-Acuña R G and Chaparro-Romo H A 2020 *Stigmatic Optics* (Bristol: IOP Publishing)
González-Acuña R G and Chaparro-Romo H A 2021 General mirror formula for adaptive optics *Appl. Opt.* **60** 375–9
González-Acuña R G and Guitiérrez-Vega J C 2018 Generalization of the axicon shape: the gaxicon *J. Opt. Soc. Am.* A **35** 1915–8
González-Acuña R G and Gutiérrez-Vega J C 2019a Analytic formulation of a refractive-reflective telescope free of spherical aberration *Opt. Eng.* **58** 085105
González-Acuña R G and Gutiérrez-Vega J C 2019b General formula for aspheric collimator lens design free of spherical aberration *Proc. SPIE* **11104** 111040P
González-Acuña R G and Gutiérrez-Vega J C 2019c General formula to eliminate spherical aberration produced by an arbitrary number of lenses *Opt. Eng.* **58** 085106
González Acuña R G and Gutiérrez-Vega J C 2019d General formula of the refractive telescope design free spherical aberration *Proc. SPIE* **11** 111050Z
González-Acuña R G and Gutiérrez-Vega J C 2020a Analytic design of a spherochromatic singlet *J. Opt. Soc. Am.* A **37** 149–53
González-Acuña R G and Gutiérrez-Vega J C 2020b The entrance pupil of an on-axis stigmatic singlet lens *Proc. SPIE* **11482** 1148205
González-Acuña R G and Gutiérrez-Vega J C Sr 2020c Analytic formulation of a spherochromatic collimator lens *Proc. SPIE* **11482** 114820Y
González-Acuña R G and Thibault S 2021 *The general equation of the stigmatic lenses: its history and what we have learned from it OSA Optical Design and Fabrication 2021 (Flat Optics, Freeform, IODC, OFT)* (Optica Publishing Group) 1207803
González-Acuña R G and Gutiérrez-Vega J C 2021 Exact equations to design a stigmatic singlet that meets the Herschel's condition *Opt. Commun.* **485** 126727
González-Acuña R G, Avendaño-Alejo M and Gutiérrez-Vega J C 2019a Singlet lens for generating aberration-free patterns on deformed surfaces *J. Opt. Soc. Am.* A **36** 925–9
González-Acuña R G, Chaparro-Romo H A and Gutiérrez-Vega J C 2019b General formula to design freeform singlet free of spherical aberration and astigmatism *Appl. Opt.* **58** 1010–5
González-Acuña R G, Chaparro-Romo H A and Gutiérrez-Vega J C 2020a Analytic solution of the eikonal for a stigmatic singlet lens *Phys. Scr.* **95** 085201
González-Acuña R G, Chaparro-Romo H A and Gutiérrez-Vega J C 2020b *Analytical Lens Design* (Bristol: IOP Publishing)
González-Acuña R G, Chaparro-Romo H A and Gutiérrez-Vega J C 2020c Uniqueness of stigmatic solutions *J. Opt. Soc. Am.* A **37** 1832–8
Gonzalez Acuña R G, Chaparro-Romo H A and Gutiérrez-Vega J C 2020d Analytic aplanatic singlet lens: setting and design for three-point objects and images in the meridional plane *Opt. Eng.* **59** 055104
González-Acuña R G, Gutiérrez-Vega J C and Rocha M C 2020e Freeform axicon with azimuthal variation *J. Mod. Opt.* **67** 1170–5
González-Acuña R G, Chaparro-Romo H A and Gutiérrez-Vega J C 2021a General stigmatic surfaces *J. Opt. Soc. Am.* A **38** 298–302

González-Acuña R G, Chaparro-Romo H A and Gutiérrez-Vega J C 2021b Analytic equations to design optical systems with three stigmatic pairs in the meridional plane *Opt. Eng.* **60** 051202

Gonzalez-Acuna R G, Chaparro-Romo H A and Gutierrez-Vega J C 2021c Exact equations for stigmatic singlet design meeting the Abbe sine condition *Opt. Commun.* **479** 126415

González-Acuña R G, Chaparro-Romo H A, Gutiérrez-Vega J C and Thibault S 2021d Bidirectional wavefront transfer function lens *Opt. Commun.* **498** 127215

Herschel J F W 1828 On the Theory of Light *Encyclopedia Metropolitana* (London: Rest Fenner) p 676

Hogendijk J P 2002 The burning mirrors of Diocles: reflections on the methodology and purpose of the history of pre-modern science *Early Sci. Med.* **7** 181–97

Huygens C 1690 *Traité de la lumière* (Leiden: Pieter van der Aa)

Kingslake R and Johnson R B 2009 *Lens Design Fundamentals* (New York: Academic)

Levi-Civita T 1900 Complementi al teorema di Malus–Dupin: nota *Atti della Reale Accademia del Lincei* (Rome: Tipografia della R. Accademia del Lincei) pp 185–9

Malacara D 2007 *Optical Shop Testing* vol 59 (New York: Wiley)

Rashed R 1990 A pioneer in anaclastics: Ibn Sahl on burning mirrors and lenses *ISIS* **81** 464–91

Scott J F 2016 *The Scientific Work of René Descartes: 1596–1650* (Milton Park: Routledge)

Shafer D 1992 Null lens design techniques *Appl. Opt.* **31** 2184–7

Southall J P C 1922 Aplanatic (or Cartesian) optical surfaces *J. Franklin Inst.* **193** 609–26

Tobin R 1990 Ancient perspective and Euclid's optics *J. Warburg Courtauld Inst.* **53** 14–41

Toomer G J 2012a *Apollonius: Conics Books V to VII: The Arabic Translation of the Lost Greek Original in the Version of the Banū Mūsā* vol 9 (Berlin: Springer)

Toomer G J 2012b *Diocles, On Burning Mirrors: The Arabic Translation of the Lost Greek Original* vol 1 (Berlin: Springer)

Tyson R 1999 *Adaptive Optics Engineering Handbook* vol 67 (Boca Raton, FL: CRC Press)

Tyson R K 1996 Adaptive optics and ground-to-space laser communications *Appl. Opt.* **35** 3640–6

Tyson R K 2000 Introduction to Adaptive Optics *Introduction to Adaptive Optics* vol 41 (Bellingham, WA: SPIE Optical Engineering Press)

Tyson R K 2015 *Principles of Adaptive Optics* (Boca Raton, FL: CRC Press)

www.ingramcontent.com/pod-product-compliance
Ingram Content Group UK Ltd.
Pitfield, Milton Keynes, MK11 3LW, UK
UKHW051859100926
8556IPUK00005B/694